Nonlinear Systems and Complexity

Volume 40

The Nonlinear Systems and Complexity series provides a platform to systematically summarize recent developments, practical applications, and overall advances in all aspects of nonlinearity, chaos, complexity, and simulation. It complements the primary research literature by presenting cohesive works that go beyond isolated novel findings. The series aims to publish cutting-edge theories and techniques in nonlinear systems and complexity; stimulate further research in nonlinearity, synchronization, complexity, and the modeling of nonlinear problems; and rapidly disseminate new knowledge to scientists, engineers, and students in the field. It features authoritative monographs covering a wide range of topics, including both fundamental advances and innovative applications. Topical areas include, but are not limited to: nonlinear dynamics; complexity, nonlinearity, and chaos; stability, bifurcation, chaos, and fractals; fractional dynamics and applications; discontinuity, synchronization, and control; computational methods for nonlinear science and engineering problems; and nonlinear physical, chemical, and biological phenomena.

Hansong Tang

Multiblock Method for Fluid Flow

Concepts, Algorithms, and Applications

Springer

Hansong Tang
Civil Engineering Department
The City College of New York
New York, NY, USA

ISSN 2195-9994 ISSN 2196-0003 (electronic)
Nonlinear Systems and Complexity
ISBN 978-3-031-78567-2 ISBN 978-3-031-78568-9 (eBook)
https://doi.org/10.1007/978-3-031-78568-9

This Springer imprint is published by the registered company Springer Nature Switzerland AG
The registered company address is: Gewerbestrasse 11, 6330 Cham, Switzerland

Dedication

To my parents,
who always, with quiet care, placed a few tasty snacks and drinks on my desk;
and to my wife and son,
who—despite their occasional complaints—patiently endured my long and monotonous work, with little explanation.

To those who strive to bridge theory and application,
and to those I cannot yet imagine—
who may one day find something of worth in this book.

To the memories of the past,
to the future that lies ahead,
and, as a leaf, to eternity beyond this life:
a timeline that, mercifully, is a longer deadline than most.

Preface

Nowadays, the multiblock method has become prevalent in the computation of fluid flows. The method is also referred to as the domain decomposition method, the overset-grid method, the zonal method, the composite-mesh method, etc., depending on the problems of under study and the authors' preferences. The multiblock method for fluid flows started in the 1980s, and, at that time, the method began to attract attention in the simulation of compressible flows in aerospace engineering. Still, I remember that, in the early 1990s, I read a review article by NASA scientists, which predicted that the method would become the future trend in computational aerodynamics. A primary motivation for the multiblock method was to overcome the difficulty in generating meshes for complex geometry. For instance, a mesh wrapping the tail of an aircraft and another covering its wake region were generated, and the multiblock method was adopted to simulate the flows in them separately, while incorporating their mutual interactions. Since then, the multiblock method has emerged as a popular topic, and it has spread to the simulation of various flow problems and evolved into an indispensable approach for parallel computation and, as illustrated in this book, for multiscale and multiphysics computation.

The author is fortunate to have witnessed and experienced a good portion of the evolution in fluid-flow simulations and multiblock techniques. Professor Bangmin Zheng, at Wuhan Institute of Hydraulic and Electric Engineering, once said that calculating the Bernoulli equation was a main computational problem in fluid flows during his college time in the 1950s. When the author was with him in the early 1980s, as a challenging computational task of that time, he worked on the simulation of a two-dimensional flow with a free surface over a dam by solving the Laplace equation. The author was first exposed to the multiblock method when working with Professor Chun-Hian Lee at Beijing University of Aeronautics and Astronautics in the 1990s. The author simulated flows over the tail and in the wake of an aircraft by solving the Euler equations via the zonal method. He then pursued simulations of incompressible laminar flows using the full Navier-Stokes equations and the multiblock method while studying with Professor Sotiropoulos at Georgia Institute of Technology around the turn of the century. Since about 2010, the author has been expanding the multiblock method to coastal ocean flows. Now, after many years, the computational capabilities have advanced to the levels of solving the three-dimensional, unsteady, full Navier-Stokes equations in applications, coupling these equations with others for multiscale, multiphysics phenomena, and even conducting machine learning on fluid flows. Such advances are infeasible without the development of multiblock techniques. This book reflects the author's journey with the multiblock method and its applications to fluid flows, and it may resonate with readers on similar journeys.

This book intends to present a study of the multiblock method across multiple flow-related disciplines in a cohesive manner. It discusses the concepts, algorithms, verifications, and applications toward capturing physical phenomena and solving engineering problems. The book starts with basic mathematical concepts and algorithms, and then proceeds to analyses of computing model equations and simplified problems. Following these, it presents the method for fluid flows in various backgrounds. Furthermore, the book discusses some special topics on the multiblock method. The multiblock method for fluid flows is rich in content. However, this book does not intend to discuss the method comprehensively; instead, it primarily describes our past work and limited observations in this direction. Additionally, the references cited in this book serve as a starting point for interested readers to study in detail. The author apologizes to colleagues and researchers whose work he might have missed mentioning and citing due to space constraints or unintentional oversight. The author also welcomes comments and discussions on the content of this book.

This instructional, research-oriented text is intended for graduate students, researchers, and practitioners across the multiple disciplines related to computer modeling and fluid flows. These disciplines include computational mathematics, computational science and engineering, aerodynamics, hydrodynamics, civil engineering, hydraulic engineering, ocean engineering, etc. It also intends to serve as a bridge among people in these disciplines and assist them in crossing disciplinary boundaries and understanding concepts and methods outside their own fields. The book is readable to people with basic knowledge of calculus, differential equations, numerical methods, and fluid mechanics, for instance, those who have taken a course in computational fluid dynamics. Readers may read the book sequentially or consult chapters and sections of particular interest.

This book is a compilation and extension of work by the author and coworkers over the past many years on the multiblock method. The author thanks the coworkers for their contribution and help, especially for the data, figures, and tables adopted as numerical examples in this book. These coworkers are advisors, collaborators, and students of the author, including Chun-Hian Lee, Fotis Sotiropoulos, Tie Zhou, Yingjie Liu, Michael Grossberg, Casey Jones, Joongcheol Paik, Xiuguang Wu, Wenlong Chen, Simon Kraatz, Ke Qu, Wenbin Dong, Lin Li, and others. The author expresses his heartfelt gratitude to Professor Dun Huang at Peking University, who shaped his interest and preparation in computation and simulation. Special thanks go to his former colleague at Stevens Institute of Technology, Dr. M. Malik, for proofreading the book and suggesting improvements. The author acknowledges the help and assistance from the editor, Dr. Albert Luo, and Springer staff. Finally, the author is indebted to his parents, Zemin Tang and Degui Zeng, his wife, Yan Qin, and his son, Da Tang, for their unwavering patience and support during the research and writing of this book.

New York, NY, USA
May 2024

Hansong Tang

Introduction

The multiblock method is a central, fundamental approach to solving complex problems. The idea of the multiblock method is natural and simple, and it may even appear in our daily life. As illustrated in Fig. 1, essentially, the idea is to split a large, complicated problem into small, manageable subproblems, solve them individually in consideration of their connections, and then present an answer to the whole problem. The connections are arranged in such a way that the solutions of the subproblems assemble into the answer to the original problem once the subproblems are solved. The multiblock method has become indispensable in simulating fluid flows in various backgrounds, including space, rivers, and oceans. In general, such flows exhibit high complexity and strong nonlinearity, and their simulation involves intensive computation of complex systems.

As a particular example, here is a situation in which the multiblock method is utilized in solving a complex flow problem. Figure 2 depicts a high-speed flow past an object. The flow field next to the wall is dominated by boundary layers and influenced strongly by viscosity, while that away from it may exhibit shocks and experience little effects from the viscosity. As a result, the computation needs to consider the viscosity effects in the region near the wall, and it can ignore the effects in the region away from the wall. The multiblock method divides the flow domain into two subdomains, see Fig. 2. The computation adopts the Navier-Stokes equations with a fine-resolution grid in the subdomain next to the wall. Such a computation is expensive in terms of computational cost. However, the computation may become less expensive in the subdomain away from the wall when it adopts simpler governing equations, such as the Euler equations or the Laplace equation, together with a coarser-resolution grid. Such a multiblock-method approach has become common in practice, especially when there is a lack of enough computing power [1, 2]. This approach faces some key questions, such as what boundary conditions are appropriate at the interfaces of the two subdomains and how to speed up the overall computation. These questions are critical to the success of the simulation. For instance, a crude, straightforward algorithm at the interfaces may lead to artifacts in the numerical solution or even blow-up of the computation. These questions have been the topics of research in the multiblock method.

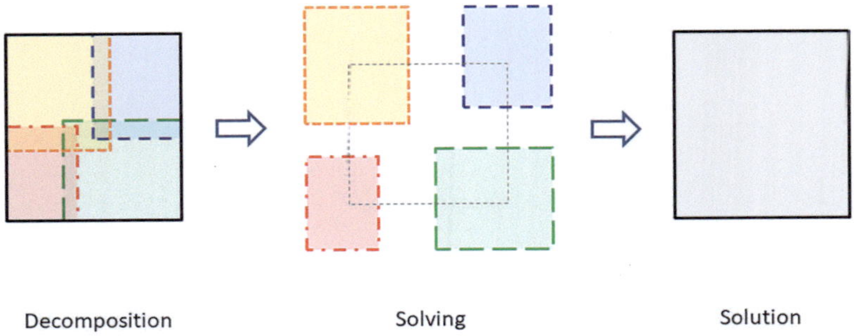

Fig. 1 Diagram of the multiblock method

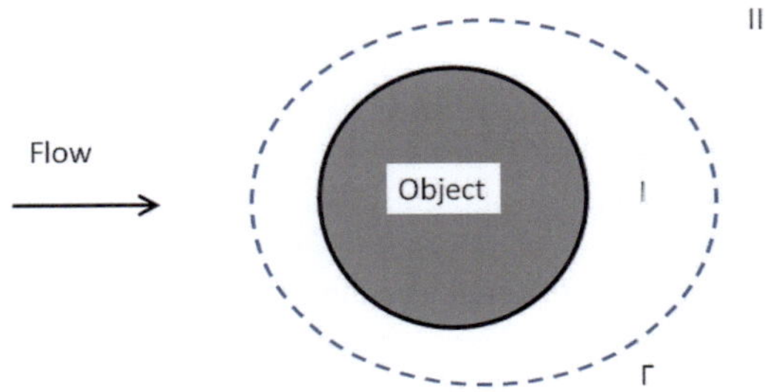

Fig. 2 A schematic representation of the multiblock method to simulate a flow past an object. The flow field is split into subdomains I and II, in which viscosity effects are included and ignored, respectively. Γ is the interface between the two subdomains

In this book, the multiblock method refers to solving problems as described above, along with the relevant solution techniques. Though "multi-block" is commonly used also, we use "multiblock" to emphasize the method's cohesive structure and application across diverse domains. In practice, the multiblock method is frequently referred to as the domain decomposition method, the zonal method, the composite-mesh method, etc., depending on the study's context and the researchers' preferences [3–5]. For instance, the method commonly refers to the domain decomposition method in the mathematics community, and it means the zonal method in the aerospace engineering community. The book uses the term multiblock method and its alternative names throughout the text.

The multiblock method has a long history, and, under the name of the domain decomposition method, traces back to the idea of German mathematician H. A. Schwarz to study partial differential equations (PDEs) over one hundred years ago [6]. Since the early 1980s, research on the multiblock method, e.g., under the name of the zonal method, grew quickly in the aerospace engineering community, and it was adopted as a measure to overcome the difficulty in dealing with complex geometry and mesh generation for compressible flows [4, 7–9]. Then, starting from the late 1980s, research on the multiblock method, also under the name of the domain-splitting method, etc., for incompressible flows has quickly become a central topic in environmental engineering, civil engineering, and other areas [3, 10].

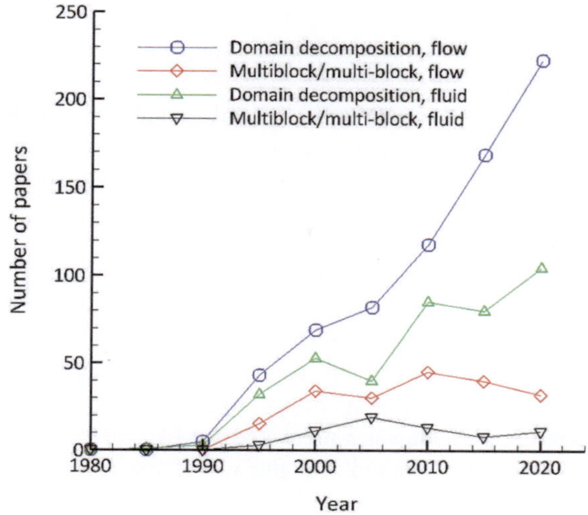

Fig. 3 Number of journal papers on the multiblock method published in the past years. The numbers result from a search within the Web of Science. They are the numbers of all papers whose titles and abstracts contain one or more of the four pairs of phrases: "domain decomposition, flow," "multiblock/multi-block, flow," "domain decomposition, fluid," and "multiblock/multi-block, fluid". "multiblock/multi-block" means "multiblock" or "multi-block"

Since then, the method has seen tremendous and continuous growth, evidenced by the publication numbers in past years in Fig. 3. As seen in the figure, the research on the topic is still expanding rapidly, particularly with the fast ongoing increase in the number of publications with the term domain decomposition, which is the technical core of the multiblock method. Now, the multiblock method has become an indispensable platform for the parallel computation of fluid flows, and it provides an avenue for simulating many emerging problems, including multiscale, multiphysics flows in the real world [11, 12].

Despite its long history, the multiblock method remains a vibrant and expanding area of research, as evidenced by its prominence in programs of various conferences. A major conference is the International Conference on Domain Decomposition Methods, which focuses on the mathematics of the method. Its first meeting was held in Paris in 1987, and since then, it has taken place approximately every two years with the publication of peer-reviewed proceedings. Other conferences also provide a platform for research on the multiblock method for fluid flows, such as the International Conference on Computational Fluid Dynamics, the AIAA SciTech Forum, and the SIAM Conference on Computational Science & Engineering.

This book intends to present a study of the multiblock method in multiple areas, but in a unified, consistent way. It goes from fundamental studies to real-world applications, crossing multiple disciplines and manifesting their connections. The book starts with concepts and algorithms in the domain decomposition method and presents the basics of computational mathematics. Then, with these concepts and algorithms, it analyzes the computation of model problems using the multiblock method. After this, the book discusses the computation of actual flow problems, from compressible flows to incompressible flows and then to ocean flows, focusing on practical algorithms and techniques. Finally, it presents two special topics: one concerns a classical numerical issue, and the other focuses on an emerging machine learning method.

The book's content is arranged to make its reading easier; the book proceeds in the order from simple to complicated problems, and it presents the content roughly in the time order of the evolution of the multiblock method. Particularly, the book is structured as follows:

Chapter 1 introduces the domain decomposition method's ideas and its concepts, followed by a brief discussion of computational algorithms and basic mathematics. The chapter deals with the Poisson equation and the diffusion equation, and it sets up basic computational frameworks for the later chapters.

As a discussion on model problems, Chap. 2 deals with the computation of advection–diffusion–reaction equations coupled at subdomain interfaces. The convergence rates are presented for two methods of computation, particularly, the conventional method and the waveform relaxation method. Optimized interface algorithms are also presented to speed up the convergence of the computation.

Chapter 3 starts with conservation laws, to which the Euler equations for compressible flows belong, and their computational methods. Then, analysis on their computation via the multiblock method is presented. Both conservative and nonconservative interface algorithms are discussed. The conservation errors and converged solutions are analyzed when nonconservative interface algorithms are adopted. Numerical experiments are presented to illustrate the effects of interface algorithms, including non-physical solutions and the prevention of their onset.

Chapter 4 presents a study on the computation of incompressible flows, and it solves the Navier-Stokes equations on overset, curvilinear grids. An interface method is presented to enforce mass conservation at grid interfaces, and its performance is illustrated on speedup of convergence, mitigation of artifacts, and reduction of conservation errors. Numerical experiments are provided to demonstrate the overset-grid methods.

Chapter 5 extends the overset-grid methods to coupling the Navier-Stokes equations and the geophysical fluid dynamics equations to simulate multiscale, multiphysics ocean flows. Relevant techniques are presented to achieve a seamless transition of numerical solutions across grid interfaces. Then, numerical experiments are made to validate the methods and techniques, and applications of the coupling system are presented to real-world problems. As an extension, the coupling of other types of PDEs is also presented for ocean flows.

Chapter 6 discusses two special topics related to the multiblock method. The first topic is artificial oscillations in numerical solutions. Based on a simple model equation, an analysis is presented on how such oscillations are generated, and conditions are presented for their onset. The second topic deals with machine learning to couple numerical solutions for differential equations as a potential new computational approach. The description lays the concepts, methods, and algorithms for the paradigm, together with numerical experiments, to illustrate its performance.

As a general pattern in presenting its contents, the book begins with the basics of concepts, methods, and algorithms, and then it proceeds with examples and applications. The book provides references for the readers to trace down to detailed techniques if they are not presented. If needed, it could be a good idea for the reader to read this book with other relevant topics, such as fluid mechanics and numerical methods for PDEs. To aid comprehension, each chapter concludes with a descriptive summary and discussion, followed by a set of review questions. Additionally, an index is presented at the end of the book to make content navigation easier. The index indicates where concepts, algorithms, and theories are introduced, defined, and primarily applied.

References

1. H.C. Chen, S.K. Lee, RANS/Laplace calculations of nonlinear waves induced by surface-piercing bodies. J. Eng. Mech. **125**, 1231–1242 (1999)
2. S.V. Guvernyuk, F.A. Maksimov, Hysteresis in supersonic flow past a plane cascade of cylindrical rods. Fluid Dyn. **56**, 886–896 (2021)
3. J.C. Strikwerdat, C.D. Scarbnick, A domain decomposition method for incompressible viscous flow. SIAM J. Sci. Comput. **14**(1), 49–67 (1993)
4. K.A. Hessenius, T.H. Pulliam, A zonal approach to solution of the Euler equations. AIAA paper 82-0969, 1982
5. W.D. Henshaw, G. Chesshire, Multigrid on composite meshes. SIAM J. Sci. Stat. Comput. **8**, 914–923 (1987)
6. H.A. Schwarz, Uber einige abbildungsaufgaben. Ges. Abh. **11**, 65–83 (1869)
7. J.A. Benek, J.L. Steger, F.C. Dougherly, A flexible grid embedding technique with application to the Euler equations. AIAA paper 83–1944, 1983
8. M.M. Rai, A conservative treatment of zonal boundary for Euler equation calculations. AIAA Paper 84-0164, 1984
9. P.G. Buning, I.T. Chiu, S. Obayashi, Y.M. Rizk, J.L. Steger, Numerical simulation of the integrated space shuttle vehicle in ascent. AIAA Paper 88-4359, 1988
10. R.L. Meakin, R.L. Street, Simulation of environmental flow problems in geometrically complex domains. Part 2: A domain-splitting method. Comput. Meths. Appl. Mech. Eng. **68**, 311–331 (1988)
11. D.E. Keyes, Multiphysics simulations: Challenges and opportunities. Int. J. High Performance Comput. Appl. **27**(1), 4–48 (2012)
12. H.S. Tang, R.D. Haynes, G. Houzeaux, A review of domain decomposition methods for simulation of fluid flows: Concepts, algorithms, and applications. Arch. Comput. Methods Eng. **28**, 841–873 (2021)

Contents

About the Author

Hansong Tang studies fluid mechanics, numerical methods, and computational physics. His work focuses on algorithm development and modeling in areas such as shock wave phenomena, non-Newtonian fluid behaviors, and coastal ocean flows. Throughout his career, Dr. Tang has held research and teaching positions at Changsha Communication College, Beijing University of Aeronautics and Astronautics, Stevens Institute of Technology, Pacific Northwest National Laboratory (Battelle), and the Naval Research Laboratory. He is currently a professor in the Department of Civil Engineering at the City College of New York, CUNY.

Chapter 1
Fundamentals of Domain Decomposition

1.1 Schwarz Method

Overlapping Subdomains The multiblock method traces back to the method of domain decomposition. The idea of domain decomposition originated from German mathematician Hermann Amandus Schwarz in the nineteenth century [1]. In his study, a boundary value problem of a partial differential equation (PDE), particularly the Laplace equation, is considered:

$$\begin{cases} \Delta u = 0, & \mathbf{x} \in \Omega \\ u = g, & \mathbf{x} \in \partial\Omega \end{cases} \tag{1.1}$$

in which $\mathbf{x}$ is the coordinate in space, $\mathbf{x} \in \mathbb{R}^n$ $(n = 2, 3, \ldots)$, u: $\mathbb{R}^n \to \mathbb{R}$ is an unknown scalar function of $\mathbf{x}$, and g: $\mathbb{R}^n \to \mathbb{R}$ is a prescribed scalar function of $\mathbf{x}$. Here and throughout this work, bold indicates a vector. In the study $(n = 2)$, the domain was decomposed into two zones, specifically subdomains 1 and 2, Ω_1 and Ω_2, respectively (Fig. 1.1). Problem (1.1) was transferred into two boundary value problems and solved with an iteration algorithm ($m = 0, 1, 2, \ldots$, and m is the iteration index):

$$\begin{cases} \Delta u_1^{m+1} = 0, & \mathbf{x} \in \Omega_1 \\ u_1^{m+1} = g, & \mathbf{x} \in \partial\Omega_1 \backslash \Gamma_1 \\ u_1^{m+1} = u_2^m, & \mathbf{x} \in \Gamma_1 \end{cases} \qquad \begin{cases} \Delta u_2^{m+1} = 0, & \mathbf{x} \in \Omega_2 \\ u_2^{m+1} = g, & \mathbf{x} \in \partial\Omega_2 \backslash \Gamma_2 \\ u_2^{m+1} = u_1^{m+1}, & \mathbf{x} \in \Gamma_2 \end{cases} \tag{1.2}$$

In Fig. 1.1, the two subdomains, Ω_1 and Ω_2, overlap with each other in a common area, and they are called overlapping subdomains. Γ_1 and Γ_2 are not boundaries of original problem (1.1) but artificial boundaries, and they are called subdomain interfaces. The conditions on the interfaces, Γ_1 and Γ_2, are referred to as interface conditions or transmission conditions. The interface conditions of the two subdomains in (1.2) are Dirichlet conditions, referred to as the Dirichlet–Dirichlet

H. Tang, *Multiblock Method for Fluid Flow*, Nonlinear Systems and Complexity 40,
https://doi.org/10.1007/978-3-031-78568-9_1

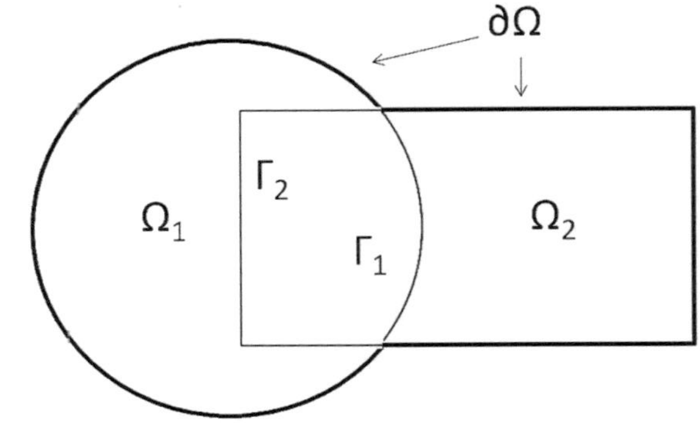

Fig. 1.1 Partition of a domain into overlapping subdomains, one being a circle and one being a rectangle

interface condition, and they exchange solutions in the two subdomains. Algorithm (1.2) is a Dirichlet–Dirichlet algorithm for domain decomposition. The algorithm is about the computation of differential equations, whose variables, i.e., u_1^m and u_2^m, are continuous functions, and it is domain decomposition at the continuous level.

Algorithm (1.2) is called the Schwarz method, particularly, the alternating Schwarz method. The method is an iteration algorithm, and it is frequently referred to as Schwarz iteration. In the iteration, the computation starts from subdomain 1 with an initial guess for the solution in subdomain 2, u_2^0 (particularly, its value on the interface of subdomain 1). After finishing in subdomain 1, the computation proceeds to subdomain 2 with u_1^1 on its own interface. After finishing in subdomain 2, it goes back to subdomain 1, and so on. With the aid of the maximum principle for a boundary value problem of the Laplace equation, Schwarz proved that, as $m \to \infty$, u_1^m and u_2^m converged to u_1 and u_2, respectively, and $u_1 \cup u_2$ equaled u, which was the solution to problem (1.1) [1]. After about 100 years, Lions provided a correction to Schwarz's proof [2].

There are variants of the Schwarz method. In alternating Schwarz method (1.2), just changing the interface condition of subdomain 2 to be $u_2^{m+1} = u_1^m$, one has $(m = 0, 1, 2, \ldots)$

$$\begin{cases} \Delta u_1^{m+1} = 0, & \mathbf{x} \in \Omega_1 \\ u_1^{m+1} = g, & \mathbf{x} \in \partial\Omega_1 \backslash \Gamma_1 \\ u_1^{m+1} = u_2^m, & \mathbf{x} \in \Gamma_1 \end{cases} \qquad \begin{cases} \Delta u_2^{m+1} = 0, & \mathbf{x} \in \Omega_2 \\ u_2^{m+1} = g, & \mathbf{x} \in \partial\Omega_2 \backslash \Gamma_2 \\ u_2^{m+1} = u_1^m, & \mathbf{x} \in \Gamma_2 \end{cases} \tag{1.3}$$

In this method, the computations in subdomains 1 and 2 start from u_1^0 and u_2^0, respectively, and, while marching from m to $m + 1$, they proceed simultaneously in the iteration. This method was introduced by Lions [2], and it is referred to as the parallel Schwarz method.

Patched Subdomains Besides overlapping subdomains, a domain can also be split into patched subdomains, or, non-overlapping subdomains aligning with each other at their interface, Γ (Fig. 1.2). In the case of two patched subdomains, a Schwarz method reads as $(m = 0, 1, 2, \ldots)$

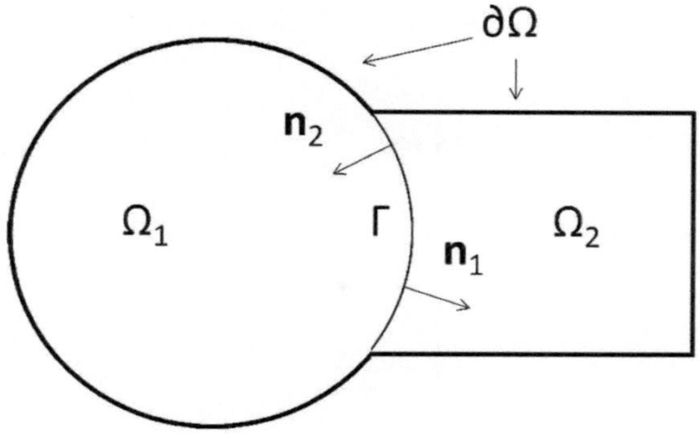

Fig. 1.2 Partition of a domain into patched subdomains

$$\begin{cases} \Delta u_1^{m+1} = 0, & \mathbf{x} \in \Omega_1 \\ u_1^{m+1} = g, & \mathbf{x} \in \partial\Omega_1 \backslash \Gamma \\ (\alpha_1 \frac{\partial}{\partial n_1} + \beta_1) u_1^{m+1} = (\alpha_1 \frac{\partial}{\partial n_1} + \beta_1) u_2^m, & \mathbf{x} \in \Gamma \end{cases}$$

$$\begin{cases} \Delta u_2^{m+1} = 0, & \mathbf{x} \in \Omega_2 \\ u_2^{m+1} = g, & \mathbf{x} \in \partial\Omega_2 \backslash \Gamma \\ (\alpha_2 \frac{\partial}{\partial n_2} + \beta_2) u_2^{m+1} = (\alpha_2 \frac{\partial}{\partial n_2} + \beta_2) u_1^{m+1}, & \mathbf{x} \in \Gamma \end{cases} \quad (1.4)$$

$\alpha_1, \alpha_2, \beta_1, \beta_2$ are prescribed parameters. A frequently referred situation is $\alpha_1 = 0$, $\beta_1 = 1$, $\alpha_2 = 1$, and $\beta_2 = 0$. It can be proven that the Schwarz method converges when $\alpha_1, \alpha_2 = 1$ [3].

More algorithms for domain decomposition by the Schwarz method have been discussed among researchers [4]. For instance, regarding non-overlapping subdomains, one of them is the Dirichlet–Neumann algorithm ($m = 0, 1, 2, \ldots$):

$$\begin{cases} \Delta u_1^{m+1/2} = 0, & \mathbf{x} \in \Omega_1 \\ u_1^{m+1/2} = g, & \mathbf{x} \in \partial\Omega_1 \backslash \Gamma \\ u_1^{m+1/2} = u_\Gamma^m, & \mathbf{x} \in \Gamma \end{cases}$$

$$\begin{cases} \Delta u_2^{m+1} = 0, & \mathbf{x} \in \Omega_2 \\ u_2^{m+1} = g, & \mathbf{x} \in \partial\Omega_2 \backslash \Gamma \\ \frac{\partial u_2^{m+1}}{\partial n_2} = -\frac{\partial u_1^{m+1/2}}{\partial n_1}, & \mathbf{x} \in \Gamma \end{cases} \quad (1.5)$$

$$u_\Gamma^{m+1} = (1-\theta) u_\Gamma^m + \theta u_2^{m+1}, \ \ \mathbf{x} \in \Gamma$$

in which θ is a constant, and $\theta \in (0, \theta_{max})$. Here, θ_{max} is a parameter. This algorithm applies a Dirichlet-Neumann interface condition. The computation starts in subdomain 1 with an initial guess for u_Γ^0, and then proceeds to subdomain 2.

Another algorithm is the Neumann–Neumann algorithm ($m = 0, 1, 2, \ldots$):

$$k = 1, 2: \begin{cases} \Delta u_k^{m+1/2} = 0, & \mathbf{x} \in \Omega_k \\ u_k^{m+1/2} = g, & \mathbf{x} \in \partial\Omega_k \backslash \Gamma \\ u_k^{m+1/2} = u_\Gamma^m, & \mathbf{x} \in \Gamma \end{cases}$$

$$k = 1, 2: \begin{cases} \Delta \Phi_k^{m+1} = 0, & \mathbf{x} \in \Omega_k \\ \Phi_k^{m+1} = g, & \mathbf{x} \in \partial\Omega_k \backslash \Gamma \\ \dfrac{\partial \Phi_k^{m+1}}{\partial n_k} = \dfrac{\partial u_1^{m+1/2}}{\partial n_1} + \dfrac{\partial u_2^{m+1/2}}{\partial n_2}, & \mathbf{x} \in \Gamma \end{cases} \tag{1.6}$$

$$u_\Gamma^{m+1} = u_\Gamma^m - \theta(\Phi_1^{m+1} + \Phi_2^{m+1}), \quad \mathbf{x} \in \Gamma$$

where, θ is again a constant, and $\vartheta \in (0, \theta_{max})$.

1.2 Time-Dependent Problem

Computational Problem The alternating Schwarz method was originally proposed for a boundary value problem in space, i.e., problem (1.1). Since then, various Schwarz methods have been developed and investigated for other types of problems [4]. For a time-dependent problem of a PDE , a Schwarz method may be built on those for a boundary value problem in space. Consider the computation of the following initial-boundary value problem of a parabolic equation:

$$\begin{cases} u_t = \Delta u, & \mathbf{x} \in \Omega, & t \in (0, T] \\ u = g, & \mathbf{x} \in \partial\Omega, & t \in (0, T] \\ u = u_0, & \mathbf{x} \in \Omega, & t = 0 \end{cases} \tag{1.7}$$

in which t is the time, T is either a constant or infinity, $\mathbf{x}$ is the coordinate in space, and $\mathbf{x} \in \mathbb{R}^n$ ($n = 1, 2, \ldots$). u: $\mathbb{R}^n \times [0, T] \rightarrow \mathbb{R}$ is an unknown scalar function of $(\mathbf{x}, t)$, g: $\mathbb{R}^n \times [0, T] \rightarrow \mathbb{R}$ is a prescribed scalar function of $(\mathbf{x}, t)$, and u_0: $\mathbb{R}^n \rightarrow \mathbb{R}$ is a prescribed scalar function of $\mathbf{x}$. Again, bold indicates a vector. By domain decomposition, the above problem is split into two subproblems:

$$\begin{cases} u_{1t} = \Delta u_1, & \mathbf{x} \in \Omega_1, & t \in (0, T] \\ u_1 = g, & \mathbf{x} \in \partial\Omega_1 \backslash \Gamma_1, & t \in (0, T] \\ u_1 = u_2, & \mathbf{x} \in \Gamma_1, & t \in (0, T] \\ u_1 = u_0, & \mathbf{x} \in \Omega_1, & t = 0 \end{cases}$$

$$\begin{cases} u_{2t} = \Delta u_2, & \mathbf{x} \in \Omega_2, & t \in (0, T] \\ u_2 = g, & \mathbf{x} \in \partial\Omega_2 \backslash \Gamma_2, & t \in (0, T] \\ u_2 = u_1, & \mathbf{x} \in \Gamma_2, & t \in (0, T] \\ u_2 = u_0, & \mathbf{x} \in \Omega_2, & t = 0 \end{cases} \tag{1.8}$$

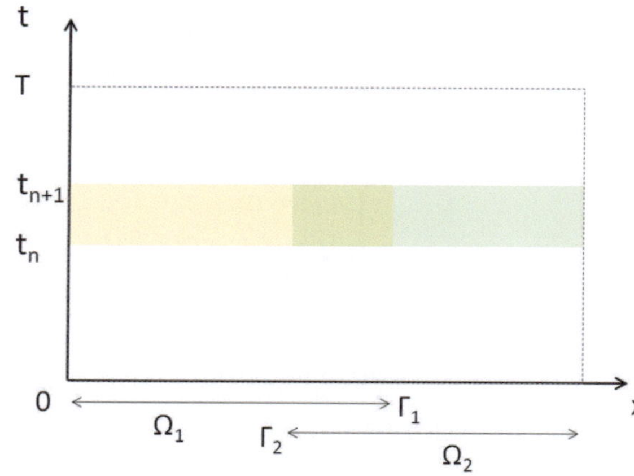

Fig. 1.3 Domain division for the conventional method

where $\Omega_1 \cup \Omega_2 = \Omega$, and Ω_1 and Ω_2 overlap with each other (Fig. 1.1). Note that an initial-boundary value problem and overlapping subdomains are discussed in this subsection. However, similar discussions, for example, on computational algorithms, may also apply to other situations, such as initial value problems and patched subdomains.

There are two categories of methods to carry out computational problem (1.8). In the first category, the computations in the two subdomains proceed simultaneously in time. Particularly, when marching in time, they exchange subdomain solutions at interfaces at each time step. This category of methods is a natural approach, and it is widely adopted in practical computation [5–7]. Hereafter, this category of methods is called the conventional Schwarz method. In the second category, the computation in each subdomain marches from the beginning to the end in time. Then, after they finish, the computations in the subdomains exchange their solutions at the interfaces. This category of methods is called the Schwarz waveform relaxation method. The waveform relaxation method dates back over three decades, for instance, in solving systems of algebraic equations and differential equations [8–10].

Conventional Schwarz Method Now, we compute problem (1.8) using the conventional Schwarz method. The computation marches in time, step by step: $[t_0, t_1]$, $[t_1, t_2]$, $\ldots$, $[t_{N-1}, t_N]$, $t_n = n\Delta t$, $n = 0, 1, 2, \ldots, N$. $t_0 = 0$, Δt is the time step, and $\Delta t = T/N$ (Fig. 1.3). The conventional method reads as

$$u_1^0 = u_0, \ \mathbf{x} \in \Omega_1$$

$$u_2^0 = u_0, \ \mathbf{x} \in \Omega_2$$

$$\mathit{Do}\ n = 0, 1, \ldots, N-1$$

$$\begin{cases} u_{1t} = \Delta u_1, & \mathbf{x} \in \Omega_1, & t \in (t_n, t_{n+1}] \\ u_1 = g, & \mathbf{x} \in \partial\Omega_1 \backslash \Gamma_1, & t \in (t_n, t_{n+1}] \\ u_1 = u_2, & \mathbf{x} \in \Gamma_1, & t \in (t_n, t_{n+1}] \\ u_1 = u_1^n & \mathbf{x} \in \Omega_1, & t = t_n \end{cases} \tag{1.9}$$

$$\begin{cases} u_{2t} = \Delta u_2, & \mathbf{x} \in \Omega_2, & t \in (t_n, t_{n+1}] \\ u_2 = g, & \mathbf{x} \in \partial\Omega_2 \backslash \Gamma_2, & t \in (t_n, t_{n+1}] \\ u_2 = u_1, & \mathbf{x} \in \Gamma_2, & t \in (t_n, t_{n+1}] \\ u_2 = u_2^n, & \mathbf{x} \in \Omega_2, & t = t_n \end{cases}$$

End Do

in which the first two lines set up the initial value condition, and superscript n indicates $t = t_n$. Essentially, by the conventional method, the computations of the initial-boundary value problems in the two subdomains are conducted simultaneously over time. Additionally, when marching from time step n to $n + 1$, they exchange solutions of the subdomains at the interfaces.

When the conventional method is discretized by finite difference, an explicit finite difference scheme in time of the equations may be adopted ($k = 1, 2$):

$$\frac{v_k^{n+1} - v_k^n}{\Delta t} = \Delta v_k^n$$

Here, v_k^n is an approximation to u_k at $t = t_n$. Utilizing the aforementioned semi-discretization, one can formulate the domain decomposition computation as

$$\begin{aligned} v_1^0 &= u_0, & \mathbf{x} \in \Omega_1 \\ v_2^0 &= u_0, & \mathbf{x} \in \Omega_2 \end{aligned}$$

Do $n = 0, 1, \ldots, N - 1$

$$\begin{cases} v_1^{n+1} = v_1^n + \Delta t \Delta v_1^n, & \mathbf{x} \in \Omega_1 \\ v_1^{n+1} = g^{n+1}, & \mathbf{x} \in \partial\Omega_1 \backslash \Gamma_1 \\ v_1^{n+1} = v_2^{n+1}, & \mathbf{x} \in \Gamma_1 \end{cases} \tag{1.10}$$

$$\begin{cases} v_2^{n+1} = v_2^n + \Delta t \Delta v_2^n, & \mathbf{x} \in \Omega_2 \\ v_2^{n+1} = g^{n+1}, & \mathbf{x} \in \partial\Omega_2 \backslash \Gamma_2 \\ v_2^{n+1} = v_1^{n+1}, & \mathbf{x} \in \Gamma_2 \end{cases}$$

End Do

The computation can be carried out numerically after discretization in space. It first predicts the solutions in the two subdomains and then exchanges their interface values. Since the temporal discretization and the interface algorithm are explicit, the computation does not need iteration between the solutions from the two subdomains, and the Schwarz iteration becomes unnecessary.

The domain decomposition computation via the conventional Schwarz method may also be performed via an implicit finite difference scheme ($k = 1, 2$):

$$\frac{v_k^{n+1} - v_k^n}{\Delta t} = \Delta v_k^{n+1}$$

With this discretization, the computation is formulated as

$$\begin{aligned} v_1^0 &= u_0, \quad \mathbf{x} \in \Omega_1 \\ v_2^0 &= u_0, \quad \mathbf{x} \in \Omega_2 \end{aligned}$$

Do $n = 0, 1, \ldots, N-1$

$$\begin{cases} \Delta v_1^{n+1} - \dfrac{v_1^{n+1}}{\Delta t} = -\dfrac{v_1^n}{\Delta t}, & \mathbf{x} \in \Omega_1 \\ v_1^{n+1} = g^{n+1}, & \mathbf{x} \in \partial\Omega_1 \backslash \Gamma_1 \\ v_1^{n+1} = v_2^{n+1}, & \mathbf{x} \in \Gamma_1 \end{cases} \tag{1.11}$$

$$\begin{cases} \Delta v_2^{n+1} - \dfrac{v_2^{n+1}}{\Delta t} = -\dfrac{v_2^n}{\Delta t}, & \mathbf{x} \in \Omega_2 \\ v_2^{n+1} = g^{n+1}, & \mathbf{x} \in \partial\Omega_2 \backslash \Gamma_2 \\ v_2^{n+1} = v_1^{n+1}, & \mathbf{x} \in \Gamma_2 \end{cases}$$

End Do

Now, the aforementioned computation becomes implicit and needs iteration between the solutions in the two subdomains.

The computation for problem (1.11), particularly the computation between two consecutive time steps, is actually the same as that for a boundary value problem of a PDE. Therefore, a domain decomposition method for a boundary value problem, such as the alternating Schwarz method and the parallel Schwarz method for problem (1.1), in the last subsection may be adopted for problem (1.11), or, actually time-dependent problem (1.7). By the alternating Schwarz method, as in (1.2), the computation of (1.11) may be carried out via a conventional method:

$v_1^0 = u_0, \ \mathbf{x} \in \Omega_1$

$v_2^0 = u_0, \ \mathbf{x} \in \Omega_2$

Do $n = 0, 1, \ldots, N-1$

$v_1^{n+1,0} = v_1^n, \ \mathbf{x} \in \Omega_1$

$v_2^{n+1,0} = v_2^n, \ \mathbf{x} \in \Omega_2$

Do $m = 0, 1, 2, \ldots$

$$\begin{cases} \Delta v_1^{n+1,m+1} - \dfrac{v_1^{n+1,m+1}}{\Delta t} = -\dfrac{v_1^n}{\Delta t}, & \mathbf{x} \in \Omega_1 \\ v_1^{n+1,m+1} = g^{n+1}, & \mathbf{x} \in \partial\Omega_1 \backslash \Gamma_1 \\ v_1^{n+1,m+1} = v_2^{n+1,m}, & \mathbf{x} \in \Gamma_1 \end{cases}$$
$$\begin{cases} \Delta v_2^{n+1,m+1} - \dfrac{v_2^{n+1,m+1}}{\Delta t} = -\dfrac{v_2^n}{\Delta t}, & \mathbf{x} \in \Omega_2 \\ v_2^{n+1,m+1} = g^{n+1}, & \mathbf{x} \in \partial\Omega_2 \backslash \Gamma_2 \\ v_2^{n+1,m+1} = v_1^{n+1,m+1}, & \mathbf{x} \in \Gamma_2 \end{cases} \tag{1.12}$$

End Do

End Do

in which m is the index for Schwarz iteration between the solutions in the two subdomains, and the two lines right before the do loop for m set up the initial guesses for the iteration. At convergence, $v_k^{n+1,m} \rightarrow v_k^{n+1}$. In this situation, the computational method and procedure between two time steps, i.e., from t_n to t_{n+1}, are the same as those for computation (1.2). As a result, time-dependent problem (1.8), can be computed using the methods developed for the computation of a boundary value problem in the last subsection.

Schwarz Waveform Relaxation Method Now, let us compute problem (1.8) using the Schwarz waveform relaxation method. As shown in Fig. 1.4, the space-time domain $\Omega \times [0, T]$ is partitioned into two overlapping space-time subdomains $\Omega_1 \times [0, T]$ and $\Omega_2 \times [0, T]$. The Schwarz waveform relaxation method solves the problem as follows:

$$u_1^0 = g_1, \ \mathbf{x} \in \Gamma_1, \ t \in (0, T]$$
$$u_2^0 = g_2, \ \mathbf{x} \in \Gamma_2, \ t \in (0, T]$$

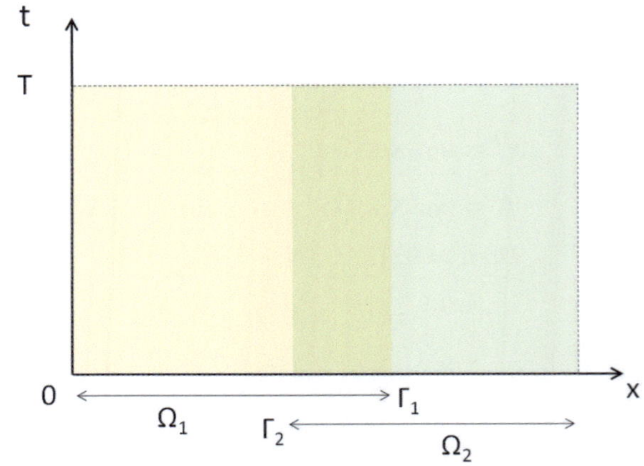

Fig. 1.4 Domain division for the waveform relaxation method

Do $m = 0, 1, 2, \ldots$

$$
\begin{cases}
{u_1}_t^{m+1} = \Delta u_1^{m+1}, & \mathbf{x} \in \Omega_1, \quad t \in (0, T] \\
u_1^{m+1} = g, & \mathbf{x} \in \partial\Omega_1 \backslash \Gamma_1, \ t \in (0, T] \\
u_1^{m+1} = u_2^m, & \mathbf{x} \in \Gamma_1, \quad t \in (0, T] \\
u_1^{m+1} = u_0, & \mathbf{x} \in \Omega_1, \quad t = 0
\end{cases}
$$

$$
\begin{cases}
{u_2}_t^{m+1} = \Delta u_2^{m+1}, & \mathbf{x} \in \Omega_2, \quad t \in (0, T] \\
u_2^{m+1} = g, & \mathbf{x} \in \partial\Omega_2 \backslash \Gamma_2, \ t \in (0, T] \\
u_2^{m+1} = u_1^{m+1}, & \mathbf{x} \in \Gamma_2, \quad t \in (0, T] \\
u_2^{m+1} = u_0, & \mathbf{x} \in \Omega_2, \quad t = 0
\end{cases}
\tag{1.13}
$$

End Do

Here, g_1 and g_2 are prescribed functions of $(\mathbf{x}, t)$, as initial guesses for u_1 and u_2, respectively, at the interfaces. The alternating Schwarz method is adopted in the aforementioned computation, but another method, such as the parallel Schwarz method, may be applied. The computation may be carried out via either an explicit or an implicit discretization. For instance, if an implicit discretization is adopted, an algorithm for the Schwarz waveform relaxation method is formulated as

Do $n = 0, 1, \ldots, N-1$

$$v_1^{n+1,0} = g_1, \quad \mathbf{x} \in \Gamma_1, \ t \in (0, T]$$

$$v_2^{n+1,0} = g_2, \quad \mathbf{x} \in \Gamma_2, \ t \in (0, T]$$

End Do

Do $m = 0, 1, 2, \ldots$

$$v_1^0 = u_0, \quad \mathbf{x} \in \Omega_1$$

$$v_2^0 = u_0, \quad \mathbf{x} \in \Omega_2$$

Do $n = 0, 1, \ldots, N-1$

$$
\begin{cases}
\Delta v_1^{n+1,m+1} - \dfrac{v_1^{n+1,m+1}}{\Delta t} = -\dfrac{v_1^n}{\Delta t}, & \mathbf{x} \in \Omega_1 \\
v_1^{n+1,m+1} = g^{n+1}, & \mathbf{x} \in \partial\Omega_1 \backslash \Gamma_1 \\
v_1^{n+1,m+1} = v_2^{n+1,m}, & \mathbf{x} \in \Gamma_1
\end{cases}
$$

$$
\begin{cases}
\Delta v_2^{n+1,m+1} - \dfrac{v_2^{n+1,m+1}}{\Delta t} = -\dfrac{v_2^n}{\Delta t}, & \mathbf{x} \in \Omega_2 \\
v_2^{n+1,m+1} = g^{n+1}, & \mathbf{x} \in \partial\Omega_2 \backslash \Gamma_2 \\
v_2^{n+1,m+1} = v_1^{n+1,m+1}, & \mathbf{x} \in \Gamma_2
\end{cases}
\tag{1.14}
$$

End Do

End Do

In the aforementioned, the solution at the interfaces is assumed/guessed before the do loop for m starts, and the initial condition is applied after the do loop for m starts. At convergence, $v_k^{n,m} \to v_k^n$ $(k = 1, 2)$.

As shown in (1.12) and (1.14), in both the conventional method and the waveform relaxation method, there is Schwarz iteration between solutions of the two subdomains when an implicit discretization is adopted for the PDEs and interface conditions. However, it is worth noting that a main difference between them is that in the former, the Schwarz iteration is within two physical time steps, i.e., n and $n + 1$. Whereas, in the latter, the iteration is over all time steps. More discussion on the difference will be given, e.g., in Sect. 2.3.3.

In general, for both conventional and waveform relaxation methods, the convergence of the Schwarz iteration depends on the transmission conditions and algorithms, the size of the overlapping zone (if overlapping partitions are used), the nature of the PDEs, the length of the time window, etc., and discussion is available in the literature [11–14].

1.3 Domain Decomposition at Discrete Level

Discrete-Level Domain Decomposition Consider the discretization of alternating Schwarz method (1.2). For illustration, consider the two-dimensional (2D) Laplace equation within Ω in the (x, y) plane. Ω is covered by two rectangle grids. The grids are overlapping grids, and they correspond to Ω_1 and Ω_2 (Fig. 1.5). The Laplace equation is discretized with a finite difference method, particularly the central finite difference method. $v_{1,i,j}$ and $v_{2,i,j}$ denote the discretized values for u_1 and u_2, respectively, at grid node $(x, y) = (x_i, y_j)$. Then, the discretized version of alternating Schwarz method (1.2) becomes $(m = 0, 1, 2, \ldots)$

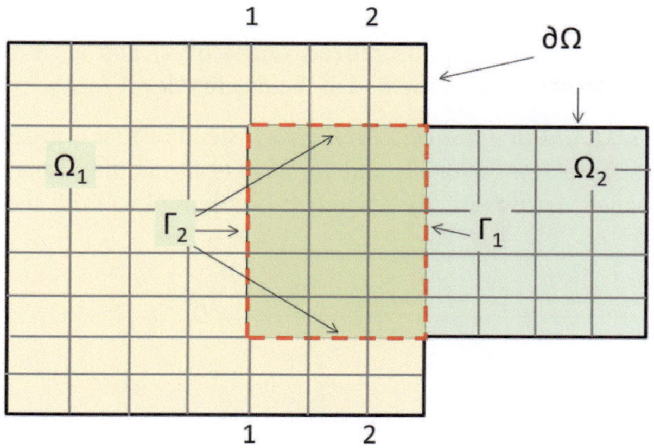

Fig. 1.5 Domain division and grids for the 2D Laplace equation

$$\begin{cases} \dfrac{v_{1,i-1,j}^{m+1}+v_{1,i+1,j}^{m+1}-2v_{1,i,j}^{m+1}}{\Delta x^2}+\dfrac{v_{1,i,j-1}^{m+1}+v_{1,i,j+1}^{m+1}-2v_{1,i,j}^{m+1}}{\Delta y^2}=0, \quad in\ \Omega_1 \\ v_{1,i,j}^{m+1}=g_{i,j}, \quad in\ \partial\Omega_1\backslash\Gamma_1; \quad v_{1,i,j}^{m+1}=v_{2,i,j}^{m}, \quad on\ \Gamma_1 \end{cases}$$

$$\begin{cases} \dfrac{v_{2,i-1,j}^{m+1}+v_{2,i+1,j}^{m+1}-2v_{2,i,j}^{m+1}}{\Delta x^2}+\dfrac{v_{2,i,j-1}^{m+1}+v_{2,i,j+1}^{m+1}-2v_{2,i,j}^{m+1}}{\Delta y^2}=0, \quad in\ \Omega_2 \\ v_{2,i,j}^{m+1}=g_{i,j}, \quad in\ \partial\Omega_2\backslash\Gamma_2; \quad v_{2,i,j}^{m+1}=v_{1,i,j}^{m+1}, \quad on\ \Gamma_2 \end{cases}$$

where Δx and Δy are grid spacing in the x-(horizontal) and y-(vertical) directions, respectively. In the aforementioned, the PDE is fully discretized into algebraic equations, and the variables are no longer continuous but discrete. This subsection discusses Schwarz methods in such discretized forms, and the domain decomposition method is referred to as being at the discrete level.

Let V_1 and V_2 stand for the vectors consisting of variables $v_{1,i,j}^{m+1}$ and $v_{2,i,j}^{m+1}$, respectively, at all inner nodes of Ω_1 and Ω_2, respectively. Then, the aforementioned discretization can be formulated as

$$\begin{bmatrix} A_{11}V_1^{m+1} \\ A_{22}V_2^{m+1} \end{bmatrix}=\begin{bmatrix} b_1-A_{12}V_2^{m} \\ b_2-A_{21}V_1^{m+1} \end{bmatrix}$$

which, when A_{11} and A_{22} are invertible, gives rise to

$$\begin{bmatrix} V_1^{m+1} \\ V_2^{m+1} \end{bmatrix}=\begin{bmatrix} A_{11}^{-1}(b_1-A_{12}V_2^{m}) \\ A_{22}^{-1}(b_2-A_{21}V_1^{m+1}) \end{bmatrix} \tag{1.15}$$

In the above, A_{11} and A_{22} are matrices resulting from the discretization at inner nodes in Ω_1 and Ω_2, respectively, A_{12} and A_{21} are matrices corresponding to interface nodes on Γ_1 and Γ_2, respectively, and b_1 and b_2 are two vectors of constants. Now, (1.15) involves discretized values of u_1 and u_2 in problem (1.2), so it is a domain decomposition method at the discrete level for the problem.

Multiplicative and Additive Schwarz Methods Instead of first transferring problem (1.1) into two subproblems over Ω_1 and Ω_2 as in (1.2) and then solving them as in the aforementioned, let us directly discretize problem (1.1) with the central finite difference in Ω:

$$\frac{v_{i-1,j}+v_{i+1,j}-2v_{i,j}}{\Delta x^2}+\frac{v_{i,j-1}+v_{i,j+1}-2v_{i,j}}{\Delta y^2}=0,\ \ in\ \Omega;\qquad v_{i,j}=g_{i,j},\ \ on\ \partial\Omega$$

which leads to a linear system

$$AV=b \tag{1.16}$$

where A is the coefficient matrix, V is the vector for solution $v_{i,j}$ at inner nodes of Ω, and b is a vector of constants. Linear system (1.16) can be solved using the so-called multiplicative Schwarz method (e.g., [15]). The method solves the system iteratively using a domain decomposition scheme at the discrete level. In particular, the coefficient matrix A is split into two smaller, local matrices A_k that are formed by a so-called restriction matrix R_k:

$$A_k=R_kAR_k^T$$

where $k=1,2$, and

$$R_1=\begin{bmatrix} 1 & 0 & \cdots & & & & & 0\\ 0 & 1 & \cdots & & & & & 0\\ \vdots & \vdots & \ddots & & & & & \\ 0 & 0 & & 1 & 0 & 0 & \cdots & 0\\ 0 & 0 & & 0 & 1 & 0 & \cdots & 0 \end{bmatrix},\qquad R_2=\begin{bmatrix} 0 & \cdots & 0 & 1 & 0 & & 0 & 0\\ 0 & \cdots & 0 & 0 & 1 & & 0 & 0\\ & & & & & \ddots & \vdots & \vdots\\ 0 & & & & & \cdots & 1 & 0\\ 0 & & & & & \cdots & 0 & 1 \end{bmatrix}$$

The multiplicative Schwarz method reads as

$$\begin{cases} V^{m+1/2}=V^m+R_1^TA_1^{-1}R_1(b-AV^m)\\ V^{m+1}=V^{m+1/2}+R_2^TA_2^{-1}R_2(b-AV^{m+1/2}) \end{cases} \tag{1.17}$$

In this method, because of the restriction matrices, the iteration is split into two steps, and it is performed between the solution associated with A_1 and that with A_2.

In the multiplicative Schwarz method, when R_1 and R_2 are non-overlapping with each other, i.e., $R_1^TR_1+R_2^TR_2=I$ (I is the identity matrix), the matrices in system (1.16) can be split as

$$A=\begin{bmatrix} A_{11} & A_{12}\\ A_{21} & A_{22} \end{bmatrix},\qquad V=\begin{bmatrix} V_1\\ V_2 \end{bmatrix},\qquad b=\begin{bmatrix} b_1\\ b_2 \end{bmatrix}$$

Here, since restriction matrices R_1 and R_2 are non-overlapping, the solution associated with A_1 does not overlap with that associated with A_2 in space. Then, the first step in (1.17) becomes

$$\begin{aligned}
\begin{bmatrix} V_1^{m+1/2} \\ V_2^{m+1/2} \end{bmatrix} &= \begin{bmatrix} V_1^m \\ V_2^m \end{bmatrix} + R_1^T A_{11}^{-1} R_1 (b - AV^m) \\
&= \begin{bmatrix} V_1^m \\ V_2^m \end{bmatrix} + R_1^T A_{11}^{-1} (b_1 - A_{11} V_1^m - A_{12} V_2^m) \\
&= \begin{bmatrix} V_1^m \\ V_2^m \end{bmatrix} + R_1^T (A_{11}^{-1} b_1 - V_1^m - A_{11}^{-1} A_{12} V_2^m) \\
&= \begin{bmatrix} A_{11}^{-1}(b_1 - A_{12} V_2^m) \\ V_2^m \end{bmatrix}
\end{aligned} \tag{1.18}$$

In the aforementioned, the terms with the restriction matrix actually only compute the rows corresponding to the rows of $V_1^{m+1/2}$ due to the action of the matrix, and thus the first step only updates V_1. Similarly, it can readily check that the second step in (1.17) only updates V_2. As a result, $V_1^{m+1} = V_1^{m+1/2}$. These, together with the derivation in the second step lead to

$$\begin{bmatrix} V_1^{m+1} \\ V_2^{m+1} \end{bmatrix} = \begin{bmatrix} A_{11}^{-1}(b_1 - A_{12} V_2^m) \\ A_{22}^{-1}(b_2 - A_{21} V_1^{m+1}) \end{bmatrix} \tag{1.19}$$

This shows that the multiplicative Schwarz method first updates V_1 and then, with the aid of the updated V_1, it updates V_2. Consequently, the multiplicative Schwarz method is actually a block Gauss–Seidel method. Additionally, it is also seen that (1.19), actually (1.17), is identical to (1.15), which is the discretized version of alternating Schwarz method (1.2). Therefore, the multiplicative Schwarz method, i.e., (1.17), could be considered as the discretized version of the alternating Schwarz method at the continuous level (1.2). This is indeed true if we consider the one-dimensional (1D) space and that the two grids overlap only at two nodes, similarly as if Γ_2 shrinks from cross-section 1-1 to 2-2 in Fig. 1.5.

To solve system (1.16) via parallel computation, the additive Schwarz method is available. In the case of two subdomains, the method reads as follows [16]:

$$V^{m+1} = V^m + (R_1^T A_1^{-1} R_1 + R_2^T A_2^{-1} R_2)(b - AV^m) \tag{1.20}$$

When R_k $(k = 1, 2)$ are non-overlapping, making a derivation similar to (1.18), one has

$$\begin{bmatrix} V_1^{m+1} \\ V_2^{m+1} \end{bmatrix} = \begin{bmatrix} A_{11}^{-1}(b_1 - A_{12} V_2^m) \\ A_{22}^{-1}(b_2 - A_{21} V_1^m) \end{bmatrix} \tag{1.21}$$

It can be shown that (1.21) is the discrete version of (1.3). Particularly, if R_k are non-overlapping, the additive Schwarz method is equivalent to the discretization of the parallel Schwarz method at the continuous level (1.3). However, as long as R_k are overlapping, the additive Schwarz method does not converge with iteration (e.g., [4, 17]). Accidentally, it was found that after a slight modification of the additive Schwarz method as

$$V^{m+1} = V^m + (\overline{R}_1^T A_1^{-1} R_1 + \overline{R}_2^T A_2^{-1} R_2)(b - AV^m) \tag{1.22}$$

convergence is achieved even when R_k are overlapping [18]. Here $\overline{R}_k$ result from replacing all ones at overlapping by zeros in R_k, and $\overline{R}_1^T R_1 + \overline{R}_2^T R_2 = I$. The method in (1.22) is called the restricted additive Schwarz method.

Schur Complement System Discretization of non-overlapping problem (1.4) will lead to a linear system in the form of (1.16). Moreover, it is easy to show that the system can be expressed as [15, 19]

$$A = \begin{bmatrix} A_1 & 0 & A_{1\Gamma} \\ 0 & A_2 & A_{2\Gamma} \\ A_{\Gamma 1} & A_{\Gamma 2} & A_{\Gamma\Gamma} \end{bmatrix}, \quad V = \begin{bmatrix} V_1 \\ V_2 \\ V_\Gamma \end{bmatrix}, \quad b = \begin{bmatrix} b_1 \\ b_2 \\ b_\Gamma \end{bmatrix}$$

in which, as before, subscripts 1 and 2 stand for the interior of Ω_1 and Ω_2, respectively, and subscript Γ denotes an interface between the two subdomains, see Fig. 1.2. After the elimination of V_1 and V_2, a so-called Schur complement system for the solution on the interface is derived as

$$\begin{cases} SV_\Gamma = g_\Gamma \\ S = A_{\Gamma\Gamma} - A_{\Gamma 1} {A_1}^{-1} A_{1\Gamma} - A_{\Gamma 2} {A_2}^{-1} A_{2\Gamma} \\ g_\Gamma = b_\Gamma - A_{\Gamma 1} {A_1}^{-1} b_1 - A_{\Gamma 2} {A_2}^{-1} b_2 \end{cases} \tag{1.23}$$

The Schur complement system is smaller in terms of the coefficient matrices' dimensions than the original system, i.e., (1.16). Once the former is solved, the solution for the latter is readily obtained. The Schur complement system serves as a basis for the study of a number of domain decomposition algorithms, including the Dirichlet–Neumann algorithm and the Neumann–Neumann algorithm [4].

It is noted that there are more techniques for domain decomposition computation, and the readers may read about them in the literature [4, 20, 21]. The rest chapters in the book will apply the techniques, particularly the concepts and algorithms, in this section.

Summary and Question

Summary This chapter discusses the fundamentals of the domain decomposition method and lays the groundwork for the computations discussed in the subsequent chapters. The discussion begins with a boundary value problem for the

Laplace equation. It introduces the alternating Schwarz method and basic concepts, including overlapping subdomains, patched subdomains, interface/transmission conditions, and Schwarz iterations. A brief note is on the convergence of the computation. Other relevant topics are also mentioned, including the parallel Schwarz method, the Dirichlet-Neumann algorithm, and the Neumann-Neumann algorithm.

The chapter then presents the frameworks for computing time-evolution problems using parabolic equations that are coupled at their subdomain interfaces. The computational methods are categorized into two types: (1) the conventional Schwarz method and (2) the Schwarz waveform relaxation method. The conventional Schwarz method is presented, which allows for either explicit or implicit time integration schemes. When an implicit scheme is adopted, iteration between subdomains is necessary. In this case, as the solution advances over time, the Schwarz methods developed for boundary value problems can be utilized. The waveform relaxation method is also presented, which adopts a different approach to solving the coupling problem. A comparative discussion is provided, highlighting that while the two methods may share the same discretizations and techniques, their computational procedures are fundamentally different.

The chapter further extends to domain decomposition at the discrete level. To illustrate this, a boundary value problem for the 2D Laplace equation is discretized into a system of algebraic equations. The multiplicative Schwarz method is first discussed, serving as a discrete-level domain decomposition. Through a derivation, it is shown that the multiplicative Schwarz method corresponds to a discretized version of the alternating Schwarz method at the continuous level. The additive Schwarz method is also introduced. Under certain conditions, it is equivalent to the discretization of the parallel Schwarz method at the continuous level. However, when these conditions are not satisfied, the method may fail to converge. To address this issue, the restricted additive Schwarz method is introduced, which overcomes the convergence problem. Finally, the chapter presents the ideas, concepts, and formulation of the so-called Schur complement system, which transforms the original system of equations into a reduced-size system.

This chapter presents the fundamental ideas, concepts, theories, and algorithms of domain decomposition. Its content has been extensively studied and remains an active area of research within the mathematics community. A challenging yet important direction lies in bridging the theoretical foundations and practical applications to real-world problems, such as those discussed in the following chapters.

Questions The following questions are provided to aid in understanding and discussing the content. Their answers and clues are embedded within the chapter.

1. Review the alternating Schwarz method, which is the original method proposed by Schwarz, to solve boundary problem (1.1). Discuss its underlying idea, procedure, and the relationship between its converged solutions and the solution of the boundary value problem.

2. Review the basic concepts, including subdomain, interface, overlapping subdomains, patched subdomains, and interface/transmission conditions.
3. In addition to the alternating Schwarz method (the original Schwarz method), this chapter introduces several other Schwarz methods, which are either variants of the original or are designed for different situations. List these methods.
4. Review the interface conditions and their algorithms in the Schwarz methods listed in the above question.
5. For the computation of a time-evolution problem, what are the two categories of Schwarz methods introduced?
6. How can the Schwarz method developed for a boundary value problem be applied to solve an initial value problem?
7. For an initial-boundary value problem, what are the similarities and differences between the conventional Schwarz method and the Schwarz waveform relaxation method in terms of the discretizations and procedures?
8. Step by step, derive Eq. (1.15) from the discretization of the boundary value problem for the 2D Laplace equation.
9. Following the derivation in the chapter, and filling in any missing steps if needed, prove that the multiplicative Schwarz method is a discrete version of the alternating Schwarz method at the continuous level.
10. Let A be an $n \times n$ coefficient matrix, R_1 be a $i \times n$ restriction matrix, and R_2 be a $j \times n$ restriction matrix, $n > i, j$. Construct a set of these matrices with numbers and then compute $A_k = R_k A R_k^T$ $(k = 1, 2)$.
11. Consider a square domain covered by two patched grids, each occupying half of the domain. Discretize the Laplace equation on these grids using parameters of your choice (e.g., sizes of square, numbers of nodes), and then derive the Schur complement system, Eq. (1.23).

References

1. H.A. Schwarz, Uber einige abbildungsaufgaben. Ges. Abh. **11**, 65–83 (1869)
2. P.L. Lions, On the Schwarz alternating method, I, in *First International Symposium on Domain Decomposition Methods for Partial Differential Equations*, ed. by R. Glowinski, G.H. Golub, G.A. Meurant, J. Periaux (SIAM, Philadelphia, 1988), pp. 1–42
3. P.L. Lions, On the Schwarz alternating method, III: a variant for nonoverlapping subdomains, in *Third International Symposium on Domain Decomposition Methods for Partial Differential Equations*, ed. by T. Chan, R. Glowinski, J. Periaux, O. Widlund (SIAM, Philadelphia, 1990), pp. 202–223
4. A. Toselli, O. Widlund, *Domain Decomposition Methods - Algorithms and Theory* (Springer, 2005)
5. G.A. Meurant, A domain decomposition method for parabolic problems. Appl. Numer. Math. **8**, 427–441 (1991)
6. H.S. Tang, C. Jones, F. Sotiropoulos, An overset - grid method for 3D unsteady incompressible flows. J. Comput. Phys. **191**, 567–600 (2003)
7. E. Blayo, About interface conditions for coupling hydrostatic and nonhydrostatic Navier-Stokes flows. Discrete Contin. Dynam. Syst. S **9**(5), 1565–1574 (2016)

8. E. Lelarasmee, A.E. Ruehli, A.L. Sangiovanni-Vincentelli, The waveform relaxation method for time-domain analysis of large scale integrated circuits. IEEE Trans. Comput. Aided Design Integr. Circuits Syst. **1**(3), 131–145 (1982)
9. M.L. Crow, M.D. Ilić, The waveform relaxation method for systems of differential/algebraic equations. Math. Comput. Modell. **19**, 67–84 (1994)
10. M. BjØrhus, *On domain decomposition, subdomain iteration and waveform relaxation*. Ph.D. thesis, University of Trondheim, 1995
11. V. Martin, An optimized Schwarz waveform relaxation method for the unsteady convection diffusion equation in two dimensions. Comput. Fluids **33**(5-6), 829–837 (2004)
12. V. Martin, An optimized Schwarz waveform relaxation method for the unsteady convection diffusion equation in two dimensions. Appl. Numer. Math. **52**(4), 401–428 (2005)
13. W.B. Dong, H.S. Tang, Y.J. Liu, Convergence analysis on computation of coupled advection-diffusion-reaction problems. Appl. Math. Comput. **420**, 126876 (2022)
14. W.B. Dong, H.S. Tang, Convergence analysis of waveform relaxation method to compute coupled advection-diffusion-reaction equations. Math. Comput. Simul. **218**, 462–481 (2024)
15. T.F. Chan, T.P. Mathew, Domain decomposition algorithms. Acta Numer. **3**, 61–143 (1994)
16. M. Dryja, O.B. Widlund, Some domain decomposition algorithms for elliptic problems, in *Iterative Methods for Large Linear Systems*, ed. by D. Kincaid, L. Hayes (Academic Press, 1990), pp. 273–291
17. M.J. Gander, Schwarz methods over the course of time. Electronic Trans. Numer. Anal. **31**, 228–255 (2008)
18. X.C. Cai, M. Sarkis, A restricted additive Schwarz preconditioner for general sparse linear systems. SIAM J. Sci. Comput. **21**, 239–247 (1999)
19. L.F. Pavarino, Domain decomposition methods and scientific computing applications. Bollettino dell'Unione Matematica Italiana, series 8, **8-B**, 21–54 (2005)
20. D.E. Keyes, Multiphysics simulations: Challenges and opportunities. Int. J. High Performance Comput. Appl. **27**(1), 4–48 (2012)
21. H.S. Tang, R.D. Haynes, G. Houzeaux, A review of domain decomposition methods for simulation of fluid flows: Concepts, algorithms, and applications. Arch. Comput. Methods Eng. **28**, 841–873 (2021)

Chapter 2
Advection–Diffusion–Reaction Equation

2.1 Problem of Computation

Advection, diffusion, and reaction are fundamental processes underlying many real-world flow problems, and these processes and their interactions are described by advection–diffusion–reaction equations as well as coupled advection–diffusion–reaction equations [1]. These equations and their couplings have been solved numerically to understand these real-world problems, such as the migration of pollutants in a porous medium, the chemical reaction in a nuclear reactor, and the propagation of tsunamis in the ocean [2–4]. Besides, advection–diffusion–reaction equations and their couplings are commonly used as model equations, or simplified equations, to study complicated PDEs and their couplings, such as the Navier–Stokes equations, their hydrostatic versions, and their coupling, as to be discussed in Sect. 5.3. This chapter presents a study on the computation of coupled advection–diffusion–reaction equations, intending to illustrate the multiblock method and also lay out frameworks for the simulation problems in the later chapters.

The problem of computation in this chapter concerns an initial value problem for 1D advection–diffusion–reaction equations coupled at subdomain interfaces:

$$
\begin{cases} u_{1t} + a_1 u_{1x} = b_1 u_{1xx} - c_1 u_1, & x < x_1,\ t > 0 \\ u_1 = u_2, & x = x_1,\ t > 0 \\ u_1 = u_0, & x < x_1,\ t = 0 \end{cases}
$$
$$
\begin{cases} u_{2t} + a_2 u_{2x} = b_2 u_{2xx} - c_2 u_2, & x > x_2,\ t > 0 \\ u_2 = u_1, & x = x_2,\ t > 0 \\ u_2 = u_0, & x > x_2,\ t = 0 \end{cases}
\tag{2.1}
$$

H. Tang, *Multiblock Method for Fluid Flow*, Nonlinear Systems and Complexity 40,
https://doi.org/10.1007/978-3-031-78568-9_2

In the aforementioned, t is the time, and x is the space coordinate. $a_k, b_k, c_k=const.$ $(k = 1, 2)$. Subdomain 1 is on the left, $x \leq x_1$, and subdomain 2 is on the right, $x \geq x_2$. The two subdomains overlap with each other, $x_2 < x_1$. Locations x_1 and x_2 are the interfaces of subdomains 1 and 2, respectively. Interface conditions apply at the interfaces as boundary conditions, and they fulfill the coupling of the PDEs in the subdomains. Moreover, boundary conditions may be added at the left and right ends of subdomains 1 and 2, respectively. u_0 is a given function of x, Without loss of generality, it is assumed that u_0 has finite support, that is, $u_0 = 0$, as long as $|x|$ is sufficiently large. Consistently, for convenience of discussion, it is considered that $u_1,\ u_2 = 0$, when $|x|$ and t are large enough.

It is noted that in (2.1), a_k, b_k, and c_k $(k = 1, 2)$ may take different values in the two subdomains. As a result, problem (2.1) contains several coupling scenarios. The examples are the coupling between two advection–diffusion–reaction equations with different coefficients, between an advection–diffusion–diffusion equation $(a_1, b_1, c_1 \neq 0)$ and an advection equation $(b_2, c_2 = 0)$, and between an advection equation $(b_1, c_1 = 0)$ and a diffusion-reaction equation $(a_2 = 0)$. Moreover, unlike the common approach, which adopts patched subdomains, we consider overlapping subdomains. Such overlapping treatment yields flexibility, e.g., in mesh generation, interface location selection, and application of transmission conditions. Thus, it is frequently adopted in practical problems, such as ocean flow simulations in Chap. 5. When the equations are distinct, in general their solutions differ in the overlapping region of the two subdomains. Thus, the well-posedness of the problem deserves discretion; for instance, it may permit different solutions or multiple solutions in the overlapping region.

In this chapter, the analysis is primarily on the computation of linear problem (2.1). However, its extension and application to nonlinear equations will also be considered. The problem of computation for the nonlinear equations concerns the following initial value problem:

$$
\begin{cases}
u_{1t} + a_1(u_1)u_{1x} = b_1 u_{1xx} - c_1(u_1)u_1, & x < x_1,\ t > 0 \\
u_1 = u_2, & x = x_1,\ t > 0 \\
u_1 = u_0, & x < x_1,\ t = 0
\end{cases}
$$

$$
\begin{cases}
u_{2t} + a_2(u_2)u_{2x} = b_2 u_{2xx} - c_2(u_2)u_2, & x > x_2,\ t > 0 \\
u_2 = u_1, & x = x_2,\ t > 0 \\
u_2 = u_0, & x < x_2,\ t = 0
\end{cases}
\tag{2.2}
$$

In the equations, $a_k(u_k)$ and $c_k(u_k)$ are functions of u_k $(k = 1, 2)$. Two situations will be considered: (1) $c_k(u_k) = const.$, (2) $c_k(u_k) = const. \cdot u_k$. The two situations correspond to the first- and second-order reactions, respectively [5]. The nonlinearity of the equations comes from the advection and source terms. When $a_k(u_k), c_k(u_k) = const.$, the equations reduce to the ones in problem (2.1).

This chapter applies two categories of Schwarz methods for time-dependent problems introduced in Sect. 1.2. The first category of Schwarz methods is the conventional method, as illustrated in (1.9). In this method, the computations in the two subdomains march together step by step in time: $(t_0, t_1]$, $(t_1, t_2]$, ..., $(t_{N-1}, t_N]$, with time step $\Delta t = T/N$. Here, $t_n = n\Delta t$, and $n = 0, 1, 2, \ldots, N$. T and N are prescribed. The method exchanges subdomains' solutions at their interfaces at each time step. For problem (2.1), the conventional method reads as

$$u_1^0 = u_0,\ x < x_2,\ t = 0$$
$$u_2^0 = u_0,\ x > x_1,\ t = 0$$

Do $n = 0, 1, 2, \ldots, N-1$

$$\begin{cases} u_{1t} + a_1 u_{1x} = b_1 u_{1xx} - c_1 u_1, & x < x_1,\ t_n < t \le t_{n+1} \\ u_1 = u_2, & x = x_1,\ t_n < t \le t_{n+1} \\ u_1 = u_1^n, & x < x_1,\ t = t_n \end{cases} \tag{2.3}$$

$$\begin{cases} u_{2t} + a_2 u_{2x} = b_2 u_{2xx} - c_2 u_2, & x > x_2,\ t_n < t \le t_{n+1} \\ u_2 = u_1, & x = x_2,\ t_n < t \le t_{n+1} \\ u_2 = u_2^n, & x < x_2,\ t = t_n \end{cases}$$

End Do

Here, superscript n indicates $t = t_n$. The conventional method has been analyzed concerning different aspects, such as preconditioners, algorithms, and error estimates [6–8].

On the grid $(\ldots, i-1, i, i+1, \ldots)$ of subdomain k $(k = 1, 2)$, discretization of the advection–diffusion–reaction equations by a finite difference method, particularly the forward finite difference in time and the central finite difference in space, leads to an explicit scheme:

$$\frac{v_{k,i}^{n+1} - v_{k,i}^n}{\Delta t} + a_k \frac{v_{k,i+1}^n - v_{k,i-1}^n}{2\Delta x} = b_k \frac{v_{k,i+1}^n - 2v_{k,i}^n + v_{k,i-1}^n}{\Delta x^2} - c_k v_{k,i}^n \tag{2.4}$$

where $v_{k,i}^n$ is the numerical solution to u_k at (x_i, t_n), Δt and Δx are the time step and the grid spacing, respectively. Instead of an explicit scheme, an implicit finite difference may also compute the equations. Particularly, the equations are discretized with the backward finite difference in time and the central finite difference in space, and this leads to an implicit scheme:

$$\frac{v_{k,i}^{n+1} - v_{k,i}^n}{\Delta t} + a_k \frac{v_{k,i+1}^{n+1} - v_{k,i-1}^{n+1}}{2\Delta x} = b_k \frac{v_{k,i-1}^{n+1} + v_{k,i+1}^{n+1} - 2v_{k,i}^{n+1}}{\Delta x^2} - c_k v_{k,i}^{n+1} \tag{2.5}$$

The second category of Schwarz methods is the Schwarz waveform relaxation method, as illustrated in Eq. (1.13). In this category, Schwarz iteration is made between solutions in the two subdomains simultaneously at all time steps, rather than individually at each time step as in the aforementioned conventional method. The waveform relaxation method reads as

$$u_1^0 = g_1, \; x = x_1, \; 0 < t \le T$$
$$u_2^0 = g_2, \; x = x_2, \; 0 < t \le T$$

Do $m = 0, 1, 2, \ldots$

$$\begin{cases} {u_1}_t^{m+1} + a_1 {u_1}_x^{m+1} = b_1 {u_1}_{xx}^{m+1} - c_1 u_1^{m+1}, & x < x_1, \; 0 < t \le T \\ u_1^{m+1} = u_2^m, & x = x_1, \; 0 < t \le T \\ u_1^{m+1} = u_0, & x < x_1, \; t = 0 \end{cases} \tag{2.6}$$

$$\begin{cases} {u_2}_t^{m+1} + a_2 {u_2}_x^{m+1} = b_2 {u_2}_{xx}^{m+1} - c_2 u_2^{m+1}, & x > x_2, \; 0 < t \le T \\ u_2^{m+1} = u_1^m, & x = x_2, \; 0 < t \le T \\ u_2^{m+1} = u_0, & x > x_2, \; t = 0 \end{cases}$$

End do

In this category of methods, an initial guess for the iteration, i.e., $g_1(t)$ and $g_2(t)$, is needed at interfaces $x = x_1, x_2$. The initial guess may be selected in a way that assures the iteration converges. The waveform relaxation method for advection–diffusion–reaction equations has been intensively studied [9].

For the computation by waveform relaxation method (2.6), let $v_{k,i}$ be a numerical approximation to u_k at x_i on the grid of subdomain k ($k = 1, 2$). A semi-discretization of the advection–diffusion–reaction equation, or discretization of the equation in space only, gives rise to

$$\frac{dv_{k,i}(t)}{dt} + a_k \frac{v_{k,i+1}(t) - v_{k,i-1}(t)}{2\Delta x} = b_k \frac{v_{k,i-1}(t) + v_{k,i+1}(t) - 2v_{k,i}(t)}{\Delta x^2} - c_k v_{k,i}(t) \tag{2.7}$$

in which $t \in [0, T]$. Such equations at different grid nodes form a system of ordinary differential equations.

By Taylor expansion, it can be verified that the truncation of schemes (2.4) and (2.5) are $\mathcal{O}(\Delta t, \Delta x^2)$, or their discretization accuracy is first-order in time and second-order in space. Additionally, the truncation scheme of (2.7) is $\mathcal{O}(\Delta x^2)$. If needed, higher-order schemes are possible. For instance, truncation error of $\mathcal{O}(\Delta t^2, \Delta x^2)$ can be achieved by discretizing the time derivative with a second-order accurate, three-time-level backward difference (e.g., [10]), or, discretizing the whole equation with the Crank–Nicholson scheme (e.g., [11]). However, the analysis in the following sections of this chapter remains the same or similar when such higher-order discretization is adopted. The content of this chapter is derived from the studies in [12–14], and the reader may refer to them for relevant information.

2.2 Conventional Method

2.2.1 *Explicit Scheme*

Consider the computation of problem (2.1) by conventional method (2.3). The computation is on two grids: $1 \le i \le I_1$ for subdomain 1, $1 \le i \le I_2$ for subdomain 2, as shown in Fig. 2.1. The two grids overlap at two nodes. Node I_1 of the left grid is its interface node, and node 1 of the right grid is its interface node. The coordinates of these two nodes correspond to the points x_1 and x_2, respectively, in (2.3). Let the equations in (2.3) be discretized by explicit scheme (2.4) on the two grids. Then, the computation between two time steps becomes

$$\begin{cases} \dfrac{v_{1,i}^{n+1} - v_{1,i}^{n}}{\Delta t} + a_1 \dfrac{v_{1,i+1}^{n} - v_{1,i-1}^{n}}{2\Delta x} \\ \quad = b_1 \dfrac{v_{1,i+1}^{n} - 2v_{1,i}^{n} + v_{1,i-1}^{n}}{\Delta x^2} - c_1 v_{1,i}^{n}, \quad i \le I_1 - 1 \\ v_{1,I_1}^{n+1} = v_{2,2}^{n+1} \end{cases}$$

$$\begin{cases} \dfrac{v_{2,i}^{n+1} - v_{2,i}^{n}}{\Delta t} + a_2 \dfrac{v_{2,i+1}^{n} - v_{2,i-1}^{n}}{2\Delta x} \\ \quad = b_2 \dfrac{v_{2,i+1}^{n} - 2v_{2,i}^{n} + v_{2,i-1}^{n}}{\Delta x^2} - c_2 v_{2,i}^{n}, \quad i \ge 2 \\ v_{2,1}^{n+1} = v_{1,I_1-1}^{n+1} \end{cases} \tag{2.8}$$

This computation follows the algorithm illustrated in (1.10). In actual computation, boundary values are required for the solutions at node $i = 1$ of the left grid and $i = I_2$ of the right grid. Here and hereafter, in view of their finite support, values such as zero will be assumed as the default values for the solutions there. By rearranging the terms, the computation is formulated as

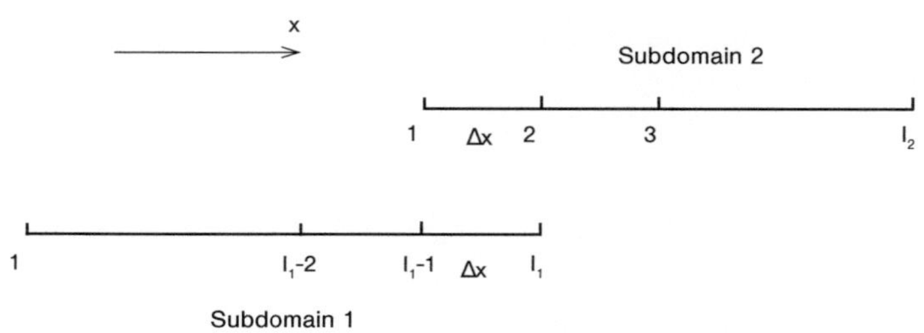

Fig. 2.1 Subdomains and grids. The grid nodes for subdomain 1 are indexed as $(1, 2, \ldots, I_1)$, and those for subdomain 2 are indexed as $(1, 2, \ldots, I_2)$. The total number of grid nodes is $I_1 + I_2 - 2$ (two overlapping nodes count as one node)

$$\begin{cases} v_{1,i}^{n+1} &= (\epsilon_1+\eta_1)v_{1,i-1}^n + (1-2\epsilon_1-\gamma_1)v_{1,i}^n \\ &\quad +(\epsilon_1-\eta_1)v_{1,i+1}^n, \quad i \le I_1 - 1 \\ v_{1,I_1}^{n+1} &= v_{2,2}^{n+1} \\ v_{2,i}^{n+1} &= (\epsilon_2+\eta_2)v_{2,i-1}^n + (1-2\epsilon_2-\gamma_2)v_{2,i}^n \\ &\quad +(\epsilon_2-\eta_2)v_{2,i+1}^n, \quad i \ge 2 \\ v_{2,1}^{n+1} &= v_{1,I_1-1}^{n+1} \end{cases} \tag{2.9}$$

in which ($k = 1, 2$)

$$\begin{aligned} \epsilon_k &= b_k \frac{\Delta t}{\Delta x^2} \\ \eta_k &= a_k \frac{\Delta t}{2\Delta x} \\ \gamma_k &= c_k \Delta t \end{aligned}$$

In (2.9), a Dirichlet condition is adopted at node I_1 and node 1, the interface nodes of the left and right grids, respectively. The computation is made first at the inner nodes in both grids and then at the interface nodes. Therefore, when an explicit discretization is adopted in the subdomains in the framework of the conventional method, there is no need for Schwarz iteration between the solutions in the two subdomains, and the computation becomes explicit.

In computational problem (2.9), the interface algorithm corresponds to that in (1.2), and it adopts a Dirichlet–Dirichlet interface condition

$$\begin{cases} u_1 = u_2, \ x = x_1 \\ u_2 = u_1, \ x = x_2 \end{cases} \tag{2.10}$$

Moreover, in view of the two overlapping nodes, the algorithm yields

$$\begin{aligned} v_{1,I_1}^{n+1} &= v_{2,2}^{n+1} \\ \frac{v_{1,I_1}^{n+1} - v_{1,I_1-1}^{n+1}}{\Delta x} &= \frac{v_{2,2}^{n+1} - v_{2,1}^{n+1}}{\Delta x} \end{aligned}$$

This is an approximation (in terms of first-order accuracy, $\mathcal{O}(\Delta x)$) to the interface condition

$$\begin{cases} u_1 = u_2, & x = x_1 \\ \dfrac{\partial u_2}{\partial n_2} = -\dfrac{\partial u_1}{\partial n_1}, & x = x_2 \end{cases} \tag{2.11}$$

which is actually a Dirichlet–Neumann interface condition, similar to that in (1.5). Therefore, when the two grids overlap at two nodes only, the Dirichlet–Dirichlet interface condition for overlapping grids is actually also a Dirichlet–Neumann condition for non-overlapping grids.

Computation (2.9) shows how its solution will change with time. To examine this, the computation is formulated as

$$V^{n+1} = AV^n \tag{2.12}$$

in which A is a matrix, and V is a solution vector. Particularly,

$$A = \left[\begin{array}{cccc|ccc} \ddots & \ddots & \ddots & & & & \\ & r_1 & s_1 & t_1 & & & \\ & & r_1 & s_1 & t_1 & & \\ \hline & & & r_2 & s_2 & t_2 & \\ & & & & r_2 & s_2 & t_2 \\ & & & & \ddots & \ddots & \ddots \end{array}\right], \quad V = \left[\begin{array}{c} \vdots \\ v_{1,I_1-2} \\ v_{1,I_1-1} \\ \hline v_{2,2} \\ v_{2,3} \\ \vdots \end{array}\right] \tag{2.13}$$

in which $r_k = \epsilon_k + \eta_k$, $s_k = 1 - 2\epsilon_k - \gamma_k$, $t_k = \epsilon_k - \eta_k$ $(k = 1, 2)$. Note that zero values are assigned to the solutions at the left and right nodes of left and right grids, respectively. The aforementioned linear system only involves the computation at the inner nodes of the two grids, and the interface conditions are already included. Additionally, the computation of (2.12) and (2.13) is equivalent to the computation on a single grid consisting of all inner nodes. Particularly, consider a single grid whose nodes consist of all nodes of the two grids, except node I_1 of the left grid and node 1 of the right grid. Then, the computation on the single grid will also be in the form of linear system (2.12) and (2.13).

Let us consider a more general situation of an iteration algorithm

$$V^{n+1} = AV^n + d$$

where n may be considered as iteration index, $A = (a_{ij})$, and $d = (d_i)$, with $i, j = 1, 2, \ldots$. Let

$$K := \max_i \left\{ \sum_j |a_{ij}| \right\} = ||A||_\infty \tag{2.14}$$

Since a sufficient condition for the convergence of the iteration is [15]

$$||A||_\infty < 1 \tag{2.15}$$

the computation converges if $K < 1$, or

$$\max_i \left\{ \sum_j |a_{ij}| \right\} < 1$$

For the computation of (2.12), the above becomes $\max_k\{|r_k| + |s_k| + |t_k|\} < 1$ $(k = 1, 2)$. All of these lead to the following conclusion:

Proposition 2.2.1 *For iteration (2.12), a sufficient condition of convergence is (2.15). Particularly, the condition is*

$$\max_k \left\{ |\epsilon_k + \eta_k| + |1 - 2\epsilon_k - \gamma_k| + |\epsilon_k - \eta_k| \right\} < 1 \tag{2.16}$$

Let V^* be the converged value of V^n as $n \to \infty$, also let $E^n = V^n - V^*$. Then, (2.12) leads to $E^n = AE^{n-1}$, and one has

$$||E^n|| = ||AE^{n-1}|| \leq ||A||||E^{n-1}|| \leq \ldots \leq ||A||^n||E^0||$$

Under condition (2.15), the above indicates that $||E^n||_\infty = 0$ when $n \to \infty$. Since $||V^n - V^*||_\infty = ||E^n||_\infty = 0$, the numerical solution becomes steady (independent of n) as $n \to \infty$, or $t \to \infty$. Note that the convergence of the computation implies its stability. According to the Lax equivalence theorem, for a consistent finite difference scheme applied to a well-posed linear initial value problem, stability is a necessary and sufficient condition for convergence [16].

2.2.2 *Implicit Scheme Solved Iteratively*

Now, let the equations be discretized by implicit scheme (2.5) on the two grids as follows (Fig. 2.1):

$$\begin{cases} \dfrac{v_{1,i}^{n+1} - v_{1,i}^n}{\Delta t} + a_1 \dfrac{v_{1,i+1}^{n+1} - v_{1,i-1}^{n+1}}{2\Delta x} \\ \quad = b_1 \dfrac{v_{1,i+1}^{n+1} - 2v_{1,i}^{n+1} + v_{1,i-1}^{n+1}}{\Delta x^2} - c_1 v_{1,i}^{n+1}, \quad i \leq I_1 - 1 \\ v_{1,I_1}^{n+1} = v_{2,2}^{n+1} \end{cases}$$
$$\begin{cases} \dfrac{v_{2,i}^{n+1} - v_{2,i}^n}{\Delta t} + a_2 \dfrac{v_{2,i+1}^{n+1} - v_{2,i-1}^{n+1}}{2\Delta x} \\ \quad = b_2 \dfrac{v_{2,i+1}^{n+1} - 2v_{2,i}^{n+1} + v_{2,i-1}^{n+1}}{\Delta x^2} - c_2 v_{2,i}^{n+1}, \quad i \geq 2 \\ v_{2,1}^{n+1} = v_{1,I_1-1}^{n+1} \end{cases} \tag{2.17}$$

Here, Dirichlet–Derichlet interface condition (2.10) for overlapping grids is applied. As indicated in the last subsection, it is also an approximation to Dirichlet–Neumann interface condition (2.11) between patched grids. The aforementioned computation follows the algorithm described in (1.11).

In algorithm (2.17), the solutions on the two grids exchange their values at the interfaces, grid nodes I_1 and 1 of subdomain 1 and 2, respectively, at the time level

$n+1$, identical to the time level of the solution at the inner nodes. Therefore, the entire algorithm is implicit at both inner and interface nodes, and it cannot be solved directly. A method of computation can be an iterative, explicit scheme:

$$
\begin{array}{l}
\text{Do } n = 0, 1, 2, \ldots, N-1 \\
\quad \text{Do } m = 0, 1, 2, \ldots \\
\quad\quad \begin{cases}
\dfrac{v_{1,i}^{n+1,m+1} - v_{1,i}^{n}}{\Delta t} + a_1 \dfrac{v_{1,i+1}^{n+1,m} - v_{1,i-1}^{n+1,m}}{2\Delta x} \\
\quad = b_1 \dfrac{v_{1,i+1}^{n+1,m} - 2v_{1,i}^{n+1,m+1} + v_{1,i-1}^{n+1,m}}{\Delta x^2} - c_1 v_{1,i}^{n+1,m+1}, \quad i \le I_1 - 1 \\
v_{1,I_1}^{n+1,m+1} = v_{2,2}^{n+1,m+1}
\end{cases} \\
\quad\quad \begin{cases}
\dfrac{v_{2,i}^{n+1,m+1} - v_{2,i}^{n}}{\Delta t} + a_2 \dfrac{v_{2,i+1}^{n+1,m} - v_{2,i-1}^{n+1,m}}{2\Delta x} \\
\quad = b_2 \dfrac{v_{2,i+1}^{n+1,m} - 2v_{2,i}^{n+1,m+1} + v_{2,i-1}^{n+1,m}}{\Delta x^2} - c_2 v_{2,i}^{n+1,m+1}, \quad i \ge 2 \\
v_{2,1}^{n+1,m+1} = v_{1,I_1-1}^{n+1,m+1}
\end{cases} \\
\quad \text{End Do} \\
\text{End Do}
\end{array}
\tag{2.18}
$$

Here, the initial condition of the problem and initial value of Schwarz iteration are omitted, but they may be referenced to those in algorithm (1.12). As previously explained, the numerical solution is set to zero at node $i = 1$ of the left grid and $i = I_2$ of the right grid. The computation is first carried out at the inner nodes, $i \le I_1 - 1$ for subdomain 1 and $i \ge 2$ for subdomain 2, then it proceeds to the interface nodes, $i = I_1$ for subdomain 1 and $i = 1$ for subdomain 2. With this order of computation, the computed values of the $(m+1)$th iteration at the inner nodes are transferred to the interface nodes as the interface conditions. At convergence, $v_{1,i}^{n+1,m+1}$ and $v_{2,i}^{n+1,m+1}$ become $v_{1,i}^{n+1}$ and $v_{2,i}^{n+1}$, respectively.

Now, let us analyze scheme (2.18). Rearranging its terms and omitting superscript $n+1$, one can express the scheme as

$$
\begin{cases}
(1 + 2\epsilon_1 + \gamma_1) v_{1,i}^{m+1} = (\eta_1 + \epsilon_1) v_{1,i-1}^{m} + (\epsilon_1 - \eta_1) v_{1,i+1}^{m} + v_{1,i}^{n}, \quad i \le I_1 - 1 \\
v_{1,I_1}^{m+1} = v_{2,2}^{m+1} \\
(1 + 2\epsilon_2 + \gamma_2) v_{2,i}^{m+1} = (\eta_2 + \epsilon_2) v_{2,i-1}^{m} + (\epsilon_2 - \eta_2) v_{2,i+1}^{m} + v_{2,i}^{n}, \quad i \ge 2 \\
v_{2,1}^{m+1} = v_{1,I_1-1}^{m+1}
\end{cases}
\tag{2.19}
$$

in which ϵ_k, η_k, and γ_k are defined as before. Furthermore, (2.19) is formulated as

$$V^{m+1} = (I - D^{-1}A)V^m + D^{-1}b \tag{2.20}$$

in which A is a matrix, V is a vector, and they are defined as in (2.13). However, in the present case, $r_k = -\epsilon_k - \eta_k$, $s_k = 1 + 2\epsilon_k + \gamma_k$, $t_k = \eta_k - \epsilon_k$ $(k = 1, 2)$. Moreover, $b = (\ldots, v^n_{1,I_1-2}, v^n_{1,I_1-1}, v^n_{2,2}, v^n_{2,3}, \ldots)^T$, I is the identity matrix, and D is the diagonal matrix with the diagonal elements of A as its diagonal elements.

Actually, scheme (2.20) is the Jacobi iteration algorithm to solve [17]

$$AV = b$$

Here, $A = (a_{ij})$, $b = (b_i)$, and $a_{ii} \neq 0$, $i, j = 1, 2, \ldots$. Scheme (2.20) is simple but permits parallel computation, and it is frequently in use for practical problems. The following condition is a sufficient condition for the convergence of (2.20):

$$\left(\sum_{j,\,\neq i} |a_{ij}|\right)/|a_{ii}| \begin{cases} \leq 1, \; i = 1, 2, \ldots \\ < 1, \; at\ least\ one\ i \end{cases} \tag{2.21}$$

Such a condition is easy to check and commonly used in practical computation, e.g., for heat and mass transfer [18, 19]. Scarborough proposed this condition [20, 21], and it is referred to as the Scarborough criterion [18, 19]. When first proposed [20, 21], the criterion was proved but not strictly. A strict proof is now available in [13]. According to Scarborough criterion (2.21), the following is concluded.

Proposition 2.2.2 *For iteration (2.19), a sufficient condition of convergence is*

$$K := \max_k \left\{ \frac{|\epsilon_k + \eta_k| + |\eta_k - \epsilon_k|}{|1 + 2\epsilon_k + \gamma_k|} \right\} < 1 \tag{2.22}$$

For Jacobi iteration (2.20), a necessary and sufficient condition for convergence is that the spectral radius, ρ, of its iteration matrix, $I - D^{-1}A$, is less than 1 [17]:

$$\rho(I - D^{-1}A) < 1 \tag{2.23}$$

Note that actually $\rho(I - D^{-1}A) \leq ||I - D^{-1}A||_\infty = K$. As a result, in comparison to condition (2.23), condition (2.22) is a less restrictive condition for convergence. However, the latter is easier to use in practical computation since the former needs computation of the spectral radius.

Example 2.2.1 A numerical experiment is made with two cases, as shown in Table 2.1. In both cases, the equations are different in the two subdomains. The length of the space domain (from the left end of subdomain 1 to the right end of subdomain 2) is 2, and $\Delta x = 2/(I - 1)$, $I = I_1 + I_2 - 2$. I_1 and I_2 are the numbers of grid nodes in subdomains 1 and 2, respectively, and here I is the total number of all grid nodes (two overlapping nodes count as one). Additionally, the

Table 2.1 Cases for numerical experiments with advection–diffusion–reaction equations

Case	a_1	b_1	c_1	a_2	b_2	c_2
1	0.5	0.002	0.0	0.25	0.001	0.0
2	0.25	0.01	0.25	−0.25	0.01	0.5

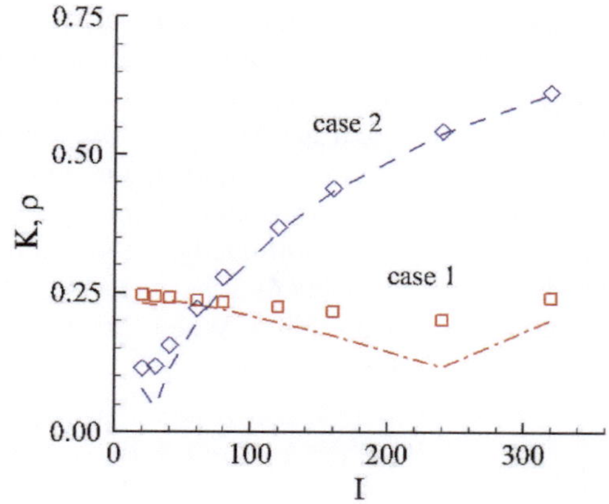

Fig. 2.2 Values of K and ρ at different mesh resolution. Line—ρ, symbol—K

time step is set as $\Delta t = 0.5\Delta x$. The experiment uses a series of meshes: $(I_1, I_2) = (10, 10), (10, 20), (20, 20), (20, 40), (40, 40), (40, 80), (80, 80), (80, 160),$ and $(160, 160)$. The computed values for K and ρ for the two cases are plotted in Fig. 2.2. It is seen that K is larger than but quite close to ρ in value, indicating that, indeed, K is a good indicator of convergence.

2.2.3 Implicit Scheme Solved Directly

Instead of iteratively, let us solve scheme (2.17) directly, e.g., by matrix inversion. The scheme can be solved directly via the conventional Schwarz method as follows:

$$
\begin{aligned}
&\text{Do } n = 0, 1, 2, \ldots, N-1\\
&\quad \text{Do } m = 0, 1, 2, \ldots\\
&\quad\left\{\begin{array}{l}
\dfrac{v_{1,i}^{n+1,m+1} - v_{1,i}^{n}}{\Delta t} + a_1 \dfrac{v_{1,i+1}^{n+1,m+1} - v_{1,i-1}^{n+1,m+1}}{2\Delta x}\\
\quad = b_1 \dfrac{v_{1,i+1}^{n+1,m+1} - 2v_{1,i}^{n+1,m+1} + v_{1,i-1}^{n+1,m+1}}{\Delta x^2} - c_1 v_{1,i}^{n+1,m+1}, \quad i \le I_1 - 1\\
v_{1,I_1}^{n+1,m+1} = v_{2,2}^{n+1,m}\\
\dfrac{v_{2,i}^{n+1,m+1} - v_{2,i}^{n}}{\Delta t} + a_2 \dfrac{v_{2,i+1}^{n+1,m+1} - v_{2,i-1}^{n+1,m+1}}{2\Delta x}\\
\quad = b_2 \dfrac{v_{2,i+1}^{n+1,m+1} - 2v_{2,i}^{n+1,m+1} + v_{2,i-1}^{n+1,m+1}}{\Delta x^2} - c_2 v_{2,i}^{n+1,m+1}, \quad i \ge 2\\
v_{2,1}^{n+1,m+1} = v_{1,I_1-1}^{n+1,m}
\end{array}\right.\\
&\quad \text{End Do}\\
&\text{End Do}
\end{aligned}
\tag{2.24}
$$

Again, the initial condition of the problem and the initial guess of Schwarz iteration for the computation may be referenced to those in algorithm (1.12). Moreover, the numerical solution is set to zero at the left and right ends of the left and right grids, respectively. It is seen that, in the computation, Dirichlet–Dirichlet interface condition (2.10) is directly implemented as an iterative algorithm. This algorithm will be referred to as a classical interface algorithm for the Dirichlet–Dirichlet interface condition. Additionally, the solution of a linear system is involved in each subdomain at every iteration m, and it may be performed via various techniques, including matrix inversion. When marching from time level n to $n+1$, one can formulate the computation as

$$\begin{cases} -(\eta_1+\epsilon_1)v_{1,i-1}^{m+1}+(1+2\epsilon_1+\gamma_1)v_{1,i}^{m+1}+(\eta_1-\epsilon_1)v_{1,i+1}^{m+1}=v_{1,i}^{n}, \ \ i\le I_1-1 \\ v_{1,I+1}^{m+1}=v_{2,2}^{m} \\ -(\eta_2+\epsilon_2)v_{2,i-1}^{m+1}+(1+2\epsilon_2-\gamma_2)v_{2,i}^{m+1}+(\eta_2-\epsilon_2)v_{2,i+1}^{m+1}=v_{2,i}^{n}, \ \ i\ge 2 \\ v_{2,1}^{m+1}=v_{1,I_1-1}^{m} \end{cases} \tag{2.25}$$

Here, superscript $n+1$ is omitted. The computation, at each iteration, updates the solution first at the interfaces, and then it solves a linear system for the solution in each subdomain.

To analyze the convergence of the computation, consider its iteration residual, i.e., $e_{1,i}^{m+1}=v_{1,i}^{m+1}-v_{1,i}^{n+1}$ and $e_{2,i}^{m}=v_{2,i}^{m}-v_{2,i}^{n+1}$. Note $v_{1,i}^{n+1}$ and $v_{2,i}^{n+1}$ are the converged values of $v_{1,i}^{m+1}$ and $v_{2,i}^{m+1}$, respectively. With these, scheme (2.25) becomes

$$\begin{cases} -(\eta_1+\epsilon_1)e_{1,i-1}^{m+1}+(1+2\epsilon_1+\gamma_1)e_{1,i}^{m+1}+(\eta_1-\epsilon_1)e_{1,i+1}^{m+1}=0, \ \ i\le I_1-1 \\ e_{1,I_1}^{m+1}=e_{2,2}^{m} \\ -(\eta_2+\epsilon_2)e_{2,i-1}^{m+1}+(1+2\epsilon_2+\gamma_2)e_{2,i}^{m+1}+(\eta_2-\epsilon_2)e_{2,i+1}^{m+1}=0, \ \ i\ge 2 \\ e_{2,1}^{m+1}=e_{1,I_1-1}^{m} \end{cases} \tag{2.26}$$

The convergence speed has been derived via the mathematical induction method and residual equation (2.26) [13], and the derivation is presented here. For the convenience of the derivation, two recursive formulas are defined:

$$\begin{aligned} R_1(i) &= 1+2\epsilon_1+\gamma_1+\frac{\eta_1^2-\epsilon_1^2}{R_1(i-1)}, \ \ i=3,\ldots,I_1-1 \\ R_2(i) &= 1+2\epsilon_2+\gamma_2+\frac{\eta_2^2-\epsilon_2^2}{R_2(i+1)}, \ \ i=I_2-2,\ldots,2 \end{aligned} \tag{2.27}$$

in which $R_1(2)=1+2\epsilon_1+\gamma_1$, $R_2(I_2-1)=1+2\epsilon_2+\gamma_2$, and $I_1, I_2\ge 3$.

The derivation starts from subdomain 1. In (2.26), when $i=2, 3$, the equation for subdomain 1 respectively becomes

$$(1 + 2\epsilon_1 + \gamma_1)e_{1,2}^{m+1} + (\eta_1 - \epsilon_1)e_{1,3}^{m+1} = 0$$
$$-(\eta_1 + \epsilon_1)e_{1,2}^{m+1} + (1 + 2\epsilon_1 + \gamma_1)e_{1,3}^{m+1} + (\eta_1 - \epsilon_1)e_{1,4}^{m+1} = 0$$

In the aforementioned, without loss of generality (e.g., u_0 in (2.1) has finite support on the x-axis, or a value for $v_{1,i}^{m+1}$ is prescribed at $i = 1$), $e_{1,1}^{m+1} = 0$ is adopted. Elimination of $e_{1,2}^{m+1}$ in the aforementioned two equations leads to

$$\left(1 + 2\epsilon_1 + \gamma_1 + \frac{\eta_1^2 - \epsilon_1^2}{R_1(2)}\right) e_{1,3}^{m+1} + (\eta_1 - \epsilon_1)e_{1,4}^{m+1} = 0$$

Then, using the aforementioned equation and the equation in (2.26) when $i = 4$ in subdomain 1, and so on, one can continue the above derivation. By deductive reasoning, it is expected that such a derivation gives rise to

$$\left(1 + 2\epsilon_1 + \gamma_1 + \frac{\eta_1^2 - \epsilon_1^2}{R_1(I_1 - 3)}\right) e_{1,I_1-2}^{m+1} + (\eta_1 - \epsilon_1)e_{1,I_1-1}^{m+1} = 0$$

At the same time, equation (2.26) at $i = I_1 - 1$ in subdomain 1 is

$$-(\eta_1 + \epsilon_1)e_{1,I_1-2}^{m+1} + (1 + 2\epsilon_1 + \gamma_1)e_{1,I_1-1}^{m+1} + (\eta_1 - \epsilon_1)e_{1,I_1}^{m+1} = 0$$

From the last two equations, also noticing that the first parenthesis in the first equation is actually $R_1(I_1 - 2)$, one arrives at

$$\left(1 + 2\epsilon_1 + \gamma_1 + \frac{\eta_1^2 - \epsilon_1^2}{R_1(I_1 - 2)}\right) e_{1,I_1-1}^{m+1} + (\eta_1 - \epsilon_1)e_{1,I_1}^{m+1} = 0$$

After applying the two interface conditions in (2.26) and changing index m, the above equation yields

$$\left(1 + 2\epsilon_1 + \gamma_1 + \frac{\eta_1^2 - \epsilon_1^2}{R_1(I_1 - 2)}\right) e_{2,1}^{m+2} + (\eta_1 - \epsilon_1)e_{2,2}^{m} = 0$$

The aforementioned derivation utilizes the equations of discretization in subdomain 1. By a similar derivation from (2.26), starting from the equation in subdomain 2 when $i = I_2 - 1$, $I_2 - 2$, one can derive

$$-(\eta_2 + \epsilon_2)e_{2,1}^{m+1} + \left(1 + 2\epsilon_2 + \gamma_2 + \frac{\eta_2^2 - \epsilon_2^2}{R_2(3)}\right) e_{2,2}^{m+1} = 0$$

From the pair of the last two equations, the elimination of $e_{2,2}^{m+1}$ yields

$$\frac{e_{2,1}^{m+2}}{e_{2,1}^{m}} = \frac{(\epsilon_1 - \eta_1)(\epsilon_2 + \eta_2)}{R_1(I_1 - 1)R_2(2)}$$

which is the contraction factor of the computation, or actually the convergence speed, at node $i = 1$ in subdomain 2.

In the aforementioned pair of the two equations, by changing the index from $m + 1$ to $m + 2$ in the second equation and then eliminating $e_{2,1}^{m+2}$, it is seen that the contraction factor is the same at $i = 2$ of subdomain 2: $e_{2,2}^{m+2}/e_{2,2}^{m} = e_{2,1}^{m+2}/e_{2,1}^{m}$. Moreover, it can be further checked by derivation that actually the contraction factor is the same at all nodes of subdomain 2:

$$e_{2,1}^{m+2}/e_{2,1}^{m} = e_{2,2}^{m+2}/e_{2,2}^{m} = e_{2,3}^{m+2}/e_{2,3}^{m} = \cdots$$

In the same way, one can prove the aforementioned for subdomain 1 and derive the same contraction factor at all of its nodes. All of these lead to the following conclusion.

Theorem 2.2.1 *The following recursive expression is the contraction factor, ρ, of the computation by conventional method (2.26) associated with the classical interface algorithm:*

$$\rho = \frac{(\epsilon_1 - \eta_1)(\epsilon_2 + \eta_2)}{R_1(I_1 - 1)R_2(2)} \tag{2.28}$$

where $\rho = e_{1,i}^{m+2}/e_{1,i}^{m}$, $e_{2,i}^{m+2}/e_{2,i}^{m}$, for all i.

Remark This theorem indicates that the contraction factor of the computation is the same, regardless of grid nodes and subdomains. Because of this, it is readily derived that such a factor is global, i.e., $\rho = ||e_{1,i}^{m+2}||_2/||e_{1,i}^{m}||_2 = ||e_{2,i}^{m+1}||_2/||e_{2,i}^{m}||_2$. Note that, unlike in the last subsection, here and hereafter, ρ represents a contraction factor.

Example 2.2.2 For its validation, the value of the theoretically derived contraction factor in (2.28) is calculated for the two cases in Table 2.1. Additionally, the two cases with the following initial and boundary conditions are solved numerically by algorithm (2.24):

$$u_0 = -sin(\pi x), \ -1 < x < 1; \ u_1 = 0, x = -1; \ u_2 = 0, \ x = 1$$

Then, the value for the contraction factor is directly calculated from the numerical solutions according to $e_{1,i}^{m+2}/e_{1,i}^{m}$, or, $e_{2,i}^{m+2}/e_{2,i}^{m}$, for any i. In this computation, and in the subsequent ones in this section, the grids overlap at two nodes (Fig. 2.1), and the grid spacing is $\Delta x = 2/(I_1 + I_2 - 3) = 0.0171$). The interfaces are at $x_1 = -1/3$, and $x_2 = -1/3 - \Delta x$. Additionally, double-precision computer arithmetic is used to ensure sufficient solution accuracy. The theoretical and numerical values for the

Table 2.2 Theoretical and numerical contraction factors. $(I_1, I_2) = (40, 80)$, $\Delta t, \Delta x = 0.0171$

Case	ρ_{theo}	ρ_{num}
1	−0.01701	−0.01701
2	0.08296	0.08296

contraction factor are listed in Table 2.2. The table shows that the theoretical and numerical factors are basically identical, validating formula (2.28).

Now, let us extend the foregoing analysis to nonlinear problem (2.2). In reference to algorithm (2.24), the computation of the nonlinear problem by the conventional Schwarz method is formulated as

$$\begin{aligned}
&\text{Do } n = 0, 1, 2, \ \dots, \ N-1\\
&\quad \text{Do } m = 0, 1, 2, \ \dots\\
&\quad \begin{cases}
\dfrac{v_{1,i}^{n+1,m+1} - v_{1,i}^{n}}{\Delta t} + a_1(v_{1,i}^{n+1,m})\dfrac{v_{1,i+1}^{n+1,m+1} - v_{1,i-1}^{n+1,m+1}}{2\Delta x}\\
\quad = b_1\dfrac{v_{1,i+1}^{n+1,m+1} - 2v_{1,i}^{n+1,m+1} + v_{1,i-1}^{n+1,m+1}}{\Delta x^2} - c_1(v_{1,i}^{n+1,m})v_{1,i}^{n+1,m+1}, \quad i \le I_1 - 1\\
v_{1,I_1}^{n+1,m+1} = v_{2,2}^{n+1,m}\\
\dfrac{v_{2,i}^{n+1,m+1} - v_{2,i}^{n}}{\Delta t} + a_2(v_{2,i}^{n+1,m})\dfrac{v_{2,i+1}^{n+1,m+1} - v_{2,i-1}^{n+1,m+1}}{2\Delta x}\\
\quad = b_2\dfrac{v_{2,i+1}^{n+1,m+1} - 2v_{2,i}^{n+1,m+1} + v_{2,i-1}^{n+1,m+1}}{\Delta x^2} - c_2(v_{2,i}^{n+1,m})v_{2,i}^{n+1,m+1}, \quad i \ge 2\\
v_{2,1}^{n+1,m+1} = v_{1,I_1-1}^{n+1,m}
\end{cases}\\
&\quad \text{End Do}\\
&\text{End Do}
\end{aligned} \tag{2.29}$$

The computation remains the same as that for (2.24), and the solution of a linear system is involved in each subdomain at each iteration m.

To analyze algorithm (2.29), let us linearize it by freezing its $a_k(v_{k,i}^{n+1,m})$ and $c_k(v_{k,i}^{n+1,m})$ as $a_k(v_{k,i}^n)$ and $c_k(v_{k,i}^n)$, respectively. Then, with the same procedure, one can analyze the computation of the nonlinear problem. The analysis will lead to a contraction factor in the form of (2.28), together with ϵ_k, η_k, and γ_k being replaced by $\epsilon_{k,i}^n$, $\eta_{k,i}^n$, and $\gamma_{k,i}^n$ as follows ($k = 1, 2$)

$$\begin{aligned}
\epsilon_{k,i}^n &= b_k\frac{\Delta t}{\Delta x^2}\\
\eta_{k,i}^n &= a_k(v_{k,i}^n)\frac{\Delta t}{2\Delta x}\\
\gamma_{k,i}^n &= c_k(v_{k,i}^n)\Delta t
\end{aligned}$$

Example 2.2.3 A numerical experiment is made with two nonlinear cases as shown in Table 2.3. The nonlinearity comes from the advection and reaction terms in Case

Table 2.3 Cases for numerical experiments on nonlinear advection–diffusion–reaction equations

Case	a_1	b_1	c_1	a_2	b_2	c_2
3	u_1	0.02	$2u_1$	u_2	0.02	u_2
4	u_1	0.1	0.2	u_2	0.2	0.1

Table 2.4 Theoretical and numerical contraction factors for nonlinear equations. $(I_1, I_2) = (40, 80)$, Δt, $\Delta x = 0.0171$

t	0.085		0.170		0.255	
Case	ρ_{theo}	ρ_{num}	ρ_{theo}	ρ_{num}	ρ_{theo}	ρ_{num}
3	0.13478	0.13468	0.14108	0.14107	0.14683	0.14685
4	0.45712	0.47331	0.46213	0.47895	0.46688	0.48323

3, and it results from the advection term in Case 4. For the computation of Case 3, the initial and boundary values are

$$u_0 = 1,\ -1 < x < 0;\ u_0 = 0,\ 0 < x < 1;\ u_1 = 1,\ x = -1;\ u_2 = 0,\ x = 1$$

For Case 4, they are

$$u_0 = -sin(\pi x),\ -1 < x < 1;\ u_1 = 0,\ x = -1;\ u_2 = 0,\ x = 1$$

In these cases, the contraction factor is related to $\eta_{k,i}^n$ and $\gamma_{k,j}^n$, and thus, it changes with time. The values of the contraction factor, at three time moments, calculated from formula (2.28) and from the numerical solutions are presented in Table 2.4. The table shows that the theoretical and numerical values are close, indicating the validity of the above analysis based on linearization.

2.2.4 *Optimized Interface Algorithm*

A central issue in the multiblock computation is to speed up the convergence of Schwarz iteration if it is involved, and an approach is to adopt optimized interface algorithms. The classical interface algorithm is adopted to enforce the Dirichlet–Dirichlet interface condition in the computation of (2.24). To speed up convergence in the computation of (2.24), the following is an optimized interface algorithm to enforce the Dirichlet–Dirichlet interface condition (Fig. 2.1):

$$\begin{aligned}(v_{1,I_1}^{m+1} - v_{1,I_1-1}^{m+1}) + \alpha v_{1,I_1}^{m+1} &= (v_{2,2}^{m} - v_{2,1}^{m}) + \alpha v_{2,2}^{m} \\ (v_{2,2}^{m+1} - v_{2,1}^{m+1}) + \beta v_{2,1}^{m+1} &= (v_{1,I_1}^{m} - v_{1,I_1-1}^{m}) + \beta v_{1,I_1-1}^{m}\end{aligned} \tag{2.30}$$

in which $\alpha, \beta = const$. Such an optimal interface algorithm is first proposed for waveform relaxation of circuit problems by Gander and coworkers (e.g., [22]), and it effectively speeds up the convergence. They show that under the condition

$$(\alpha + 1)(\beta - 1) + 1 \neq 0 \tag{2.31}$$

the optimized interface algorithm recovers the Dirichlet–Dirichlet interface condition, (2.10), at convergence as $m \to \infty$. Actually, at the convergence, the optimal interface algorithm leads to

$$\begin{aligned}(1+\alpha)(v_{1,I_1}^{n+1} - v_{2,2}^{n+1}) &= -(v_{2,1}^{n+1} - v_{1,I_1-1}^{n+1})\\ (\beta - 1)(v_{2,1}^{n+1} - v_{1,I_1-1}^{n+1}) &= v_{1,I_1}^{n+1} - v_{2,2}^{n+1}\end{aligned}$$

which gives rise to

$$((1+\alpha)(\beta-1)+1)\,(v_{2,1}^{n+1} - v_{1,I_1-1}^{n+1}) = 0$$

Under condition (2.31), the above results in $v_{2,1}^{n+1} = v_{1,I_1-1}^{n+1}$, the second interface condition in (2.24). Similarly to the above, $v_{1,I_1}^{n+1} = v_{2,2}^{n+1}$, which is the first interface condition in (2.24). By the way, it is readily seen that the optimized interface algorithm returns to the classical interface algorithm in (2.24), as $\alpha, \beta \to \infty$.

Now, let us analyze the convergence of the computation. With the optimized interface algorithm, (2.26) is modified as

$$\begin{cases} -(\eta_1+\epsilon_1)e_{1,i-1}^{m+1} + (1+2\epsilon_1+\gamma_1)e_{1,i}^{m+1} + (\eta_1-\epsilon_1)e_{1,i+1}^{m+1} = 0,\ i \le I_1 - 1\\ (e_{1,I_1}^{m+1} - e_{1\,I_1-1}^{m+1}) + \alpha e_{1,I_1}^{m+1} = (e_{2,2}^{m} - e_{2,1}^{m}) + \alpha e_{2,2}^{m}\\ -(\eta_2+\epsilon_2)e_{2,i-1}^{m+1} + (1+2\epsilon_2+\gamma_2)e_{2,i}^{m+1} + (\eta_2-\epsilon_2)e_{2,i+1}^{m+1} = 0,\ i \ge 2\\ (e_{2,2}^{m+1} - e_{2,1}^{m+1}) + \beta e_{2,1}^{m+1} = (e_{1,I_1}^{m} - e_{1,I_1-1}^{m}) + \beta e_{1,I_1-1}^{m} \end{cases} \tag{2.32}$$

For this scheme, by the same procedure in the last subsection, one can derive the contraction factor and optimal values for (α, β) [13]. These are summarized as follows.

Theorem 2.2.2 *The contraction factor, ρ, for the computation by the conventional method associated with optimized interface algorithm (2.32) is*

$$\rho = -\frac{(\alpha+1)(\eta_2+\epsilon_2) - R_2(2)}{(\alpha+1)R_1(I_1-1) + \eta_1 - \epsilon_1} \cdot \frac{(\beta-1)(\eta_1-\epsilon_1) - R_1(I_1-1)}{(\beta-1)R_2(2) + \eta_2 + \epsilon_2} \tag{2.33}$$

A set of optimal values for (α, β) is

$$\alpha^* = \frac{R_2(2)}{\eta_2+\epsilon_2} - 1,\quad \beta^* = \frac{R_1(I_1-1)}{\eta_1-\epsilon_1} + 1 \tag{2.34}$$

which makes the contraction factor zero. The contraction factor has a saddle point at (α^*, β^*).

Table 2.5 Theoretical and numerical contraction factors and optimal values (α^*, β^*). $(I_1, I_2) = (40, 80)$, Δt, $\Delta x = 0.0171$

Case	α^* β^*	ρ_{theo}	ρ_{num}
1	5.1462, 10.5666	0	-6.12×10^{-15}
2	1.7496, −3.3839	0	0

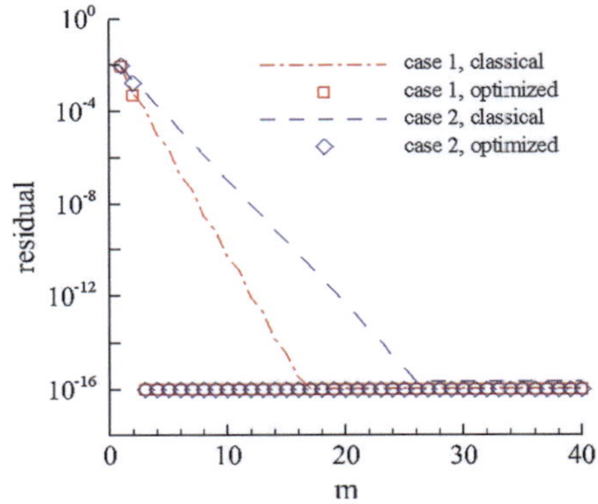

Fig. 2.3 Convergence associated with the classical and optimized interface algorithms. The residual is e^m_{1,I_1}, $e^m_{2,1}$

It is directly seen from (2.33) that the contraction factor becomes zero when either α takes the value in (2.34), i.e., $\alpha = \alpha^*$, regardless that of β, or when β takes the value in (2.34), i.e., $\beta = \beta^*$, regardless that of α. Since the contraction factor is the ratio of residuals of two consecutive times of iterations, i.e., $\rho = e^2_{1,i}/e^0_{1,i}$, $\rho = e^2_{2,i}/e^0_{2,i}$, its zero value implies that convergence is achieved within two iterations, referred to as "perfect convergence".

Example 2.2.4 This numerical example illustrates the performance of the optimized interface algorithm. Particularly, numerical solutions are made for the two cases in Table 2.1 and the initial and boundary conditions in the last subsection. The computation is formulated as (2.24) but with interface algorithms (2.30) and (2.34). The interfaces, grids, and time step are set as before. The results are shown in Table 2.5. The table shows that, at optimal values (α^*, β^*), the values for the numerical contraction factors are very small or essentially zeros in magnitude, which are consistent with the theoretical values.

To directly illustrate the effectiveness of the optimized interface algorithm, the convergence of the actual computation of the two cases is shown in Fig. 2.3. It is seen that in comparison to the classical algorithm, the optimal algorithm substantially speeds up the convergence. For the two cases, it does lead to the "perfect convergence," achieving convergence in about two iterations.

It is also interesting to find out the performance of optimized interface algorithm (2.30) for speeding up the computation of the nonlinear equations. In application of the algorithm, the optimal values (α^*, β^*) in (2.34) will be computed by replacing its ϵ_k, η_k, and γ_k by $\epsilon^n_{k,i}$, $\eta^n_{k,i}$, and $\gamma^n_{k,i}$, respectively, as defined before.

Table 2.6 Theoretical and numerical contraction factors and optimal values (α^*, β^*) for nonlinear equations. $(I_1, I_2) = (40, 80)$, Δt, $\Delta x = 0.0171$

Case	t	α^*	β^*	ρ_{theo}	ρ_{num}
3	0.085	8.45139×10^{-1}	-3.11444×10^{0}	0	0.00×10^{0}
	0.170	8.44148×10^{-1}	-3.10923×10^{0}	0	5.25×10^{-14}
	0.255	8.42956×10^{-1}	-3.10358×10^{0}	0	0.00×10^{0}
4	0.085	7.58689×10^{-1}	-4.29258×10^{-1}	0	9.44×10^{-2}
	0.170	7.62619×10^{-1}	-1.64360×10^{-1}	0	7.72×10^{-2}
	0.255	7.67611×10^{-1}	-9.94238×10^{-1}	0	6.31×10^{-2}

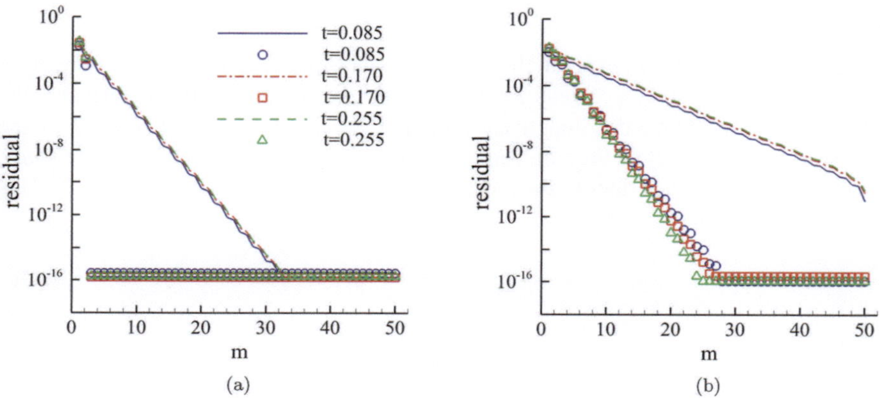

Fig. 2.4 Convergence associated with the classical and optimized interface algorithms for nonlinear equations. Line—classical interface algorithm, symbol—optimized interface algorithm. The residual is e^m_{1,I_1}, $e^m_{2,1}$. **(a)** Case 3. **(b)** Case 4

Example 2.2.5 Numerical results for the two nonlinear cases in Table 2.3 are presented in Table 2.6. In the computation, the optimal values (α^*, β^*) are calculated at each time step since they are solution dependent. The table shows the results at three time instants. It shows that when the optimized interface algorithm is applied, the actual contraction factor in Case 3 is almost zero or the theoretical value. It is not zero, but still small, in Case 4. These indicate that the algorithm is effective in speeding up the convergence.

The convergence history is plotted in Fig. 2.4 to illustrate the speedup clearly. The figure shows that the computation of Case 3 again exhibits the "perfect convergence"; that is, it converges within about two iterations. In Case 4, convergence speed slows down, and the computation with the optimal interface algorithm consumes many iterations. However, its associated convergence is still much faster than that associated with the classical interface algorithm.

2.3 Waveform Relaxation Method

2.3.1 Convergence Analysis

Solution Formulation Now, let us analyze the computation of coupled advection–diffusion–reaction equations by the waveform relaxation method, particularly (2.6). The analysis will be made through semi-discretization (2.7). Instead of directly on the variable $v_{k,i}(t)$, the analysis is made on its Laplace transform:

$$\hat{v}_{k,i} = \int_0^\infty e^{-st} v_{k,i}(t)dt \tag{2.35}$$

in which the hat "∧" stands for the Laplace transform, and $s = \sigma + \mathbf{i}\omega$, being a complex number. σ (≥ 0) and ω are the real and imaginary parts, respectively, and $\mathbf{i} = \sqrt{-1}$, being the imaginary unit. For simplicity, let $v_{k,i}(0) = 0$ in the transform. As a result of the Laplace transform, the analysis on (2.7) will be made in the s-domain rather than in the t-domain. By the Laplace transform, Eq. (2.7) is transformed as

$$\epsilon_k \hat{v}_{k,i-1} + (\eta_k - s)\hat{v}_{k,i} + \gamma_k \hat{v}_{k,i+1} = 0 \tag{2.36}$$

in which

$$\begin{aligned}
\epsilon_k &= \frac{a_k}{2\Delta x} + \frac{b_k}{\Delta x^2} \\
\eta_k &= -\frac{2b_k}{\Delta x^2} - c_k \\
\gamma_k &= -\frac{a_k}{2\Delta x} + \frac{b_k}{\Delta x^2}
\end{aligned}$$

Note here ϵ_k, η_k, and γ_k are defined differently from those in the last section, right after (2.9). In this analysis, assume $a_k \neq 0$, b_k, $c_k > 0$, which lead to $\epsilon_k > 0$ (true as long as Δx is sufficiently small), $\eta_k < 0$. Additionally, $\gamma_k \neq 0$, unless otherwise specified.

Equation (2.36) comprises a linear system when i takes different values ($-\infty < i < \infty$), and it has a solution in the form of [15]:

$$\hat{v}_{k,i} = Ar_-{}^i + Br_+{}^i \tag{2.37}$$

as long as

$$|r_-| < 1, \qquad |r_+| > 1 \tag{2.38}$$

In the aforementioned, A, $B = const.$, and

$$r_{\pm} = \frac{s - \eta_k \pm \sqrt{(s - \eta_k)^2 - 4\epsilon_k \gamma_k}}{2\gamma_k} \tag{2.39}$$

Here, $r_{\pm}$ are the roots for the characteristic equation of Eq. (2.36):

$$\gamma_k r^2 + (\eta_k - s) r + \epsilon_k = 0$$

From (2.39), it is readily to check that $r_- r_+ = \epsilon_k / \gamma_k$, and

$$\begin{aligned} & \epsilon_k {r_+}^{-1} + \eta_k - s \\ = & \gamma_k r_- + \eta_k - s \\ = & \gamma_k \frac{s - \eta_k - \sqrt{(s - \eta_k)^2 - 4\epsilon_k \gamma_k}}{2\gamma_k} + \eta_k - s \\ = & -\gamma_k r_+ \\ = & -\epsilon_k {r_-}^{-1} \end{aligned} \tag{2.40}$$

Condition (2.38) is the base of the analysis for the waveform relaxation method in this section. This condition is guaranteed as in the following proposition, which has been proved along with a discussion of its applicable scenarios in [14].

Proposition 2.3.1 *The roots in (2.39) are analytical. Furthermore, $|r_+| > 1$, and, under condition*

$$\left| \mathbf{i}\omega - \eta_k + \sqrt{(\mathbf{i}\omega - \eta_k)^2 - 4\epsilon_k \gamma_k} \right| > 2\epsilon_k$$

$|r_-| < 1$ holds.

Convergence Speed Now, consider the computation of the waveform relaxation method in two subdomains, particularly (2.6). For convenience, the grid for the subdomain 1 is indexed with ($i = -I_1, \ldots - 2, -1, 0$), and that for the subdomain 2 is indexed with ($i = 0, 1, 2, \ldots, I_2$), see Fig. 2.5. By discretization (2.7), the

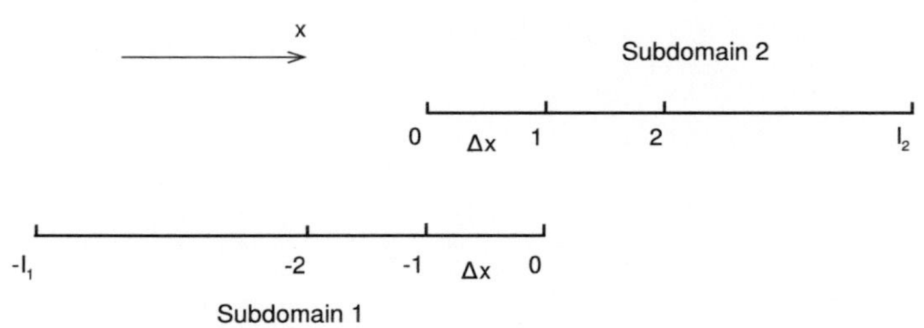

Fig. 2.5 Two grids overlap at two nodes

Schwarz waveform relaxation method becomes ($m = 0, 1, 2, \ldots$):

$$\begin{cases} dv_{1,i}^{m+1}/dt = \epsilon_1 v_{1,i-1}^{m+1} + \eta_1 v_{1,i}^{m+1} + \gamma_1 v_{1,i+1}^{m+1}, & i \le -1 \\ v_{1,0}^{m+1} = v_{2,1}^{m} \end{cases}$$
$$\begin{cases} dv_{2,i}^{m+1}/dt = \epsilon_2 v_{2,i-1}^{m+1} + \eta_2 v_{2,i}^{m+1} + \gamma_2 v_{2,i+1}^{m+1}, & i \ge 1 \\ v_{2,0}^{m+1} = v_{1,-1}^{m} \end{cases} \tag{2.41}$$

in which $t \in [0, T]$. Node 0 of subdomain 1 and node 0 of subdomain 2 are the interface nodes, corresponding to the points x_1 and x_2, respectively, in (2.6). The classical interface algorithm is adopted to enforce the Dirichlet-Dirichlet interface condition at the interfaces. Additionally, in view of their finite support, zero is assumed as the default value for the solution at $i = -I_1$ and $i = I_2$ of the left and right grids, respectively.

In (2.41), eliminating $v_{1,0}^{m+1}$ and $v_{2,0}^{m+1}$ according to the interface condition, and then taking the Laplace transform, in reference of (2.36), one obtains two linear algebraic systems:

$$\begin{bmatrix} \ddots & \vdots & \vdots & \vdots & \vdots \\ \cdots & \epsilon_1 & \eta_1 - s & \gamma_1 & 0 \\ \cdots & 0 & \epsilon_1 & \eta_1 - s & \gamma_1 \\ \cdots & 0 & 0 & \epsilon_1 & \eta_1 - s \end{bmatrix} \begin{bmatrix} \vdots \\ \hat{v}_{1,-3}^{m+1} \\ \hat{v}_{1,-2}^{m+1} \\ \hat{v}_{1,-1}^{m+1} \end{bmatrix} = \begin{bmatrix} \vdots \\ \vdots \\ 0 \\ -\gamma_1 \hat{v}_{2,1}^{m} \end{bmatrix}$$
$$\begin{bmatrix} \eta_2 - s & \gamma_2 & 0 & 0 & \cdots \\ \epsilon_2 & \eta_2 - s & \gamma_2 & 0 & \cdots \\ 0 & \epsilon_2 & \eta_2 - s & \gamma_2 & \cdots \\ \vdots & \vdots & \vdots & \vdots & \ddots \end{bmatrix} \begin{bmatrix} \hat{v}_{2,1}^{m+1} \\ \hat{v}_{2,2}^{m+1} \\ \hat{v}_{2,3}^{m+1} \\ \vdots \end{bmatrix} = \begin{bmatrix} -\epsilon_2 \hat{v}_{1,-1}^{m} \\ 0 \\ \vdots \\ \vdots \end{bmatrix} \tag{2.42}$$

The solutions to the systems are in the form of (2.37). For the first system, the magnitude of ${r_-}^i$ goes to infinity as i tends to $-\infty$ since $|r_-| < 1$. Therefore, in the solution of the first system, the coefficient A has to be zero. Similarly, for the solution to the second system, B becomes zero. Therefore, one has

$$\begin{aligned} \hat{v}_{1,i}^{m+1} &= B^{m+1} {r_+}^i, \quad i = -1, -2, \ldots \\ \hat{v}_{2,i}^{m+1} &= A^{m+1} {r_-}^i, \quad i = 1, 2, \ldots \end{aligned} \tag{2.43}$$

in which

$$r_- = \frac{s - \eta_2 \pm \sqrt{(s - \eta_2)^2 - 4\epsilon_2\gamma_2}}{2\gamma_2}$$
$$r_+ = \frac{s - \eta_1 \pm \sqrt{(s - \eta_1)^2 - 4\epsilon_1\gamma_1}}{2\gamma_1}$$

That is, r_- and r_+ are evaluated by $(\epsilon_2, \eta_2, \gamma_2)$ and $(\epsilon_1, \eta_1, \gamma_1)$, respectively.

Now, let us analyze the convergence of the Schwarz iteration in (2.42) in terms of m. In (2.42), the last equation in the first system leads to

$$\epsilon_1 \hat{v}_{1,-2}^{m+1} + (\eta_1 - s)\hat{v}_{1,-1}^{m+1} + \gamma_1 \hat{v}_{2,1}^{m} = 0$$

Plugging (2.43) into the aforementioned equation yields

$$B^{m+1} = -\frac{\gamma_1 r_-}{\epsilon_1 {r_+}^{-2} + (\eta_1 - s){r_+}^{-1}} A^m$$

Similarly, one has

$$A^{m+1} = -\frac{\epsilon_2 {r_+}^{-1}}{\gamma_2 {r_-}^2 + (\eta_2 - s)r_-} B^m$$

Given the aforementioned two equations, (2.43) yields

$$\begin{aligned} \hat{v}_{1,i}^{m+1} &= \rho \hat{v}_{1,i}^{m-1}, \quad i = 0, -1, -2, \ldots \\ \hat{v}_{2,i}^{m+1} &= \rho \hat{v}_{2,i}^{m-1}, \quad i = 0, 1, 2, \ldots \end{aligned}$$

in which

$$\rho = \frac{\epsilon_2 \gamma_1}{(\gamma_2 r_- + \eta_2 - s)(\epsilon_1 {r_+}^{-1} + \eta_1 - s)}$$

Here, ρ is the contraction factor of the computation, and it reflects the convergence speed. Note that, due to the linearity of (2.41), the above contraction factor is actually the contraction factor of the difference of $\hat{v}_{k,i}^m$ $(k = 1, 2)$ to its converged value. This difference tends to zero as the iteration converges. When $|\rho| < 1$, waveform relaxation method (2.41) converges in the s-space. In view of (2.40), the contraction factor becomes

$$\begin{aligned} \rho &= \frac{\epsilon_2 \gamma_1}{(-\epsilon_2 {r_-}^{-1})(-\gamma_1 r_+)} \\ &= \frac{r_-}{r_+} \end{aligned} \tag{2.44}$$

which is less than 1 in magnitude. Moreover, the derivative of the contraction factor over s is

$$\rho' = \frac{{r_-}' r_+ + r_- {r_+}'}{{r_+}^2}$$

According to Proposition 2.3.1, $r_\pm$ are analytical and thus differentiable. Additionally, the denominator ${r_+}^2$ cannot be zero since $|r_+| > 1$. As a result, the contraction factor is differentiable.

Theorem 2.3.1 *For the computation of the waveform relaxation method associated with classical interface algorithm (2.41), the contraction factor is* $\rho = r_-/r_+$*, and it is analytical. Moreover,* $|\rho| < 1$*, that is, the computation converges.*

Remark 3.1 Since ρ is analytical, it achieves minimums and maximums only on the boundary of the s-domain. The s-domain is the left plane, and its boundary is along $\sigma = 0$, or $s = \omega\mathbf{i}$.

Remark 3.2 The aforementioned theorem includes different scenarios in terms of the equations. Particularly, it contains the coupling of two identical equations (e.g., two identical advection–diffusion equations), two equations of the same type but with different coefficients (e.g., two diffusion–reaction equations with different coefficients), and two equations of different types (e.g., an advection equation and an advection–diffusion equation). This theorem indicates that, once under its condition, the computation of all of these scenarios will converge.

Remark 3.3 In an early study on the RC-type circuits, which are governed by systems of ordinary differential equations, the conclusions in Theorem 2.3.1 have been reported [22]. The early study corresponds to a special situation with $\epsilon_1 = \epsilon_2 = \gamma_1 = \gamma_2$, $\eta_1 = \eta_2$ of this section. Thus, the theorem is an extension of the previous work.

Example 2.3.1 Let us compute the contraction factor associated with the classical interface algorithm, (2.44), in the two cases as shown in Table 2.7. The contraction factor is plotted in Fig. 2.6. As indicated in Theorem 2.3.1, ρ is analytical, so its maximums/minimums only happen at the boundary of the s-domain, and they are along $\sigma = 0$. This is seen in the two cases in the figure. Additionally, it is seen that the maximums and minimums for Cases 1 and 2 are smaller than 1 in magnitude, verifying the theorem.

To actually compute the semi-discretized systems in (2.41), they must be discretized in time. With the discretization using the backward Euler method, the overall scheme for the computation of the Schwarz waveform relaxation method becomes

Table 2.7 Cases for numerical experiments on the waveform relaxation method for advection–diffusion–reaction equations

Case	a_1	b_1	c_1	a_2	b_2	c_2
1	1.5	0.1	0	1	0.1	0
2	0.2	0.4	1	0.4	0.2	2

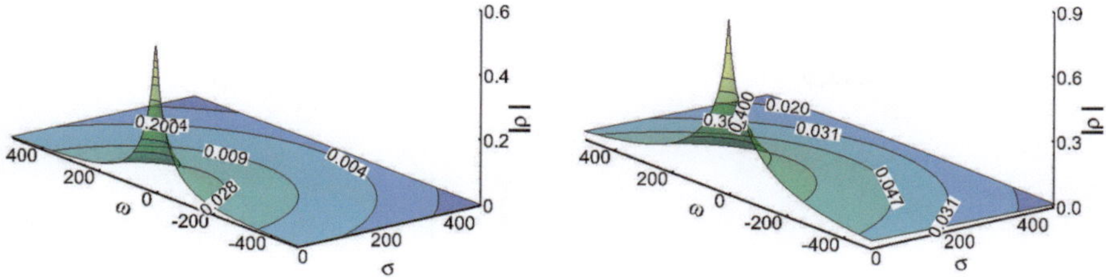

(a) (b)

Fig. 2.6 The s-domain distribution of the contraction factor associated with the classical interface algorithm. $\Delta x = 0.05$. (**a**) Case 1. The maximum of $|\rho|$ is 0.455. (**b**) Case 2. The maximum of $|\rho|$ is 0.812

$$
\begin{array}{l}
\text{Do } m = 0, 1, 2, \ldots \\
\quad \text{Do } n = 0, 1, 2, \ldots, N-1 \\
\quad \left\{
\begin{array}{l}
\dfrac{v_{1,i}^{n+1,m+1} - v_{1,i}^{n}}{\Delta t} + a_1 \dfrac{v_{1,i+1}^{n+1,m+1} - v_{1,i-1}^{n+1,m+1}}{2\Delta x} \\
\quad = b_1 \dfrac{v_{1,i+1}^{n+1,m+1} - 2v_{1,i}^{n+1,m+1} + v_{1,i-1}^{n+1,m+1}}{\Delta x^2} - c_k v_{1,i}^{n+1,m+1}, \quad i \le I_1 - 1 \\
v_{1,0}^{n+1,m+1} = v_{2,1}^{n+1,m} \\
\dfrac{v_{2,i}^{n+1,m+1} - v_{2,i}^{n}}{\Delta t} + a_2 \dfrac{v_{2,i+1}^{n+1,m+1} - v_{2,i-1}^{n+1,m+1}}{2\Delta x} \\
\quad = b_2 \dfrac{v_{2,i+1}^{n+1,m+1} - 2v_{2,i}^{n+1,m+1} + v_{2,i-1}^{n+1,m+1}}{\Delta x^2} - c_2 v_{2,i}^{n+1,m+1}, \quad i \ge 2 \\
v_{2,0}^{n+1,m+1} = v_{1,-1}^{n+1,m}
\end{array}
\right. \\
\quad \text{End Do} \\
\text{End Do}
\end{array}
\tag{2.45}
$$

In the aforementioned, the initial condition of the problem and the initial guesses of the solution at the interfaces are omitted for the computation, but they can be referred to those in algorithm (1.14). Zero is assigned to the solution at $i = -I_1$ and $i = I_2$ of the left and right grids, respectively. It can be seen that the above scheme involves solving a system of equations in each subdomain at every physical time step. A computational example is presented as follows.

Example 2.3.2 This example computes the solutions for the above two cases with a boundary condition. The space and time domains are $x \in [-1, 1]$, $t \in [0, 5]$. The initial and boundary conditions are

$$u_0 = -sin(\pi x), \; -1 < x < 1; \; u_1 = 0, \; x = -1; \; u_2 = 0, \; x = 1$$

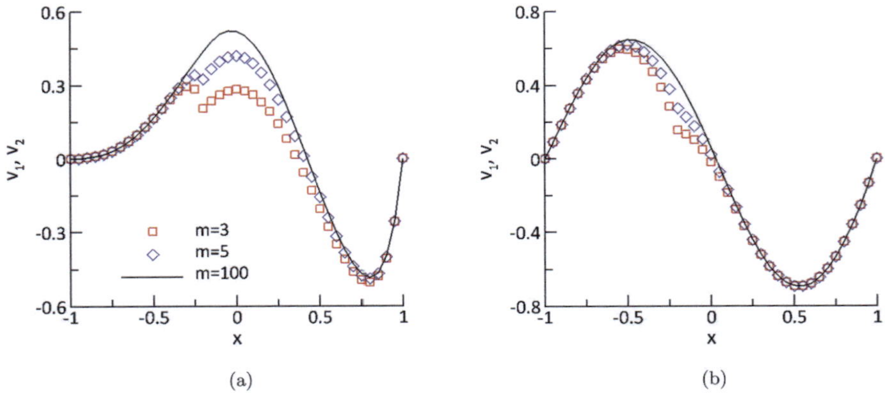

Fig. 2.7 Solutions at different Schwarz iterations associated with the classical interface algorithm. (**a**) Case 1, $t = 0.5$. (**b**) Case 2, $t = 0.15$

The left subdomain covers $-1 \leq x \leq -0.15$, and the right subdomain occupies $-0.2 \leq x \leq 1$. The two subdomains have interfaces at $x_1 = -0.15$ and $x_2 = -0.2$. The computation is conducted via (2.45), with grid spacings $\Delta x_1, \Delta x_2 = 0.05$, and time step $\Delta t = 0.05$. In the computation, initial guesses at the interfaces, i.e., $v_{1,0}^{n,0}$ and $v_{2,0}^{n,0}$, are needed to initiate the Schwarz iteration. For simplification, these initial guesses are set as zero, that is, $v_{1,0}^{n,0}, v_{2,0}^{n,0} = 0$, for all n. Note that, in correspondence to (2.6), these are equivalent to that $u_0 = -sin(\pi x)$, $g_1 = 0$, and $g_2 = 0$ are adopted in the computation.

Figure 2.7 shows the solutions at different iterations and at convergence, illustrating their convergence processes. In the figure, solutions at $m = 100$ are considered to be converged solutions (with evidence in Fig. 2.10).

2.3.2 *Convergence Speedup*

Algorithm (2.30) leads to an optimized interface algorithm to speed up the convergence of the computation in (2.41), and it is in the following form:

$$\begin{aligned} (v_{1,0}^{m+1} - v_{1,-1}^{m+1}) + \alpha v_{1,0}^{m+1} &= (v_{2,1}^{m} - v_{2,0}^{m}) + \alpha v_{2,1}^{m} \\ (v_{2,1}^{m+1} - v_{2,0}^{m+1}) + \beta v_{2,0}^{m+1} &= (v_{1,0}^{m} - v_{1,-1}^{m}) + \beta v_{1,-1}^{m} \end{aligned} \tag{2.46}$$

Again, as long as (2.31) holds, this algorithm recovers Dirichlet-Dirichlet interface condition at the convergence of the Schwarz iteration. By adoption of the above optimized interface algorithm and the discretization in (2.41), one has the following computation:

$$\begin{cases} dv_{1,i}^{m+1}/dt = \epsilon_1 v_{1,i-1}^{m+1} + \eta_1 v_{1,i}^{m+1} + \gamma_1 v_{1,i+1}^{m+1}, \; i \le -1 \\ v_{1,0}^{m+1} = v_{1,-1}^{m+1}/(1+\alpha) - v_{2,0}^{m}/(1+\alpha) + v_{2,1}^{m} \\ dv_{2,i}^{m+1}/dt = \epsilon_2 v_{2,i-1}^{m+1} + \eta_2 v_{2,i}^{m+1} + \gamma_2 v_{2,i+1}^{m+1}, \; i \ge 1 \\ v_{2,0}^{m+1} = v_{2,1}^{m+1}/(1-\beta) - v_{1,0}^{m}/(1-\beta) + v_{1,-1}^{m} \end{cases} \tag{2.47}$$

As seen in the aforementioned, the optimized interface algorithm becomes the classical interface algorithm, i.e., in (2.45), when α, β tend to infinity.

Similarly, as in the derivation of (2.42), $v_{1,0}^{m+1}$ and $v_{2,0}^{m+1}$ are eliminated in (2.47) by plugging its interface conditions into its discretization of the two subdomains. Then, by the Laplace transform, one will arrive at a linear system similar to (2.42). With the same derivation, one obtains the contraction factor with the optimized interface algorithm as

$$\rho = \frac{1+\alpha-r_-{}^{-1}}{1+\alpha-r_+{}^{-1}} \cdot \frac{1-\beta-r_+}{1-\beta-r_-} \cdot \frac{r_-}{r_+} \tag{2.48}$$

The details for the derivation are given in [14], which also proves the following conclusion.

Theorem 2.3.2 *For the waveform relaxation method associated with the optimized interface algorithm, (2.47), (1) the contraction factor in the s-domain, ρ, is (2.48), and it is an analytical function under condition*

$$\alpha > 0, \; \beta < 0$$

(2) A set of optimal (α, β) is

$$\alpha^* = -1 + r_-{}^{-1}, \;\; \beta^* = 1 - r_+ \tag{2.49}$$

which leads to $\rho = 0$.

The optimal $(\alpha^*, \; \beta^*)$ in (2.49) are in the s-domain. Consequently, obtaining their values in the t-domain needs to find their inverse Laplace transform, which is not straightforward. As a practical approach, (α^*, β^*) may be chosen as constants (real) that minimize the magnitude of ρ. Particularly, one can reduce the magnitude of the contraction factor in its whole s-domain, and choose the optimal values as

$$(\alpha^*, \beta^*) = \underset{\alpha \in R^+, \; \beta \in R^-}{argmin} \; \{ \underset{Re(s)>0}{max} |\rho(\alpha, \beta, s)| \}$$

Since it is analytical, ρ achieves its minimums and maximums on the boundary of the s-domain, i.e., along $\sigma = 0$. Therefore, one just needs to find its minimums and maximums along $\sigma = 0$. In doing this, a practical method used in [22] is to let $\beta^* = -\alpha^*$ and solve α^* from

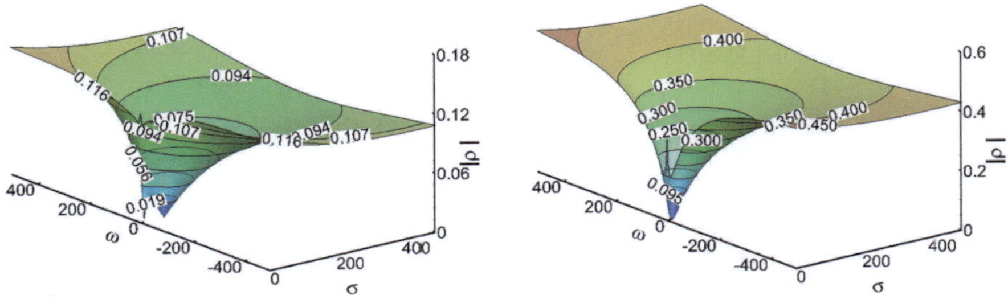

(a) (b)

Fig. 2.8 The s-domain distributions of $|\rho|$ associated with the optimized interface algorithm. $\Delta x = 0.05$. (**a**) Case 1. $\alpha^* = 1.76848$. The maximum of $|\rho|$ is 0.129. (**b**) Case 2. $\alpha^* = 0.40361$. The maximum of $|\rho|$ is 0.477

$$|\rho(\alpha^*, -\alpha^*, s)|_{s=\mathbf{i}\omega_{min}} = |\rho(\alpha^*, -\alpha^*, s)|_{s=\mathbf{i}\omega_{max}} \tag{2.50}$$

Here ω_{min} and ω_{max} may be determined as π/T and $\pi/\Delta t$, respectively, with Δt being the time step in the computation, e.g., of (2.45).

Example 2.3.3 For Cases 1 and 2 of the last subsection, the optimal value of α^* is estimated by solving (2.50). The corresponding $|\rho|$ with the optimal value is plotted in Fig. 2.8. By comparison of Figs. 2.6 and 2.8, it is seen that, indeed, the overall value of $|\rho|$, including its spikes and the troughs, has become smaller in magnitude in association with the optimized algorithm than with the classical algorithm. Additionally, the distribution of $|\rho|$ in the s-domain of the former has deviated substantially from that of the latter.

Example 2.3.4 The solutions for Cases 1 and 2 are computed using the optimized interface algorithm, particularly (2.46). The algorithm is implemented via (2.48) and (2.50), with all others as before, and the solutions are presented in Fig. 2.9. It is seen in the figure that, in comparison with that associated with the classical interface algorithm in Fig. 2.7, the convergence process of the iterated values in association with the optimized algorithm is much faster. Actually, in Case 2, the iterated values at $m = 3, 5$ become indistinct from those at convergence ($m = 100$). Additionally, it is also noted that the iterated values associated with the classical algorithm are continuous during the iteration, see Fig. 2.7. However, those associated with the optimized algorithm are not, as evidenced in Fig. 2.9a, although they become continuous at the convergence.

To further show their performance on convergence speedup, the iteration residual associated with the optimized interface algorithm, together with that associated with the classical algorithm, is plotted in Fig. 2.10. In the figure, the iteration residual is computed as the logarithm of

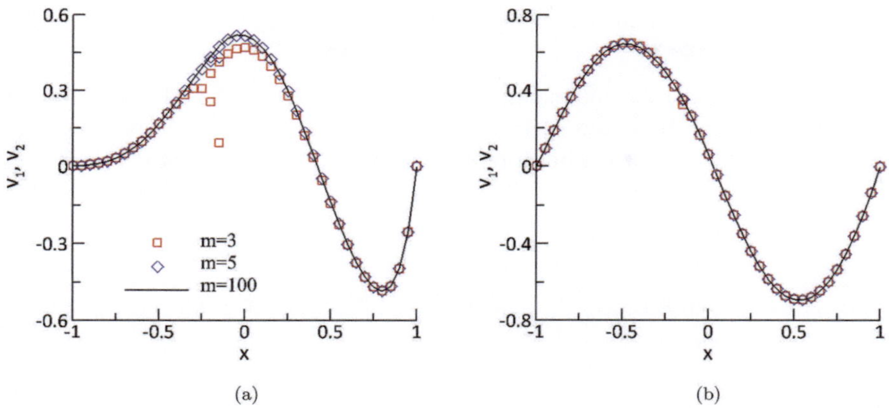

Fig. 2.9 Solutions at different Schwarz iterations associated with the optimized interface algorithm. (**a**) Case 1, $t = 0.5$. (**b**) Case 2, $t = 0.15$

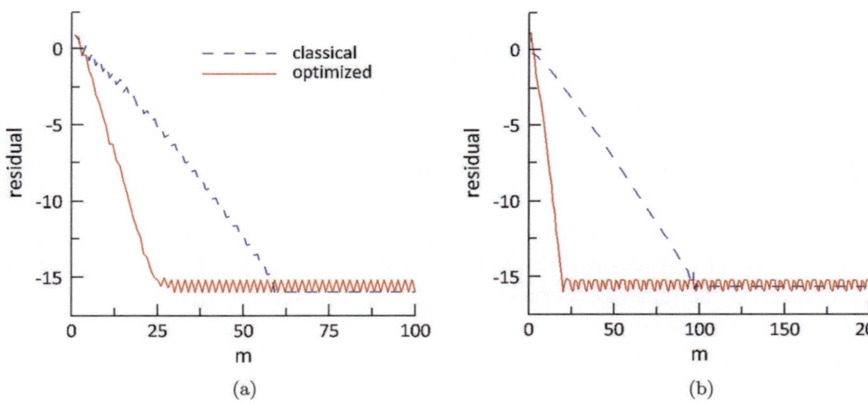

Fig. 2.10 Subdomains and of the computations with the classical interface algorithm and its optimized interface algorithm. (**a**) Case 1, $t = 0.5$. (**b**) Case 2, $t = 0.15$

$$\text{Residual}_1 = \frac{1}{N(I_1 + I_2 + 2)} \left(\sum_{n=1,i=0}^{N,-I_1} |v_{1,i}^{n,m+1} - v_{1,i}^{n,m}| + \sum_{n=1,i=0}^{N,I_2} |v_{2,i}^{n,m+1} - v_{2,i}^{n,m}| \right) \tag{2.51}$$

Here, I_1 and I_2 are the numbers of grid nodes in subdomains 1 and 2, respectively. The figure shows that, after $m = 100$, the residuals with the classical interface algorithm and its optimized interface algorithm reach machine zero, at about 10^{-16}, and then remain about flat. However, in both Cases 1 and 2, the residuals reach the machine zero for the optimized interface algorithm substantially earlier than for the classical algorithm, indicating that the former speeds up the convergence of the computation.

Now, we consider the computation of nonlinear problem (2.2). By a derivation similar to that for algorithm (2.41), the computation by the Schwarz waveform relaxation method for the nonlinear problem may be expressed as ($m = 0, 1, 2, \ldots$)

$$\begin{cases} dv_{1,i}^{m+1}/dt = \epsilon_{1,i}^m v_{1,i-1}^{m+1} + \eta_{1,i}^m v_{1,i}^{m+1} + \gamma_{1,i}^m v_{1,i+1}^{m+1}, & i \le -1 \\ v_{1,0}^{m+1} = v_{2,1}^m \end{cases}$$
$$\begin{cases} dv_{2,i}^{m+1}/dt = \epsilon_{2,i}^m v_{2,i-1}^{m+1} + \eta_{2,i}^m v_{2,i}^{m+1} + \gamma_{2,i}^m v_{2,i+1}^{m+1}, & i \ge 1 \\ v_{2,0}^{m+1} = v_{1,-1}^m \end{cases} \tag{2.52}$$

in which the coefficients are

$$\begin{aligned} \epsilon_{k,i}^m &= \frac{a_k(v_{k,i}^m)}{2\Delta x} + \frac{b_k}{\Delta x^2} \\ \eta_{k,i}^m &= -\frac{2b_k}{\Delta x^2} - c_k(v_{k,i}^m) \\ \gamma_{k,i}^m &= -\frac{a_k(v_{k,i}^m)}{2\Delta x} + \frac{b_k}{\Delta x^2} \end{aligned}$$

These coefficients are not constant but solution-dependent. Additionally, in this computation, a discretization similar to (2.45) is used, and the classical interface algorithm may be replaced by (2.46), the optimized interface algorithm.

Before further considering the computation of (2.52), let us discuss a property of the solutions to the linear equations in (2.41). From (2.43), it is derived that

$$\begin{aligned} |\hat{v}_{1,i}^{m+1}| &= |B^{m+1} r_+{}^i| = |B^{m+1} r_+{}^{i+1} r_+{}^{-1}| = |\hat{v}_{1,i+1}^{m+1}||r_+|^{-1} \\ &= \ldots = |\hat{v}_{1,i+q}^{m+1}||r_+|^{-q} \\ &= \ldots = |\hat{v}_{1,0}^{m+1}||r_+|^i, \quad i = -1, -2, \ldots \\ |\hat{v}_{2,i}^{m+1}| &= |A^{m+1} r_-{}^i| = |A^{m+1} r_-{}^{i-1} r_-| = |\hat{v}_{2,i-1}^{m+1}||r_-| \\ &= \ldots = |\hat{v}_{2,i-q}^{m+1}||r_-|^q \\ &= \ldots = |\hat{v}_{2,0}^{m+1}||r_-|^i, \quad i = 1, 2, \ldots \end{aligned}$$

where q is a positive integer. The above are actually

$$\begin{aligned} |\hat{v}_{1,i}^{m+1}| &= \hat{v}_{1,0}^{m+1}||r_+|^i, \quad i = -1, -2, \ldots \\ |\hat{v}_{2,i}^{m+1}| &= |\hat{v}_{2,0}^{m+1}||r_-|^i, \quad i = 1, 2, \ldots \end{aligned}$$

Considering condition (2.38), i.e., $|r_-| < 1$ and $|r_+| > 1$, we conclude the following proposition.

Proposition 2.3.2 *The magnitude of solution for system (2.42) decays with $|i|$, particularly,*

$$\begin{array}{l} |\hat{v}_{1,0}^{m+1}| > \cdots > |\hat{v}_{1,i+1}^{m+1}| > |\hat{v}_{1,i}^{m+1}|, \; i = -1, -2, \ldots \\ |\hat{v}_{2,0}^{m+1}| > \cdots > |\hat{v}_{2,i}^{m+1}| > |\hat{v}_{2,i+1}^{m+1}|, \; i = 1, 2, \ldots \end{array} \tag{2.53}$$

The proposition shows that the magnitude of the solutions, $|\hat{v}_{k,i}^{m+1}|$, or their errors against their converged solutions (given the linearity of system (2.42)), decreases as their location gets away from interface nodes. Such decay in magnitude is rapid because of the power $|i|$, $|r_-| < 1$, and $|r_+| > 1$ in the above derivation. In other words, the solutions at interface nodes $i = 0$ have little effect on solutions at nodes with a large $|i|$, or their influences are restricted within a small neighborhood of the interface nodes.

With the aforementioned understanding, it is conjectured that for the computation of (2.52), if we introduce linearization by freezing its coefficients $\epsilon_{k,i}^m$, $\eta_{k,i}^m$, and $\gamma_{k,i}^m$ $(k = 1, 2)$ as those at the interface nodes $i = 0$, the optimized interface algorithm could remain effective to a certain extent since the effect of the solutions at the interface nodes is restricted within their small neighborhood. Therefore, the optimized interface algorithm is computed using the coefficients evaluated by the solutions at the interfaces. Particularly, for the optimized algorithm, the roots are evaluated as

$$\begin{aligned} r_- &= \frac{s - \eta_{2,0}^m - \sqrt{(s - \eta_{2,0}^m)^2 - 4\epsilon_{2,0}^m \gamma_{2,0}^m}}{2\gamma_{2,0}^m} \\ r_+ &= \frac{s - \eta_{1,0}^m + \sqrt{(s - \eta_{1,0}^m)^2 - 4\epsilon_{1,0}^m \gamma_{1,0}^m}}{2\gamma_{1,0}^m} \end{aligned} \tag{2.54}$$

Example 2.3.5 The computation of (2.52), which is nonlinear and adopts the classical algorithm, is conducted for two cases in Table 2.8. Also, to speed up the computation, optimized algorithm (2.46) is adopted. Additionally, this algorithm is implemented via (2.48) and (2.50), which are evaluated by the solutions at the interfaces as in (2.54). The space and time domains in the experiments are $x \in [-1, 1]$, $t \in [0, 1]$. Similarly, as in the linear scenarios, the initial and boundary conditions are

Table 2.8 Cases for numerical experiments on the waveform relaxation method for nonlinear advection–diffusion–reaction equations

Case	a_1	b_1	c_1	a_2	b_2	c_2
3	$sin(\pi u_1)$	0.4	1	$sin(\pi u_2)$	0.2	2
4	u_1	0.4	u_1	u_2	0.2	$2u_2$

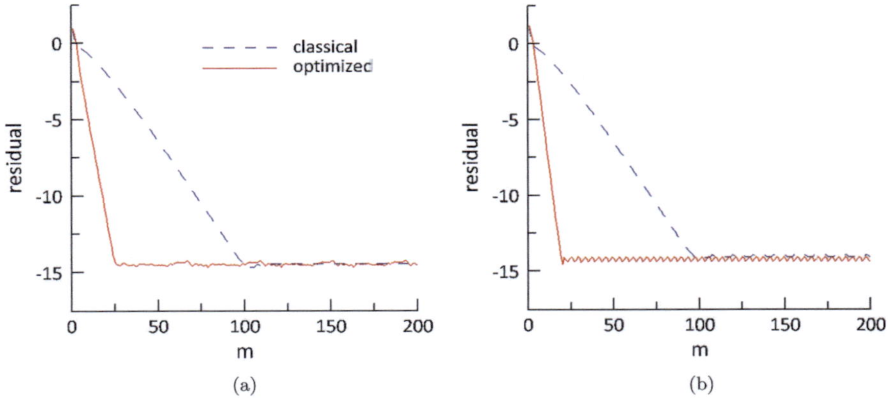

Fig. 2.11 Convergence of the computations for nonlinear equations with the classical algorithm and its optimized algorithm. $t = 0.2$. (**a**) Case 3, $\alpha^* = 0.49372$. (**b**) Case 4, $\alpha^* = 0.48292$

$$u_0 = -sin(\pi x),\ -1 < x < 1;\ u_1 = 0,\ x = -1;\ u_2 = 0,\ x = 1$$

$\Delta t = 0.05$, $\Delta x = 0.05$. The interfaces of subdomains 1 and 2 are located at $x_1 = -0.45$ and $x_2 = -0.5$, respectively. Again, for simplicity, the initial guesses for the iteration at the interfaces are set to zero, that is $v_{1,0}^{n,0}, v_{2,0}^{n,0} = 0$, for all n.

Figure 2.11 presents the convergence for the computation of the two cases. It is seen that in both cases, the computation with the classical algorithm converges as anticipated. In addition, the optimized algorithm leads to much faster convergence than the classical algorithm. The convergence speedup is comparable to that in the linear cases in Fig. 2.10, indicating the validity of the optimized algorithm for nonlinear equations. It is noted that, however, numerical experiments also show that the validity may fail due to the presence of nonlinearity [14].

2.3.3 Comparison with Conventional Method

The Schwarz waveform relaxation method and the conventional Schwarz method are two types of domain decomposition methods, and both of them are widely adopted. The former has been a focus of theoretical study [23, 24], while the latter has been popular for application problems [25, 26]. Therefore, it is interesting to compare the waveform relaxation method and the conventional method. This subsection presents a comparison of their computational expenses via numerical experiments.

First, let us look at the computational algorithms of the two methods. Algorithm (2.45) is typical for the waveform relaxation method, while algorithm (2.24) is standard for the conventional method. They show that the two methods have the same discretization. However, they differ from each other. The major difference is that their orders for the loops of m and n are opposite. In the waveform relaxation method, the inner loop is over n for marching in time, and the outer loop is over m for the Schwarz iteration. In the conventional method, the inner loop is over m for the Schwarz iteration, and the outer loop is over n for marching in time. It is noted that the discretization in (2.45) and (2.24) is implicit. But, by slightly modifying them, explicit discretization is available to compute the problem, such as that in (2.8). The following discussion is made on the basis of the implicit discretization.

As seen in (2.45) and (2.24), for both waveform relaxation and conventional methods, a linear system of algebraic equations has to be solved in each subdomain because they adopt implicit discretization. Solving the linear system is expected to comprise the primary computational expense for the two methods. For instance, to solve a linear system with a dimension of $I \times I$ matrix by the naive Gaussian elimination algorithm, the number of operation counts for the computation is at the order of $I^3/3 + I^2 + \mathcal{O}(I)$ (an operation count is one time of multiplication or division. In comparison with multiplication or division, addition and subtraction are negligible in terms of computational expense) [17]. As a result, a large number of operation counts, or high computational expense, is needed as the dimension of the matrix becomes large. Additionally, the linear systems in the two methods are the same, and they will be solved using the same techniques in the experiments. Therefore, the total number of times to compute the linear system, referred to as the computation count hereafter, roughly reflects the computational expense of the waveform relaxation and conventional methods. The computation count is computed as $N \times M$ for the two methods. Here, N is the total number of time steps, and M is the iteration number to achieve convergence, or the total number of Schwarz iteration.

To compare their computation counts, ideally, the same criterion for convergence should be used for the waveform relaxation and conventional methods. However, since their computational algorithms are not the same, different but similar forms of residuals will be applied. Particularly, the residuals of the waveform relaxation methods and the conventional methods are computed in such a way that the same values of them lead to a similar solution accuracy. For the waveform relaxation methods, in association with both the classical and optimized algorithms, the residual is computed according to

$$\text{Residual}_2 = \max_{1\le n\le N}(|v_{1,0}^{n,m+1} - v_{1,0}^{n,m}|,\ |v_{2,0}^{n,m+1} - v_{2,0}^{n,m}|) \tag{2.55}$$

which only utilizes solutions at the interface nodes. With a rough equivalence, for the conventional methods, also in association with both the classical algorithm and the optimized algorithm, the residual is calculated as

$$\text{Residual}_3 = \max(|v_{1,0}^{n,m+1} - v_{1,0}^{n,m}|,\ |v_{2,0}^{n,m+1} - v_{2,0}^{n,m}|),\ \text{any } n \tag{2.56}$$

Again, only solutions at the interface nodes are utilized.

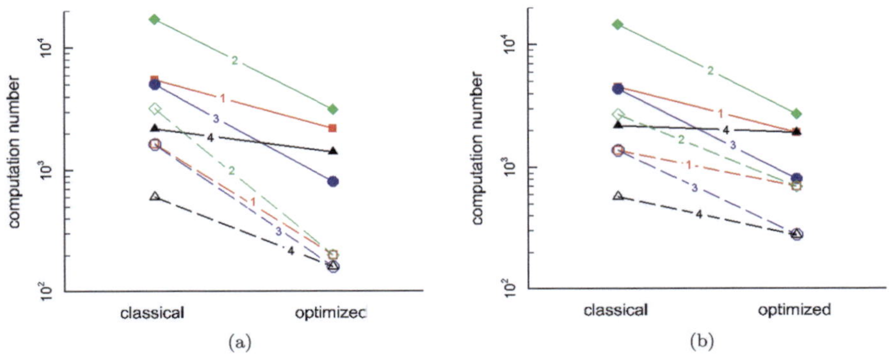

Fig. 2.12 Comparison of the computation counts. Filled symbols and solid lines — waveform relaxation Schwarz methods, open symbols and dash lines — conventional Schwarz methods. Numbers 1, 2, 3, and 4 are the case numbers. (**a**) The residual is computed using the interface nodes, according to (2.55), (2.56). (**b**) The residual is computed using all nodes, according to (2.51), (2.57)

Now, let us make experiments on the computation counts for the Schwarz waveform relaxation method, (2.45) (or its version for nonlinear equations), and the optimized Schwarz waveform relaxation method, (2.45) with its interface algorithm replaced by (2.46). The experiments also include the conventional Schwarz method, (2.24) (or its version for nonlinear equations), and the optimized conventional Schwarz method, (2.24) with its interface algorithm substituted by (2.30). The experiments are conducted via the cases in the previous subsections, with $\Delta t = 0.05$, $\Delta x = 0.05$, and $T = 5$. The tolerance of residual for the convergence is set as 10^{-10}, in terms of (2.55) or (2.56). Figure 2.12a presents the resulting computation counts for the four cases in Tables 2.7 and 2.8, including both linear and nonlinear equations. As seen in the figure, the waveform relaxation methods have higher computation counts than the conventional methods, with a ratio of at least 3:1. For instance, in Case 1, the computation count by the Schwarz waveform relaxation method is 5,500, marked by the filled squares, and that by the conventional Schwarz method is 1,648, depicted by the open squares. This trend is also true for their optimized algorithms, with an even higher ratio of at least 5:1. Note that due to its "perfect convergence" as discussed in Sect. 2.2.4, the optimized conventional method leads to convergence within two iterations when marching from time level n to $n + 1$, and thus the computation count is $2N$. For instance, for Cases 1 and 2, $2N = 2(T/\Delta t) = 2(5/0.05) = 200$, and this is shown in the lower right corner of Fig. 2.12a.

For a thorough comparison, the aforementioned experiments are repeated but with solutions at all grid nodes to evaluate the iteration residual and determine convergence. For the waveform relaxation methods, such a residual is computed according to (2.51). Similarly, for the conventional methods, the residual is calculated as

$$\text{Residual}_4 = \frac{1}{I_1+1}\sum_{i=0}^{-I_1} |v_{1,i}^{n,m+1} - v_{1,i}^{n,m}| + \frac{1}{I_2+1}\sum_{i=0}^{I_2} |v_{2,i}^{n,m+1} - v_{2,i}^{n,m}|, \text{ any } n \tag{2.57}$$

With such a way to compute the residuals, Fig. 2.12b presents the resulting computation counts. This figure shows that the computation exhibits the same patterns as in Fig. 2.12a. Particularly, the waveform relaxation method consumes more computation counts than the conventional method, and the optimized algorithm also needs more computation counts for the former than the latter.

The comparison of computational expense can be further tuned. For instance, the computation counts for the optimized waveform relaxation method may be reduced if a better estimate for optimal values (α^*, β^*) is available. Additionally, it is expected that although the computation counts for the waveform relaxation methods and the conventional methods are good indicators for their computational time, they may not be directly proportional to the computational time of the two methods, since other factors are also involved. For instance, the waveform relaxation method may need fewer communication times between the two subdomains than the conventional method, and thus, the former may consume less computational time for communication.

Summary and Question

Summary This chapter discusses the computation of advection-diffusion-reaction equations coupled at the interfaces of their subdomains. It begins by introducing the computational problem and then presents the frameworks of the conventional Schwarz method and the Schwarz waveform relaxation method. The discretized equations are presented, and both Dirichlet-Dirichlet and Dirichlet-Neumann boundary conditions are illustrated, along with a note on discretization accuracy. The chapter provides a detailed discussion of the two methods.

In the discussion of computation using the conventional Schwarz method, the chapter begins with an explicit scheme that utilizes the Dirichlet-Dirichlet interface condition. The calculation is reformulated as an iterative problem for a linear system, and it is shown that the converged solution corresponds to a steady-state solution. The calculation employs no Schwarz iteration due to its explicit nature.

In the computation of an implicit discretization and implicit interface treatment using an iterative method, the Scarborough criterion is applied as a condition for convergence. When a direct method (i.e., matrix inversion) is employed, an expression for the contraction factor is derived and numerically validated. Building on this, an optimized interface algorithm is introduced to accelerate convergence. Again, the contraction factor is derived, and "perfect convergence"—convergence within two iterations—is achieved. Numerical examples validate the derived contraction factor and the perfect convergence. Additionally, the theoretical analysis is extended to nonlinear equations, with supporting numerical experiments confirming its validity.

For the computation using the waveform relaxation Schwarz method, the governing PDEs are semi-discretized in space, resulting in a system of ordinary differential equations. These equations are then transformed into a system of algebraic equations via the Fourier transform, and the computation is reformulated as a Schwarz iteration of the system. For the classical interface algorithm, the contraction factor is derived, demonstrating that the iteration converges. A contraction factor is also derived for the case with an optimized interface algorithm. Because the analysis is conducted on the s-domain, the parameters in the optimized algorithm are approximated in the t-domain for the actual computation. Moreover, the analysis and derivation for the contraction factor are extended to the nonlinear problems. Numerical experiments illustrate and verify the theoretical results.

A comparison is made between the conventional method and the waveform relaxation method in terms of computational time. The significant computational cost in both methods comes from solving a linear system in each subdomain at every iteration. The computation unit is thus defined as the number of times such a system is solved. Numerical experiments conclude that the waveform relaxation method requires at least three times as many computation units as the conventional method, for both classical and optimized interface algorithms.

The analyses and algorithms, such as the optimized interface algorithms, in this chapter are based on linear systems for both the conventional method and the waveform relaxation method, and they remain largely effective for nonlinear systems. However, their validity may degenerate or fail in some nonlinear cases. An interesting topic for further research is to understand the reasons behind such degeneration and failure. Insights into such reasons will be valuable for designing algorithms to deal with nonlinearity. As another topic, it is also intriguing to extend the analyses and algorithms presented in this chapter to more complex equations relevant to fluid flows, such as the 2D advection-diffusion equation and the Burgers' equation.

Questions The following questions are provided to aid in understanding, discussing, and investigating the content. Their answers and clues are embedded within the chapter.

1. Derive the accuracy orders of the discretizations in Eqs. (2.4), (2.5), and (2.7).
2. Discuss the interface algorithms in Eqs. (2.10) and (2.11), and explain why they are essentially equivalent when the two grids overlap at two nodes.
3. In Section 2.2.2, it is demonstrated that K is close in value to ρ and can thus serve as an indicator of convergence. If there is, construct a case where their values differ significantly.
4. Derive condition (2.31), under which the optimized interface algorithm in Eq. (2.30) reduces to the Dirichlet-Dirichlet interface condition at convergence.
5. Explain why a zero contraction factor implies that convergence is achieved in just two iterations in computation by the conventional method.

6. Example 2.2.5 demonstrates the validity of perfect convergence in one nonlinear case but also shows its failure in another. Discuss possible reasons why perfect convergence succeeds in one case but fails in the other.
7. A cornerstone of the analysis for the waveform relaxation method is Eq. (2.38), whose validity is stated in Proposition 2.3.1. Consider whether the condition in the proposition for $|r_1| < 1$ can be removed or relaxed.
8. Explain why the contraction factor for the waveform relaxation method reaches its minimum and maximum values only along $\sigma = 0$.
9. The derivation of the contraction factor for the conventional method is based on the iteration residual, as shown in Eq. (2.26). In contrast, the derivation for the waveform relaxation method is based on the solution variable itself, as shown in Eq. (2.41). Explain why the same factor would hold if the residual were used in the latter case.
10. Equation (2.50) is used in searching for the optimal values of (α, β). Discuss this equation and whether alternative approaches for the search are available.
11. Although the conventional method and the waveform relaxation method differ in procedures, their computations share similarities. The same computational code can be used for both by simply switching the order of the loops for the Schwarz iteration and the physical time-stepping. Explain this using Eqs. (2.24) and (2.45).
12. Theorems 2.2.2 and 2.3.2 demonstrate that the optimized interface algorithm yields a zero contraction factor, indicating perfect convergence—convergence within two iterations—for both the conventional and waveform relaxation methods. Explain why, in practical calculations, the perfect convergence is achieved only for the former method and not for the latter.
13. Numerical experiments indicate that the computational cost of the waveform relaxation methods is higher than that of the conventional methods. How is this cost measured in the comparison, and can it be quantified more accurately?

References

1. G.K. Batchelor, *An Introduction to Fluid Dynamics (Cambridge Mathematical Library) 2nd edition* (Cambridge University Press, 2000)
2. S. Atis, S. Saha, H. Auradou, J. Martin, N. Rakotomalala, L. Talon, D. Salin, CHEMO-hydrodynamic coupling between forced advection in porous media and self-sustained chemical waves. Chaos **22**, 037108 (2012)
3. S. Hamilton, M. Berrill, K. Clarno, R. Pawlowski, A. Toth, C.T. Kelley, T. Evans, B. Philip, An assessment of coupling algorithms for nuclear reactor core physics simulations. J. Comput. Phys. **311**, 241–257 (2016)
4. K. Qu, H.S. Tang, A. Agrawal, Integration of fully 3D fluid dynamics and geophysical fluid dynamics models for multiphysics coastal ocean flows: Simulation of local complex free-surface phenomena. Ocean Modell. **135**, 14–30 (2019)
5. Z.-G. Ji, *Hydrodynamics and Water Quality* (Wiley, 2008)
6. G.A. Meurant, A domain decomposition method for parabolic problems. Appl. Numer. Math. **8**, 427–441 (1991)

7. C. Canuto, A.L. Giudice, A multi-timestep Robin–Robin domain decomposition method for time dependent advection-diffusion problems. Appl. Math. Comput. **363**, 124596 (2019)
8. S. Boulaaras, M.S.T. Brahim, S. Bouzenada, A posteriori error estimates for the generalized Schwarz method of a new class of advection-diffusion equation with mixed boundary condition. Math. Methods Appl. Sci. **41**(14), 5493–5505 (2018)
9. M.J. Gander, Schwarz methods over the course of time. Electronic Trans. Numer. Anal. **31**, 228–255 (2008)
10. H.S. Tang, C. Jones, F. Sotiropoulos, An overset-grid method for 3D unsteady incompressible flows. J. Comput. Phys. **191**, 567–600 (2003)
11. H.S. Tang, F. Sotiropoulos, Fractional step artificial compressibility schemes for the unsteady incompressible Navier-Stokes equations. Comput. Fluids **36**, 974–986 (2007)
12. W. Dong, *Towards simulation of ocean flows: Analysis and algorithm for computation of coupled partial differential equations*. Ph.D. thesis, Civil Eng. Dept., City College of New York, 2022
13. W.B. Dong, H.S. Tang, Y.J. Liu, Convergence analysis on computation of coupled advection-diffusion-reaction problems. Appl. Math. Comput. **420**, 126876 (2022)
14. W.B. Dong, H.S. Tang, Convergence analysis of waveform relaxation method to compute coupled advection-diffusion-reaction equations. Math. Comput. Simul. **218**, 462–481 (2024)
15. E. Kreyszig, *Advanced Engineering Mathematics, 9th edition* (Wiley, 2006)
16. G.A. Sod, *Numerical Methods in Fluid Dynamics – Initial and Initial Boundary-Value Problems* (Cambridge University Press, 1985)
17. J.F. Epperson, *An Introduction to Numerical Methods and Analysis, 2nd edition* (Wiley, 2013)
18. S.V. Patankar, *Numerical Heat Transfer and Fluid Flow* (Hemisphere Publishing Corporation, 1980)
19. H.K. Versteeg, W. Malalasekera, *An Introduction to Computational Fluid Dynamics: The Finite Volume Method* (Pearson Education Limited, 2007)
20. J.B. Scarborough, *Numerical Mathematical Analysis, 1st edition* (The Johns Hopkins Press, 1930)
21. J.B. Scarborough, *Numerical Mathematical Analysis, 5th edition* (The Johns Hopkins Press, 1958)
22. M.J. Gander, A. Ruehli, Optimized waveform relaxation methods for RC type circuits. IEEE Trans. Circuits Syst. **51**, 755–768 (2004)
23. V. Martin, An optimized Schwarz waveform relaxation method for unsteady convection diffusion equation. Appl. Numer. Math. **52**, 401–428 (2005)
24. M.J. Gander, L. Halpern, Optimized schwarz waveform relaxation methods for advection diffusion problems. SIAM J. Numer. Anal. **45**, 666–697 (2007)
25. H.S. Tang, K. Qu, X.G. Wu, An overset grid method for integration of fully 3D fluid dynamics and geophysical fluid dynamics models to simulate multiphysics coastal ocean flows. J. Comput. Phys. **273**, 548–571 (2014)
26. E. Blayo, About interface conditions for coupling hydrostatic and nonhydrostatic Navier-Stokes flows. Discrete Contin. Dynam. Syst. S **9**(5), 1565–1574 (2016)

Chapter 3
Compressible Flow

3.1 Conservation Law

Since the early 1980s, the multiblock method, under the name of the zonal method, overset-grid method, etc., has rapidly grown in the computation of compressible flows. At the early time, the primary motivation of the method was to overcome the difficulty in generating meshes for complex configurations of fluid flow boundaries and to reduce computational expense in solving the discretized equations [1–3]. The growth of the multiblock method has been accompanied by numerical methods for the system of conservation laws. The system of conservation laws forms the mathematical foundation of compressible fluid flows and, thus, the basis for their numerical algorithms. The mathematical theory for conservation laws was introduced in the 1950s [4, 5], and it has evolved into an essential foundation in computational fluid dynamics and its applications.

The initial value problem of a conservation law in 1D space reads as

$$\begin{cases} u_t + f(u)_x = 0, & t > 0 \\ u = u_0, & t = 0 \end{cases} \tag{3.1}$$

where $u(x, t)$ is an unknown scalar function, and $f(u)$ is the flux. $f(u)$ and $u_0(x)$ are prescribed scalar functions. Additionally, for convenience of discussion, it is assumed that u_0 is bounded and has finite support; $u_0 \to 0$ as $x \to \pm\infty$. Also, $f(u) \to 0$ when $x \to \pm\infty$. Without loss of generality, the discussion in this chapter is made primarily on a scalar, 1D conservation law.

When multiple conservation laws are present, they form a system of conservation laws:

$$Q_t + F(Q)_x = 0 \tag{3.2}$$

H. Tang, *Multiblock Method for Fluid Flow*, Nonlinear Systems and Complexity 40,
https://doi.org/10.1007/978-3-031-78568-9_3

where Q and F are vectors. When the Jacobian matrix $A = \partial F(Q)/\partial Q$ has real and distinct eigenvalues, system (3.2) is called a hyperbolic system of conservation laws. The governing equations for an inviscid, compressible fluid flow are the Euler equations [6], and they comprise a hyperbolic system of conservation laws. Particularly, in 1D space, the Euler equations are

$$Q = \begin{pmatrix} \rho \\ \rho u \\ E \end{pmatrix}, \qquad F(Q) = \begin{pmatrix} \rho u \\ \rho u^2 + p \\ u(E + p) \end{pmatrix} \tag{3.3}$$

where $p = (\gamma - 1)(E - u^2/2)$. Here ρ, u, p, and E are the density, velocity, pressure, and total energy, respectively, and γ is the ratio of specific heats. For an ideal gas, $\gamma = 1.4$. It can be checked that the Jacobian matrix has three eigenvalues, $u - c$, u, and $u + c$, with $c = \sqrt{\partial p/\partial \rho}$, being the speed of sound [6, 7]. As long as $c \neq 0$, these three eigenvalues are real and distinct, and the above Euler equations comprise a hyperbolic system of conservation laws. Note that the Euler equations are still a hyperbolic system of conservation laws in two-dimensional (2D) and three-dimensional (3D) spaces.

Generally, initial value problem (3.1) permits discontinuity in its solution, u. This implies that a classical (or strong) solution, which is continuous and differentiable and satisfies its equation pointwise (or almost everywhere), may fail to exist. To overcome this difficulty, the concept of a weak solution is introduced by Lax [4, 8]. The weak solution is defined as u that satisfies the weak form of the conservation law:

$$\int_0^{+\infty} \int_{-\infty}^{+\infty} (\phi_t u + \phi_x f(u))dxdt + \int_{-\infty}^{+\infty} (\phi u)|_{t=0} dx = 0 \tag{3.4}$$

where $\phi \in C_0^\infty(\Omega)$, a test function, and Ω is the upper half plane: $-\infty < x < \infty$, $t \geq 0$. Particularly, ϕ is any smooth function of x and t with finite support; it is infinitely differentiable and becomes zero as $|x|$ and t are sufficiently large. Here, "smooth" means infinitely differentiable and thus implies continuity. It is noted that the weak solution may not be unique, and it could be non-physical [9, 10].

Consider the weak solution of the initial value problem of a conservation law, particularly (3.1). Let u be a smooth function, except on a curve $x = x(t)$ (or, a finite number of curves) along which u is discontinuous, i.e., $\lim_{x \to x(t)^-} u \neq \lim_{x \to x(t)^+} u$ (Fig. 3.1). Here, $x(t)^- \equiv x(t) - 0$, and $x(t)^+ \equiv x(t) + 0$. Such a function u is called a piecewise smooth function. In addition, let [ ] represent the jump in the value of a function across the curve. For instance, $[u] = u(x^+, t) - u(x^-, t)$, with x^- and x^+ being the left and right sides, respectively, of the curve. Then, we have the following conclusion about a piecewise smooth solution to problem (3.1). Such a conclusion shows that finding a piecewise smooth solution is sufficient to obtain a weak solution. Note that, in the following discussion, instead of

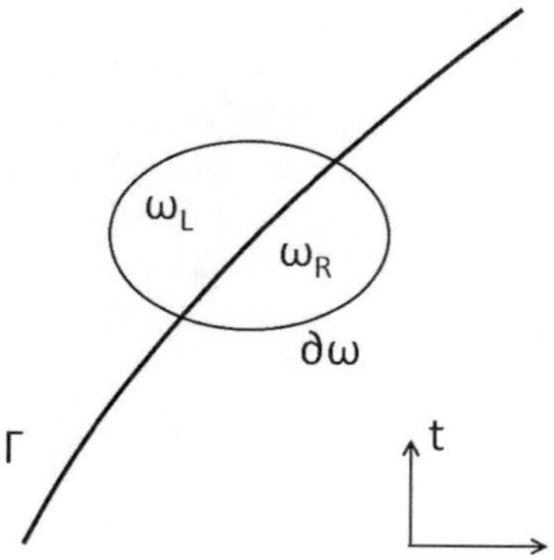

Fig. 3.1 Γ is an interface: $x = x(t)$. ω is a neighborhood of a point on the interface, $\omega = \omega_L \cup \omega_R$

being strictly piecewise smooth, it may be sufficient for u to be piecewise continuous and differentiable.

Theorem 3.1.1 *A necessary and sufficient condition for a piecewise smooth solution u to be a weak solution is (1) u satisfies Eq. (3.1) wherever it is smooth, and (2) u satisfies the Rankine–Hugoniot condition*

$$[u]s = [f(u)] \tag{3.5}$$

along curve $x = x(t)$, where u is discontinuous on the $x-t$ plane and $s = dx(t)/dt$.

Proof First, let us prove the necessary condition. Suppose u is smooth at point (x, t) in Ω, and $\phi(x, t) \in C_0^\infty(\Omega)$. According to the definition of weak solution (3.4), one has

$$\begin{aligned}
0 &= \int_0^{+\infty}\int_{-\infty}^{+\infty} (\phi_t u + \phi_x f(u))dxdt + \int_{-\infty}^{+\infty} (\phi u)|_{t=0}dx \\
&= \int_0^{+\infty}\int_{-\infty}^{+\infty} ((\phi u)_t + (\phi f(u))_x - \phi u_t - \phi f(u)_x)dxdt + \int_{-\infty}^{+\infty} (\phi u)|_{t=0}dx \\
&= \oint_{\partial\Omega} (-\phi u dx + \phi f(u)dt) + \int_0^{+\infty}\int_{-\infty}^{+\infty} (-\phi u_t - \phi f(u)_x)dxdt \\
&\quad + \int_{-\infty}^{+\infty} (\phi u)|_{t=0}dx \\
&= \int_{-\infty}^{+\infty} (-\phi u)|_{t=0}dx + \int_0^{+\infty}\int_{-\infty}^{+\infty} (-\phi u_t - \phi f(u)_x)dxdt + \int_{-\infty}^{+\infty} (\phi u)|_{t=0}dx \\
&= \int_0^{+\infty}\int_{-\infty}^{+\infty} \phi(-u_t - f(u)_x)dxdt
\end{aligned}$$

Here $\partial\Omega$ is the boundary of the upper plane and in the counterclockwise direction, and the Green formula is used. Because of the arbitrariness of ϕ, it is known that u satisfies the conservation law in Eq. (3.1).

Let u be discontinuous along curve Γ: $x = x(t)$. Consider a point (x, t) on the curve and its neighborhood of ω consisting of ω_L and ω_R, which are respectively at the left and right side of the curve, see Fig. 3.1. Suppose that ω is small enough so that u is smooth in $\omega \setminus \Gamma$, Then, the definition of weak solution (3.4) leads to

$$\begin{aligned}
0 &= \int_{\omega} (\phi_t u + \phi_x f(u))dxdt \\
&= \int_{\omega_L} (\phi_t u + \phi_x f(u))dxdt + \int_{\omega_R} (\phi_t u + \phi_x f(u))dxdt \\
&= \int_{\omega_L} ((\phi u)_t + (\phi f(u))_x)\,\phi dxdt - \int_{\omega_L} (u_t + f_x(u))\phi dxdt \\
&+ \int_{\omega_R} ((\phi u)_t + (\phi f(u))_x)\,\phi dxdt - \int_{\omega_R} (u_t + f_x(u))\phi dxdt
\end{aligned}$$

By the Green formula, the first and third area integrals after the last equal sign become closed line integrals along their boundaries, and this leads to

$$\begin{aligned}
0 &= \oint_{\partial\omega_L} (-\phi f(u)dt + \phi u dx) - \int_{\omega_L} (u_t + f_x(u))\phi dxdt \\
&+ \oint_{\partial\omega_R} (-\phi f(u)dt - \phi u dx) - \int_{\omega_R} (u_t + f_x(u))\phi dxdt \\
&= -\int_{\Gamma} \phi\,([u]dx - [f]dt) \\
&- \int_{\omega_L} (u_t + f_x(u))\phi dxdt - \int_{\omega_R} (u_t + f_x(u))\phi dxdt
\end{aligned}$$

In the aforementioned, $\phi = 0$ along $\partial\omega$ is utilized, in view $\phi \in C_0^{\infty}(\omega)$. Additionally, the closed line integrals are in counterclockwise direction, and they are in the opposite direction along Γ when running along $\partial\omega_L$ and $\partial\omega_R$. Since u satisfies the equation of (3.1) in ω_L and ω_R, the last two area integrals after the last equal sign become zero. Then

$$\begin{aligned}
0 &= \int_{\Gamma} \phi\,([u]dx - [f]dt) \\
&= \int_{\Gamma} \phi\,([u]s - [f])\,dt
\end{aligned}$$

Because of the arbitrariness of ϕ, it is known that Rankine-Hugoniot condition (3.5) holds on Γ.

Second, similarly, with the aid of the Green Formula, one can prove that conditions (1) and (2) comprise a sufficient condition for a piecewise smooth solution u to be a weak solution. This completes the proof of Theorem 3.1.1. □

Let Ω be the domain $(-\infty < x < +\infty, 0 < t' < t)$. Integration of the equation in (3.1) over Ω gives rise to

$$\begin{aligned}0 &= \int_\Omega (u_{t'} + f_x(u))dxdt' \\ &= \int_{-\infty}^{+\infty} (u - u|_{t=0})dx + \int_0^t (f(u)|_{x=-\infty} - f(u)|_{x=\infty}))dt' \\ &= \int_{-\infty}^{+\infty} (u - u_0)dx\end{aligned}$$

Here, $lim_{x\to\pm\infty} f(u) = 0$ is adopted. With the aforementioned, one reaches the following conclusion.

Proposition 3.1.1 *The solution to initial value problem (3.1) satisfies the following conservation condition:*

$$\int_{-\infty}^{+\infty} u(x,t)dx = \int_{-\infty}^{+\infty} u_0(x)dx = const. \tag{3.6}$$

Proposition 3.1.1 indicates that $\int_{-\infty}^{+\infty} udx$ is conserved over time, and this is an important property of a conservation law. It is desired that a numerical solution to (3.1), an initial value problem of a conservation law, preserves the property.

3.2 Conservative Scheme

A finite difference method in the form of a conservative scheme, introduced by Lax [4, 8], is commonly adopted to compute conservation laws. Let us compute initial value problem (3.1) on a uniform grid $(\ldots, x_{i-1}, x_i, x_{i+1}, \ldots)$, with grid spacing Δx, and a uniform time division $(t_0, t_1, \ldots, t_n, t_{n+1}, \ldots)$, with $t_0 = 0$ and time step Δt. Let v_i^n be a numerical approximation of $u(x_i, t_n)$, the conservative scheme for the conservation law in problem (3.1) reads as

$$v_i^{n+1} = v_i^n - \lambda(\overline{f}^n_{i+1/2} - \overline{f}^n_{i-1/2}) \tag{3.7}$$

where $\lambda = \Delta t/\Delta x$, and subscript $i + 1/2$ indicates the location $x = (x_i + x_{i+1})/2$. Additionally, the initial value is evaluated as

$$v_i^0 = \frac{1}{\Delta x}\int_{(i-1/2)\Delta x}^{(i+1/2)\Delta x} u_0(x)dx \tag{3.8}$$

In scheme (3.7), $\overline{f}^n_{i+1/2}$ is called the numerical flux, and

$$\overline{f}^n_{i+1/2} = \overline{f}(v^n_{i-l+1}, v^n_{i-l}, \cdots, v^n_{i+l}) \tag{3.9}$$

where l is an integer. For the conservative scheme, its numerical flux satisfies the consistency condition

$$\overline{f}^n_{i+1/2}(v, v, \cdots, v) = f(v) \tag{3.10}$$

Consistent with $\lim_{x\to\pm\infty} f(u) = 0$, it is assumed that $\lim_{i\to\pm\infty} \overline{f}^n_{i+1/2} = 0$. Scheme (3.7) is an explicit finite difference scheme, and it has $2l + 1$ stencils.

Scheme (3.7) may be written as

$$v^{n+1}_i = H(v^n_{i-l}, v^n_{i-l+1}, \cdots, v^n_{i+l}) \tag{3.11}$$

Here, H is the right-hand side of (3.7). When u is a smooth solution for the conservation law in (3.1), one has

$$\begin{aligned}
&u^{n+1}_i - H(u^n_{i-l}, u^n_{i-l+1}, \cdots, u^n_{i+l}) \\
=&\Delta t\left((u^{n+1}_i - u^n_i)/\Delta t + (\overline{f}^n_{i+1/2} - \overline{f}^n_{i-1/2})/\Delta x\right) \\
=&\Delta t\big(u_t + f(u)_x + r(u, \Delta x, \Delta t)\big) \\
=&\Delta t r(u, \Delta x, \Delta t)
\end{aligned}$$

where $r(u, \Delta x, \Delta t)$ is the truncation error of scheme (3.7), and thus a consistency condition for a conservative scheme in the form of (3.11) can also be written as

$$u^{n+1}_i - H(u^n_{i-l}, u^n_{i-l+1}, \cdots, u^n_{i+l}) = \Delta t r(u, \Delta x, \Delta t) \tag{3.12}$$

in which $\lim_{\Delta t, \Delta x\to 0} r(u, \Delta x, \Delta t) \to 0$.

There are various conservative schemes. The Lax–Friedrichs scheme is a well-known conservative scheme:

$$\frac{v^{n+1}_i - (v^n_{i-1} + v^n_{i-1})/2}{\Delta t} + \frac{f(v^n_{i+1}) - f(v^n_{i-1})}{2\Delta x} = 0 \tag{3.13}$$

In the form of the conservative scheme, its numerical flux reads as ($l = 1$)

$$\overline{f}^{LF}_{i+1/2} = \frac{1}{2}\left(f(v^n_i) + f(v^n_{i+1})\right) - \frac{\Delta x}{2\Delta t}\left(v^n_{i+1} - v^n_i\right) \tag{3.14}$$

It is easy to verify that the Lax–Friedrichs scheme satisfies consistency conditions (3.10) and (3.12). The Lax–Wendroff scheme is another popular conservative scheme, and it reads as

$$\frac{v_i^{n+1} - v_i^n}{\Delta t} + \frac{f(v_{i+1}^n) - f(v_{i-1}^n)}{2\Delta x} \\ - \frac{\Delta t}{2} \frac{a_{i+1/2}^n (f(v_{i+1}^n) - f(v_i^n)) - a_{i-1/2}^n (f(v_i^n) - f(v_{i-1}^n))}{\Delta x^2} = 0 \tag{3.15}$$

where $a = \partial f(v)/\partial v$, and

$$a_{i+1/2}^n = a\left(\frac{v_i^n + v_{i+1}^n}{2}\right) \tag{3.16}$$

The numerical flux of the Lax–Wendroff scheme can be written as ($l = 1$)

$$\overline{f}_{i+1/2}^{LW} = \frac{1}{2}\left(f(v_i^n) + f(v_{i+1}^n)\right) - \frac{\lambda}{2}\left(a_{i+1/2}^n (f(v_{i+1}^n) - f(v_i^n))\right) \tag{3.17}$$

Again, it can also be verified that the Lax–Wendroff scheme meets consistency conditions (3.10) and (3.12). The Godunov scheme is also a commonly discussed conservative scheme, and it is in the form of (3.7). Its numerical flux is evaluated using the analytical solution of a Riemann problem, $u(x/t, v_i^n, v_{i+1}^n)$. Particularly, the flux is expressed as ($l = 1$)

$$\overline{f}_{i+1/2}^{G} = f(u(x_{i+1/2}/t, v_i^n, v_{i+1}^n)) \tag{3.18}$$

and u is the solution to a Riemann problem of the conservation law [11]:

$$\begin{cases} u_t + f(u)_x = 0, \; t > 0 \\ u = v_i^n, \quad x < x_{i+1/2}, \; t = 0 \\ u = v_{i+1}^n, \quad x > x_{i+1/2}, \; t = 0 \end{cases}$$

in which $x_{i+1/2} = x_i + \Delta x/2$. A Riemann problem is an initial value problem of a conservation law, and its initial value is a piecewise constant. Here, the Riemann problem is restricted between two grid nodes in space. Therefore, it is a local problem. Particularly, it is a local Riemann problem or a local problem of the conservation law. Correspondingly, its solution is a local solution of the conservation law. Note that more details for the aforementioned schemes are available in [8].

A natural question is: why do we tend to favor a conservative scheme, or what is its advantage? The content in the following answers this question. Let us construct step functions for the numerical solution and the numerical flux of initial value problem (3.1):

$$\begin{aligned} v_\Delta(x, t) &= v_i^n, \quad x, t \in [x_{i-1/2}, x_{i+1/2}) \times [t_n, t_{n+1}) \\ f_\Delta(x, t) &= \overline{f}_{i+1/2}^n, \quad x, t \in [x_i, x_{i+1}) \times [t_n, t_{n+1}) \end{aligned} \tag{3.19}$$

which are a direct extension of the numerical solution and the numerical flux in conservative scheme (3.7). Note that f_Δ is a function of v_Δ. For the step function of the numerical flux, we have

Proposition 3.2.1 *Let $v_\Delta(x,t)$ be bounded, then $f_\Delta(x,t)$ is also bounded. Moreover, if $v_\Delta(x,t)$ converges to a function $u(x,t)$ as Δx, $\Delta t \to 0$, then, at the place where $u(x,t)$ is continuous,*

$$\lim_{\Delta x, \Delta t \to 0} f_\Delta(x,t) = f(u(x,t)) \tag{3.20}$$

At the place where it is discontinuous,

$$\lim_{\Delta x, \Delta t \to 0} f_\Delta(x,t) = f(u(x,t)) + \mathcal{O}(u(x^+,t) - u(x^-,t)) \tag{3.21}$$

Proof By definitions (3.9) and (3.19), the mean-value theorem, and consistency condition (3.10), one has

$$\begin{aligned}
f_\Delta(x,t) &= \overline{f}^n_{i+1/2}(v_\Delta(x-(l-1/2)\Delta x,t), \cdots, v_\Delta(x+(l-1/2)\Delta x,t)) \\
&= \overline{f}^n_{i+1/2}(v_\Delta(x,t), \cdots, v_\Delta(x,t)) \\
&\quad + \sum_{l'=1}^{2l} \overline{f}^n_{i+1/2_{l'}} \times (v_\Delta(x+(l'-l-1/2)\Delta x,t) - v_\Delta(x,t)) \\
&= f(v_\Delta(x,t)) \\
&\quad + \sum_{l'=1}^{2l} \overline{f}^n_{i+1/2_{l'}} \times (v_\Delta(x+(l'-l-1/2)\Delta x,t) - v_\Delta(x,t))
\end{aligned}$$

in which $\overline{f}^n_{i+1/2_{l'}}$ is the derivative of $\overline{f}^n_{i+1/2}$ over its $l'th$ variable. From the above, the proof becomes obvious. □

The aforementioned proposition leads to the main conclusion about the conservative scheme, a theorem stated as follows:

Theorem 3.2.1 *Let initial value problem (3.1) be computed by conservative scheme (3.7). Then, if the numerical solution $v_\Delta(x,t)$ is uniformly bounded and converges almost everywhere to $u(x,t)$ as Δt, $\Delta x \to 0$, the $u(x,t)$ will be a weak solution.*

Proof With the definition of the step functions, the conservative scheme reads as

$$\frac{v_\Delta(x,t+\Delta t) - v_\Delta(x,t)}{\Delta t} + \frac{f_\Delta(x+\Delta x/2,t) - f_\Delta(x-\Delta x/2,t)}{\Delta x} = 0$$

Let test function $\phi(x,t) \in C_0^\infty(\Omega)$, where $\Omega = R \times R^+$. Multiplying the equation with the test function and then integrating the resulting equation over Ω, one has

$$\int_0^\infty \int_{-\infty}^\infty \phi(x,t) \left(\frac{v_\Delta(x,t+\Delta t) - v_\Delta(x,t)}{\Delta t} \right.$$
$$\left. + \frac{f_\Delta(x+\Delta x/2,t) - f_\Delta(x-\Delta x/2,t)}{\Delta x} \right) dxdt = 0$$

By substitution of $t' = t + \Delta t$, one has

$$\begin{aligned} &\int_0^\infty \int_{-\infty}^\infty \phi(x,t) v_\Delta(x,t+\Delta t) dxdt \\ &= \int_{\Delta t}^\infty \int_{-\infty}^\infty \phi(x,t'-\Delta t) v_\Delta(x,t') dxdt' \\ &= \int_{\Delta t}^\infty \int_{-\infty}^\infty \phi(x,t-\Delta t) v_\Delta(x,t) dxdt \end{aligned}$$

Similarly, one has

$$\begin{aligned} &\int_0^\infty \int_{-\infty}^\infty \phi(x,t) f_\Delta(x \pm \Delta x/2,t) dxdt \\ &= \int_{\Delta t}^\infty \int_{-\infty}^\infty \phi(x \mp \Delta x/2,t) v_\Delta(x,t) dxdt \\ &+ \int_0^{\Delta t} \int_{-\infty}^\infty \phi(x \mp \Delta x/2,t) v_\Delta(x,t) dxdt \end{aligned}$$

Then, the aforementioned integral equation leads to

$$\begin{aligned} &\int_{\Delta t}^\infty \int_{-\infty}^\infty \left(\frac{\phi(x,t-\Delta t) - \phi(x,t)}{\Delta t} v_\Delta(x,t) \right. \\ &\left. + \frac{\phi(x-\Delta x/2,t) - \phi(x+\Delta x/2,t)}{\Delta x} f_\Delta(x,t) \right) dxdt \\ &- \int_0^{\Delta t} \int_{-\infty}^\infty \frac{\phi(x,t) v_\Delta(x,t)}{\Delta t} dxdt \\ &+ \int_0^{\Delta t} \int_{-\infty}^\infty \frac{\phi(x-\Delta x/2,t) - \phi(x+\Delta x/2,t)}{\Delta x} v_\Delta(x,t) dxdt = 0 \end{aligned}$$

As $v_\Delta(x,t)$ is uniformly bounded and it converges almost everywhere to $u(x,t)$ when Δt, $\Delta x \to 0$, the first two integrals on the left-hand side become those on the left-hand side of Eq. (3.4), and the last integral tends to zero. Here, the test function has finite support, and the mean value theorem applies. As a result, the above equation becomes Eq. (3.4), that is, $v_\Delta(x,t)$ tends to a weak solution $u(x,t)$ at the limit. This completes the proof of the theorem. □

As indicated in Theorem 3.2.1, the advantage of using a conservative scheme is that if its solution is convergent, it converges to a weak solution. A benefit of a conservative scheme can also be seen from the angle of conservation; the discrete version of conservation condition (3.6) holds:

$$\int_{-\infty}^{+\infty} v_\Delta(x,t_n)dx = const. \tag{3.22}$$

which is actually an approximation to (3.6) by the rectangle rule for numerical integration. Actually, when initial value problem (3.1) is solved using the conservative scheme, one has

$$\begin{aligned}
&\int_{-\infty}^{+\infty} v_\Delta(x,t_n)dx \\
&= \sum_{i=-\infty}^{+\infty} v_i^n \Delta x \\
&= \sum_{i=-\infty}^{+\infty} (v_i^{n-1} - \lambda(\overline{f}_{i+1/2}^{n-1} - \overline{f}_{i-1/2}^{n-1}))\Delta x \\
&= (\cdots v_{i-1}^{n-1} + v_i^{n-1} + v_{i+1}^{n-1} + \cdots)\Delta x + \\
&\quad \cdots + \Delta t(\overline{f}_{i-1/2}^{n-1} - \overline{f}_{i-3/2}^{n-1}) + \Delta t(\overline{f}_{i+1/2}^{n-1} - \overline{f}_{i-1/2}^{n-1}) + \Delta t(\overline{f}_{i+3/2}^{n-1} - \overline{f}_{i+1/2}^{n-1}) + \cdots
\end{aligned}$$

All terms of numerical fluxes cancel with each other. Moreover, assume $\lim_{i\to\pm\infty} \overline{f}_{i+1/2}^{n} = 0$. Therefore,

$$\begin{aligned}
\int_{-\infty}^{+\infty} v_\Delta(x,t_n)dx &= \sum_{i=-\infty}^{+\infty} v_i^{n-1}\Delta x \\
&= \cdots\cdots \\
&= \sum_{i=-\infty}^{+\infty} v_i^0 \Delta x \\
&= \int_{-\infty}^{+\infty} u_0(x)dx \\
&= const.
\end{aligned}$$

In the above, the initial value v_i^0 is evaluated as in (3.8). As a result, the aforementioned shows that a numerical solution of a conservative scheme satisfies discrete conservation condition (3.22). This is summarized as follows.

Proposition 3.2.2 *If initial value problem (3.1) is computed by conservative algorithm (3.7), then the numerical solution satisfies the discrete conservation condition in the form of (3.22).*

3.3 Multiblock Computation

Consider the computation of initial value problem (3.1) by the multiblock method. In this chapter, the conventional method to march in time, as illustrated in Eqs. (1.9) and (1.10), will be utilized. Explicit schemes in the form of (3.7) are adopted in the subdomains, together with explicit interface treatments. As a result, no Schwarz iteration, e.g., the Schwarz alternating iteration, is needed between the two subdomains. The computation will follow the frameworks in the previous chapters, particularly (1.10) and (2.8).

As a common practice, the space domain can be divided into two patched subdomains or two overlapping subdomains, see Fig. 3.2. Each subdomain has its own grid, which consists of inner nodes and interface nodes. Figure 3.2a shows two patched subdomains 1 and 2, which are discretized by two patched grids. The two grids have inner nodes at $i \leq I - 1$ and $i \geq I + 1$. The grids also have an interface node at $i = I$, which is right at the subdomain interface. When the subdomains overlap, their grids are overlapping grids, see Fig. 3.2b. In this situation, the grids for subdomains 1 and 2 have $i = I$ and $i = 1$, respectively, as their interface nodes. Each interface node of a grid falls into a host cell in the other grid. Particularly, interface node $i = I$ has $[x_{I''-1}, x_{I''}]$ as its host cell in the grid of subdomain 2, and interface node $i = 1$ has $[x_{I'}, x_{I'+1}]$ as its host cell in the grid of subdomain 1.

In the situation of two patched subdomains (Fig. 3.2a), let $v_{1,i}^n$, $v_{2,i}^n$, and v_I^n be the solutions in subdomain 1, in subdomain 2, and at the interface, respectively. The initial value for the computation is evaluated as

$$\begin{cases} v_{1,i}^0 = \dfrac{1}{\Delta x_1} \displaystyle\int_{x_{i-1/2}}^{x_{i+1/2}} u_0(x)dx, & i \leq I - 1 \\ v_{2,i}^0 = \dfrac{1}{\Delta x_2} \displaystyle\int_{x_{i-1/2}}^{x_{i+1/2}} u_0(x)dx, & i \geq I + 1 \\ v_i^0 = \dfrac{2}{\Delta x_1 + \Delta x_2} \displaystyle\int_{x_{i-1/2}}^{x_{i+1/2}} u_0(x)dx, & i = I \end{cases} \tag{3.23}$$

A conservative scheme is used in subdomain 1 and another in subdomain 2. In light of the conventional method to march in time, one has

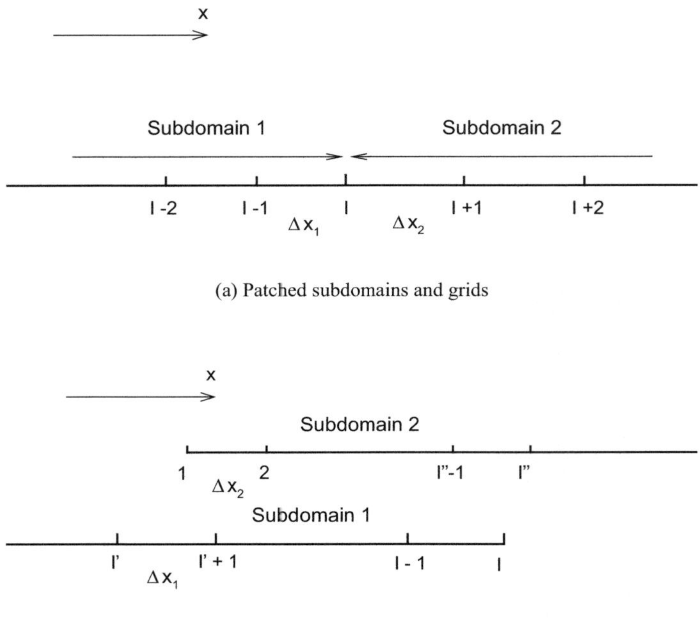

(a) Patched subdomains and grids

(b) Overlapping subdomains and grids

Fig. 3.2 Subdomains and grids for initial value problem (3.1). In (**b**), the host cells of interface nodes $i = 1$ and $i = I$ are $(x_{I'}, x_{I'+1})$ and $(x_{I''-1}, x_{I''})$, respectively

$$\begin{cases} v_{1,i}^{n+1} = v_{1,i}^{n} - \lambda_1(\overline{f}_{i+1/2}^{1} - \overline{f}_{i-1/2}^{1}), & i \leq I-1 \\ v_{2,i}^{n+1} = v_{2,i}^{n} - \lambda_2(\overline{f}_{i+1/2}^{2} - \overline{f}_{i-1/2}^{2}), & i \geq I+1 \\ v_{i}^{n+1} = q_{I}^{n+1}, & i = I \end{cases} \tag{3.24}$$

Here, subscript/superscript "1" and "2" indicate subdomains 1 and 2, respectively, and q_I^{n+1} is an explicit interface algorithm.

For the overlapping grids, one evaluates the initial value as

$$\begin{cases} v_{1,i}^{0} = \dfrac{1}{\Delta x_1}\displaystyle\int_{x_{i-1/2}}^{x_{i+1/2}} u_0(x)dx, & i \leq I-1 \\ v_{2,i}^{0} = \dfrac{1}{\Delta x_2}\displaystyle\int_{x_{i-1/2}}^{x_{i+1/2}} u_0(x)dx, & i \geq I''+1 \\ v_{1,i}^{0} = \dfrac{1}{x_{I''+1/2} - x_{I-1/2}}\displaystyle\int_{x_{I-1/2}}^{x_{I''+1/2}} u_0(x)dx, & i = I \end{cases}$$

In the aforementioned, the right ending node of subdomain 1, $i = I$, is the division point of the two subdomains. Alternatively, if the left ending node of subdomain 2, $i = 1$, is the division point, the initial value is computed as

$$\begin{cases} v_{1,i}^0 = \dfrac{1}{\Delta x_1} \displaystyle\int_{x_{i-1/2}}^{x_{i+1/2}} u_0(x)dx, & i \leq I' - 1 \\ v_{2,i}^0 = \dfrac{1}{\Delta x_2} \displaystyle\int_{x_{i-1/2}}^{x_{i+1/2}} u_0(x)dx, & i \geq 2 \\ v_{2,i}^0 = \dfrac{1}{x_{3/2} - x_{I'-1/2}} \displaystyle\int_{x_{I'-1/2}}^{x_{3/2}} u_0(x)dx, & i = 1 \end{cases}$$

For the overlapping grids, the computation is carried out as

$$\begin{cases} v_{1,i}^{n+1} = v_{1,i}^n - \lambda_1(\overline{f}^1_{i+1/2} - \overline{f}^1_{i-1/2}), & i \leq I - 1 \\ v_{2,i}^{n+1} = v_{2,i}^n - \lambda_2(\overline{f}^2_{i+1/2} - \overline{f}^2_{i-1/2}), & i \geq 2 \\ v_{1,i}^{n+1} = q_I^{n+1}, & i = I \\ v_{2,i}^{n+1} = q_1^{n+1}, & i = 1 \end{cases} \tag{3.25}$$

In the aforementioned, q_I^{n+1} and q_1^{n+1} are the interface algorithms at interface nodes $i = I$ and $i = 1$, respectively.

A key issue in the computation with the patched grids and overlapping grids, i.e., (3.24) and (3.25), respectively, is to design the interface algorithms. Generally, interface algorithms can be divided into two categories: (1) conservative interface algorithms and (2) nonconservative interface algorithms. The conservative interface algorithms are those that preserve a discretized version of conservation condition (3.6). While nonconservative interface algorithms are those that do not preserve such a condition. For the patched grids in Fig. 3.2a, a commonly adopted conservative interface algorithm for (3.24) is [12, 13]

$$q_I^{n+1} = v_I^n - \lambda_I(\overline{f}^2_{I+1/2} - \overline{f}^1_{I-1/2}) \tag{3.26}$$

where $\lambda_I = 2\Delta t/(\Delta x_1 + \Delta x_2)$. This conservative interface algorithm is actually of a numerical-flux-enforcing type; it directly enforces the numerical fluxes at the two sides of the interface node, $i = I$, by adopting those of the schemes in the two subdomains. For the overlapping grids in Fig. 3.2b, as an extension from the aforementioned numerical-flux-enforcing type of conservative interface algorithm, a conservative interface algorithm is

$$q_I^{n+1} = v_{1,I}^n - \lambda'_I(\overline{f}^2_{I''+1/2} - \overline{f}^1_{I-1/2}) \tag{3.27}$$

Here, $\lambda'_I = \Delta t/(x_{I''+1/2} - x_{I-1/2})$, and $i = I$ is the division point of subdomains. When the division point is at $i = 1$, the interface algorithm becomes

$$q_1^{n+1} = v_{2,1}^n - \lambda_1'(\overline{f}^2_{3/2} - \overline{f}^1_{I'-1/2}) \tag{3.28}$$

where $\lambda_1' = \Delta t/(x_{3/2} - x_{I'-1/2})$.

Interpolation is widely used to construct interface algorithms, and, generally speaking, such an algorithm is a nonconservative interface algorithm. For the patched grids in Fig. 3.2a, a nonconservative interface algorithm based on linear interpolation is

$$q_I^{n+1} = \frac{\Delta x_1}{\Delta x_1 + \Delta x_2}(v_{2,I+1}^{n+1} - v_{1,I-1}^{n+1}) + v_{1,I-1}^{n+1} \tag{3.29}$$

When two grids overlap, as shown in Fig. 3.2b, an interpolation-based nonconservative interface algorithm becomes

$$\begin{cases} q_I^{n+1} = \dfrac{x_I - x_{I''-1}}{\Delta x_2}(v_{2,I''}^{n+1} - v_{2,I''-1}^{n+1}) + v_{2,I''-1}^{n+1} \\ q_1^{n+1} = \dfrac{x_1 - x_{I'}}{\Delta x_1}(v_{1,I'+1}^{n+1} - v_{1,I'}^{n+1}) + v_{1,I'}^{n+1} \end{cases} \tag{3.30}$$

Note that when applying (3.29) and (3.30), the interface algorithms are computed after the computation of the schemes in the subdomains. Therefore, the computation of (3.24) and (3.25) remains explicit.

The two categories of interface algorithms are analyzed in detail in the following sections, and their advantages and disadvantages are discussed, followed by illustrations via numerical examples.

3.4 Conservative Interface Algorithm

3.4.1 Property and Performance

Numerical solutions in the form of step functions are defined to facilitate the discussion. For the patched grids in Fig. 3.2a, a numerical solution in the form of a step function is defined as

$$v_\Delta(x,t) = \begin{cases} v_{1,i}^n, & x,t \in [x_{i-1/2}, x_{i+1/2}) \times [t_n, t_{n+1}), \quad i \le I-1 \\ v_{2,i}^n, & x,t \in [x_{i-1/2}, x_{i+1/2}) \times [t_n, t_{n+1}), \quad i \ge I+1 \\ v_i^n, & x,t \in [x_{I-1/2}, x_{I+1/2}) \times [t_n, t_{n+1}), \quad i = I \end{cases} \tag{3.31}$$

Correspondingly, its numerical flux in the form of a step function is defined as

$$f_\Delta(x,t) = \begin{cases} \overline{f}^1_{i+1/2}, & x,t \in [x_i, x_{i+1}) \times [t_n, t_{n+1}), \quad i \le I-1 \\ \overline{f}^2_{i+1/2}, & x,t \in [x_i, x_{i+1}) \times [t_n, t_{n+1}), \quad i \ge I \end{cases} \tag{3.32}$$

A similar definition can be made for overlapping grids in Fig. 3.2b. When node $i = I$ is considered as the division point of the two subdomains, a numerical solution in the form of a step function is

$$v_\Delta(x,t) = \begin{cases} v_{1,i}^n, & x,t \in [x_{i-1/2}, x_{i+1/2}) \times [t_n, t_{n+1}), \quad i \le I-1 \\ v_{2,i}^n, & x,t \in [x_{i-1/2}, x_{i+1/2}) \times [t_n, t_{n+1}), \quad i \ge I''+1 \\ v_{1,i}^n, & x,t \in [x_{I-1/2}, x_{I''+1/2}) \times [t_n, t_{n+1}), \quad i = I \end{cases} \tag{3.33}$$

When node $i = 1$ is considered as the division point of the two subdomains, it is

$$v_\Delta(x,t) = \begin{cases} v_{1,i}^n, & x,t \in [x_{i-1/2}, x_{i+1/2}) \times [t_n, t_{n+1}), \quad i \le I'-1 \\ v_{2,i}^n, & x,t \in [x_{i-1/2}, x_{i+1/2}) \times [t_n, t_{n+1}), \quad i \ge 2 \\ v_{2,i}^n, & x,t \in [x_{I'-1/2}, x_{3/2}) \times [t_n, t_{n+1}), \quad i = 1 \end{cases} \tag{3.34}$$

A conservative interface algorithm strictly preserves the discretized version of conservation condition (3.6). Firstly, we consider the computation of (3.24) on patched grids in Fig. 3.2a. By interface algorithm (3.26), one has

$$\begin{aligned} &\int_{-\infty}^{+\infty} v_\Delta(x,t)dx \\ &= \sum_{i=-\infty}^{I-1} v_{1,i}^n \Delta x_1 + v_I^n \frac{\Delta x_1 + \Delta x_2}{2} + \sum_{i=I+1}^{+\infty} v_{2,i}^n \Delta x_2 \\ &= \cdots + v_{1,I-1}^{n-1} \Delta x_1 + v_I^{n-1} \frac{\Delta x_1 + \Delta x_2}{2} + v_{2,I+1}^{n-1} \Delta x_2 + \cdots \\ &\quad + \Delta t(\overline{f}_{I-1/2}^{1,n-1} - \overline{f}_{I-3/2}^{1,n-1}) + \Delta t(\overline{f}_{I+1/2}^{2,n-1} - \overline{f}_{I-1/2}^{1,n-1}) + \Delta t(\overline{f}_{I+3/2}^{2,n-1} - \overline{f}_{I+1/2}^{2,n-1}) + \cdots \end{aligned}$$

In the aforementioned, the numerical flux terms cancel each other, and they tend to zero as x goes to infinity. These, together with (3.23), lead to ($n \ge 0$)

$$\begin{aligned} \int_{-\infty}^{+\infty} v_\Delta(x,t)dx &= \sum_{i=-\infty}^{I-1} v_{1,i}^{n-1} \Delta x_1 + v_I^{n-1} \frac{\Delta x_1 + \Delta x_2}{2} + \sum_{i=I+1}^{+\infty} v_{2,i}^{n-1} \Delta x_2 \\ &= \cdots\cdots \\ &= \sum_{i=-\infty}^{I-1} v_{1,i}^0 \Delta x_1 + v_I^0 \frac{\Delta x_1 + \Delta x_2}{2} + \sum_{i=I+1}^{+\infty} v_{2,i}^0 \Delta x_2 \\ &= \int_{-\infty}^{+\infty} u_0(x)dx \\ &= const. \end{aligned}$$

These are summarized as follows.

Proposition 3.4.1 *Let initial value problem (3.1) be computed by multiblock method (3.24) in association with conservative interface algorithm (3.26). Then, the numerical solution satisfies discrete conservation condition (3.22).*

Secondly, we consider the computation of (3.25) on overlapping grids in Fig. 3.2b by conservative interface algorithms (3.27) and (3.28). In this situation, when node $i = I$ is the subdomain division point, starting from (3.33), one has

$$
\begin{aligned}
&\int_{-\infty}^{+\infty} v_\Delta(x,t)dx \\
&= \int_{-\infty}^{x_{I-1/2}} v_\Delta(x,t)dx + \int_{x_{I-1/2}}^{x_{I''+1/2}} v_\Delta(x,t)dx + \int_{x_{I''+1/2}}^{+\infty} v_\Delta(x,t)dx \\
&= \sum_{i=-\infty}^{I-1} v_{1,i}^n \Delta x_1 + v_{1,I}^n (x_{I''+1/2} - x_{I-1/2}) + \sum_{i=I''+1}^{+\infty} v_{2,i}^n \Delta x_2 \\
&= \cdots + v_{1,I-1}^{n-1}\Delta x_1 + v_{1,I}^{n-1}(x_{I''+1/2} - x_{I-1/2}) + v_{2,I''+1}^{n-1}\Delta x_2 + \cdots \\
&\quad + \Delta t(\overline{f}_{I-1/2}^{1,n-1} - \overline{f}_{I-3/2}^{1,n-1}) + \Delta t(\overline{f}_{I''+1/2}^{2,n-1} - \overline{f}_{I-1/2}^{1,n-1}) \\
&\quad + \Delta t(\overline{f}_{I''+3/2}^{2,n-1} - \overline{f}_{I''+1/2}^{2,n-1}) + \cdots
\end{aligned}
$$

in which, again, the numerical fluxes cancel each other. Then, one has

$$
\begin{aligned}
\int_{-\infty}^{+\infty} v_\Delta(x,t)dx &= \sum_{i=-\infty}^{I-1} v_{1,i}^{n-1}\Delta x_1 + v_{1,I}^{n-1}(x_{I''+1/2} - x_{I-1/2}) + \sum_{i=I''+1}^{+\infty} v_{2,i}^{n-1}\Delta x_2 \\
&= \cdots\cdots \\
&= \int_{-\infty}^{+\infty} u_0(x)dx \\
&= const.
\end{aligned}
$$

When the subdomain division is at node $i = 1$, a similar derivation can be carried out starting from (3.34). As a result, one has

Proposition 3.4.2 *Let initial value problem (3.1) be computed by multiblock method (3.25) in association with conservative interface algorithm (3.27) and (3.28). Then, the numerical solution satisfies discrete conservation condition (3.22).*

It can further be shown that a conservative interface algorithm leads to a weak solution [13]. Here, a proof is presented for the scenarios of patched grids (Fig. 3.2a). With step functions defined in (3.31) and (3.32), algorithm (3.24) in association with interface algorithm (3.26) becomes

$$\frac{v_\Delta(x,t+\Delta t)-v_\Delta(x,t)}{\Delta t}+\frac{f_\Delta(x+\Delta x_1/2,t)-f_\Delta(x-\Delta x_1/2,t)}{\Delta x_1}=0,$$
$$x\in(-\infty,x_{I-1/2})$$
$$\frac{v_\Delta(x,t+\Delta t)-v_\Delta(x,t)}{\Delta t}+\frac{f_\Delta(x+\Delta x_2/2,t)-f_\Delta(x-\Delta x_2/2,t)}{\Delta x_2}=0,$$
$$x\in[x_{I+1/2},+\infty)$$
$$\frac{v_\Delta(x,t+\Delta t)-v_\Delta(x,t)}{\Delta t}+\frac{f_\Delta(\alpha_2 x+\beta_2\Delta x_2/2,t)-f_\Delta(\alpha_1 x-\beta_1\Delta x_2/2,t)}{(\Delta x_1+\Delta x_2)/2}=0,$$
$$x\in[x_{I-1/2},x_{I+1/2}) \tag{3.35}$$

Here

$$\alpha_1=\frac{2\Delta x_1}{\Delta x_1+\Delta x_2},\quad \beta_1=\frac{4x_{I-1/2}}{\Delta x_1+\Delta x_2}-\frac{2x_{I-1}}{\Delta x_1}$$
$$\alpha_2=\frac{2\Delta x_2}{\Delta x_1+\Delta x_2},\quad \beta_2=-\frac{4x_{I+1/2}}{\Delta x_1+\Delta x_2}+\frac{2x_{I+1}}{\Delta x_2}$$

It is easy to verify that when $x_{I-1/2}<x\le x_{I+1/2}$,

$$x_{I-1}<\alpha_1 x-\beta_1\Delta x_1/2\le x_I,\quad f_\Delta(\alpha_1 x-\beta_1\Delta x_1/2,t)=f_\Delta(x_{I-1/2},t)$$
$$x_I<\alpha_2 x+\beta_2\Delta x_2/2\le x_{I+1},\quad f_\Delta(\alpha_2 x+\beta_2\Delta x_2/2,t)=f_\Delta(x_{I+1/2},t)$$

Theorem 3.4.1 *If the numerical solution $v_\Delta(x,t)$ of (3.35) is uniformly bounded and converges almost everywhere to $u(x,t)$ as the time step and grid spacing tend to zero, then $u(x,t)$ will be a weak solution.*

Proof Let $\phi(x,t)\in C_0^\infty(\Omega)$, $\Omega=R\times R^+$. Multiplying it with the three equations in (3.35) and then integrating them over their zones within Ω, one obtains

$$\int_0^\infty\int_{-\infty}^\infty\frac{v_\Delta(x,t+\Delta t)-v_\Delta(x,t)}{\Delta t}\phi(x,t)dxdt$$
$$+\int_0^\infty\int_{-\infty}^{x_{I-1/2}}\frac{f_\Delta(x+\Delta x_1/2,t)-f_\Delta(x-\Delta x_1/2,t)}{\Delta x_1}\phi(x,t)dxdt$$
$$+\int_0^\infty\int_{x_{I+1/2}}^\infty\frac{f_\Delta(x+\Delta x_2/2,t)-f_\Delta(x-\Delta x_2/2,t)}{\Delta x_2}\phi(x,t)dxdt$$
$$+\int_0^\infty\int_{x_{I-1/2}}^{x_{I+1/2}}\frac{f_\Delta(\alpha_2 x+\beta_2\Delta x_2/2,t)-f_\Delta(\alpha_1 x-\beta_1\Delta x_2/2,t)}{(\Delta x_1+\Delta x_2)/2}\phi(x,t)dxdt=0 \tag{3.36}$$

The first integral in the above is calculated as in the proof of Theorem 3.2.1. For the second integral, substitution of variables yields

$$\int_0^\infty \int_{-\infty}^{x_{I-1/2}} \frac{f_\Delta(x+\Delta x_1/2,t)-f_\Delta(x-\Delta x_1/2,t)}{\Delta x_1}\phi(x,t)dxdt$$
$$=\int_0^\infty \int_{-\infty}^{x_{I-1}} \frac{\phi(x'-\Delta x_1/2,t)-\phi(x'+\Delta x_1/2,t)}{\Delta x_1} f_\Delta(x',t)dx'dt$$
$$+\int_0^\infty \int_{x_{I-1}}^{x_I} \frac{\phi(x'-\Delta x_1/2,t)f_\Delta(x',t)}{\Delta x_1}dx'dt$$
$$=\int_0^\infty \int_{-\infty}^{x_{I-1}} \frac{\phi(x-\Delta x_1/2,t)-\phi(x+\Delta x_1/2,t)}{\Delta x_1} f_\Delta(x,t)dxdt$$
$$+\int_0^\infty \phi(\xi-\Delta x_1/2,t)f_\Delta(x_{I-1/2},t)dt$$

Here, as a step function, $f_\Delta(x',t)=f_\Delta(x_{I-1/2},t)$, $x'\in(x_{I-1},x_I)$. Also, the mean value theorem is applied. Similarly,

$$\int_0^\infty \int_{x_{I+1/2}}^{\infty} \frac{f_\Delta(x+\Delta x_2/2,t)-f_\Delta(x-\Delta x_2/2,t)}{\Delta x_2}\phi(x,t)dxdt$$
$$=\int_0^\infty \int_{x_{I+1}}^{\infty} \frac{\phi(x'-\Delta x_2/2,t)-\phi(x'+\Delta x_2/2,t)}{\Delta x_2} f_\Delta(x',t)dx'dt$$
$$-\int_0^\infty \int_{x_I}^{x_{I+1}} \frac{\phi(x'+\Delta x_2/2,t)f_\Delta(x',t)}{\Delta x_2}dx'dt$$
$$=\int_0^\infty \int_{x_{I+1}}^{\infty} \frac{\phi(x-\Delta x_2/2,t)-\phi(x+\Delta x_2/2,t)}{\Delta x_2} f_\Delta(x,t)dxdt$$
$$-\int_0^\infty \phi(\eta+\Delta x_2/2,t)f_\Delta(x_{I+1/2},t)dt$$

and

$$\int_0^\infty \int_{x_{I-1/2}}^{x_{I+1/2}} \frac{f_\Delta(\alpha_2 x+\beta_2\Delta x_2/2,t)-f_\Delta(\alpha_1 x-\beta_1\Delta x_2/2,t)}{(\Delta x_1+\Delta x_2)/2}\phi(x,t)dxdt$$
$$=\int_0^\infty \int_{x_I}^{x_{I+1}} \frac{f_\Delta(x',t)\phi((x'-\beta_2\Delta x_2)/\alpha_2,t)}{\alpha_2(\Delta x_1+\Delta x_2)/2}dx'dt$$
$$-\int_0^\infty \int_{x_{I-1}}^{x_I} \frac{f_\Delta(x',t)\phi((x'+\beta_1\Delta x_1/2)/\alpha_1,t)}{\alpha_1(\Delta x_1+\Delta x_2)/2}dx'dt$$
$$=\int_0^\infty \int_{x_{I-1/2}}^{x_I+1/2} \frac{f_\Delta(x_{I+1/2},t)\phi(x,t)-f_\Delta(x_{I-1/2},t)\phi(x,t)}{(\Delta x_1+\Delta x_2)/2}dxdt$$
$$=\int_0^\infty \phi(\zeta,t)\left(f_\Delta(x_{I+1/2},t)-f_\Delta(x_{I-1/2},t)\right)dt$$

In the above, $\xi \in (x_{I-1}, x_I)$, $\eta \in (x_I, x_{I+1})$, and $\zeta \in (x_{I-1/2}, x_{I+1/2})$. As a result, (3.36) becomes

$$\int_{\Delta t}^{\infty}\int_{-\infty}^{\infty}\frac{\phi(x,t-\Delta t)-\phi(x,t)}{\Delta t}v_{\Delta}(x,t)dxdt-\int_{0}^{\Delta t}\int_{-\infty}^{\infty}\frac{\phi(x,t)v_{\Delta}(x,t)}{\Delta t}dxdt$$
$$+\int_{0}^{\infty}\int_{-\infty}^{x_{I-1}}\frac{\phi(x-\Delta x_1/2,t)-\phi(x+\Delta x_1/2,t)}{\Delta x_1}f_{\Delta}(x,t)dxdt$$
$$+\int_{0}^{\infty}\int_{x_{I+1}}^{\infty}\frac{\phi(x-\Delta x_2/2,t)-\phi(x+\Delta x_2/2,t)}{\Delta x_2}f_{\Delta}(x,t)dxdt$$
$$+\int_{0}^{\infty}(\phi(\xi-\Delta x_1/2,t)-\phi(\zeta,t))\,f_{\Delta}(x_{I-1/2},t)dt$$
$$-\int_{0}^{\infty}(\phi(\eta+\Delta x_2/2,t)-\phi(\zeta,t))\,f_{\Delta}(x_{I+1/2},t)dt=0$$

Let Δx_1, Δx_2, and Δt tend to zero and notice that the $\phi(x,t)$ is continuous and has finite support, the last two integrals on the left-hand side go to zero, and the aforementioned equation becomes (3.4). Therefore, the numerical solution tends to a weak solution. This completes the proof of the theorem. □

In practical computation, conservative interface algorithms usually outperform nonconservative interface algorithms. They can overcome problems in solutions associated with nonconservative interface algorithms, especially when discontinuities are present in the solutions. One of such problems is inaccuracy in the location and strength of a discontinuity, such as a shock (a wave across which flow variables jump in values) crossing a grid interface [12]. Two examples are presented as follows.

Example 3.4.1 Consider the computation of a 1D inviscid flow, particularly an initial value problem of Euler equations (3.2) and (3.3) with the following initial condition:

$$\begin{cases}(\rho_l,u_l,p_l)=(1,2,1), & x<5\\(\rho_r,u_r,p_r)=(1,-2,1), & x>5\end{cases}$$

which is a piecewise constant with a discontinuity at $x = 5$. Here and hereafter, subscripts l and r indicate the left and right sides of the initial discontinuity, respectively. Also, in this chapter, if not specified, the units for time, length, velocity, density, and pressure may be considered as in s, m, m/s, kg/m^3, and Pa, respectively. This is a Riemann problem of the 1D Euler equations, or a shock tube problem, and its analytical solution is available [7, 14, 15]. In the Riemann resolution, or the solution of the Riemann problem, one shock travels to the left, and another shock propagates to the right.

The computation of the zonal method adopts patched grids, see Fig. 3.2a, and it uses the Lax–Wendroff scheme in the two grids. At the interface, both conservative interface algorithm (3.26) and nonconservative interface algorithm (3.29) are

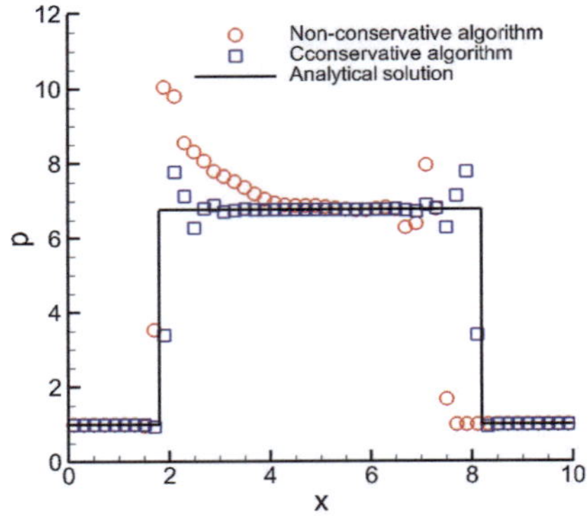

Fig. 3.3 Solution of moving shocks, $t = 3.5$. $\Delta x_1 = \Delta x_2 = 0.2$, $x_I = 6.1$, and $CFL = 0.95$

adopted The Courant–Friedrichs–Lewy (CFL) number is used to assure the stability of the computation. Particularly given grid spacing Δx_k ($k = 1, 2$), the time step is determined via $\Delta t |v^n_{k,i}|/\Delta x_k \leq CFL$, or $\Delta t = min_{i,k}(CFL \Delta x_k/|v^n_{k,i}|)$.

The computational result is presented in Fig. 3.3. Note that when the conservative interface algorithm is used, the interface algorithm becomes the Lax–Wendroff scheme, and the computation is actually that on a single grid by the Lax–Wendroff scheme. The difference in the solutions for the two shocks associated with the conservative and nonconservative interface algorithms directly shows their effects. Particularly, as seen in the figure, the simulation captures the strengths and positions of the left and right shocks, shown by the analytical solution, when the conservative interface algorithm is adopted. However, it becomes higher in strength for the left shock and slower in speed for the right shock when the neoconservative interface algorithm is adopted. It is interesting to note that when the nonconservative inerface algorithm is employed, unlike that for the right shock that passes the interface, still the numerical solution for the left shock becomes inaccurate in strength even it does not pass the interface. Additionally, the solution associated with the conservative interface algorithm exhibits oscillations at the shocks. Such oscillations are not related to the interface algorithms but artificial or numerical oscillations inherited from the Lax–Wendroff scheme [8].

Example 3.4.2 This is another example illustrating the advantages of conservative interface algorithms. Consider the following 1D inviscid flow problem:

$$\begin{cases} (\rho_l, u_l, p_l) = (0.25, 4, 0.2857142), & x < 5 \\ (\rho_r, u_r, p_r) = (1, 1, 3.2857142), & x > 5 \end{cases} \tag{3.37}$$

which is a Riemann problem. The solution exhibits a stationary shock at $x = 5$.

Two patched grids are used, and their interface is right at the shock location. The Lax–Wendroff scheme is adopted in both subdomains. As seen in Fig. 3.4, when nonconservative algorithm (3.29) is adopted, the simulated Riemann resolution presents a stationary discontinuity at the interface and a shock moving away from

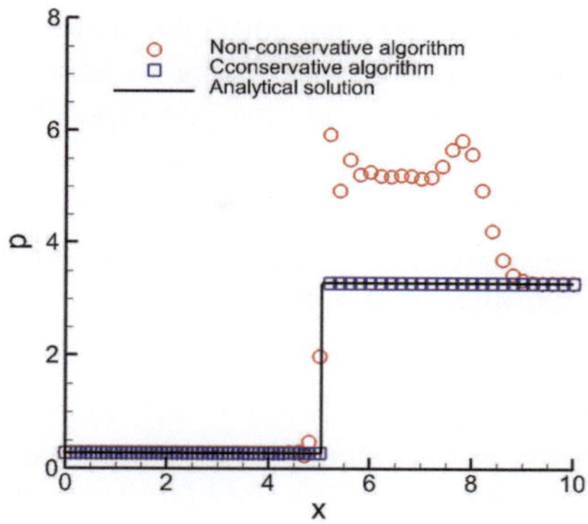

Fig. 3.4 Computation of a stationary shock, $t = 1$. $\Delta x_1 = 0.025$, $\Delta x_2 = 0.05$, $x_I = 5$, and $CFL = 0.95$

there to the right. In comparison with the analytical solution, it is seen that the stationary discontinuity at the interface is wrong in strength. Additionally, the shock is a non-physical solution, or a solution that does not occur in the physical world. However, when the conservative interface is adopted, the numerical solution correctly resolves the stationary shock and produces no non-physical solution.

3.4.2 Problem of Conservative Interface Algorithm

Computation with Different Schemes Conservative interface algorithms have been the goal in interface treatments. However, such algorithms may suffer from problems, particularly when they are constructed by enforcing numerical fluxes at interfaces, as in the form of (3.26), (3.27), and (3.28). Two situations of the problems will be discussed on the basis of early studies [16, 17]: (1) the two subdomains have the same grid spacing but different numerical schemes, and (2) the two subdomains have the same scheme but different grid spacings.

Now, let us discuss the first situation. Suppose two grids patch with each other (Fig. 3.2a), $\Delta x_1 = \Delta x_2 = \Delta x$, and algorithm (3.26) is employed at the interface. Note that a similar discussion for algorithm (3.27) and (3.28) on overlapping grids in Fig. 3.2b can also be made. Consider an interface algorithm using grid nodes from $I - l_1$ to $I + l_2$, l_1 and l_2 being positive integers. Let u be a smooth function of (x, t), in the form of Eq. (3.12), with q_I^{n+1} being its H, the interface algorithm leads to

$$
\begin{aligned}
& u_I^{n+1} - q_I^{n+1}(u_{I-l_1}^n, \cdots, u_{I+l_2}^n) \\
=& u_I^{n+1} - u_I^n + \lambda_I(\overline{f}^2_{I+1/2} - \overline{f}^1_{I-1/2}) \\
=& u_I^{n+1} - u_I^n + \lambda_I(\overline{f}^2_{I+1/2} - \overline{f}^2_{I-1/2}) + \lambda_I(\overline{f}^2_{I-1/2} - \overline{f}^1_{I-1/2}) \\
=& \Delta t(r + \tilde{r})
\end{aligned}
$$

Here and hereafter, for convenience, the location of u (also a numerical solution, e.g., v) is referred to using its node indices only, without distinguishing between the subdomains and the interface. For instance, $u^n_{I-l_1}$ and $u^n_{I+l_2}$ denote the values of u at nodes $I - l_1$ and $I + l_2$ in subdomains 1 and 2, respectively. As shown in (3.12), r is the truncation error of the scheme in subdomain 2, and $\tilde{r} = (\lambda_I/\Delta t)(\overline{f}^2_{I-1/2} - \overline{f}^1_{I-1/2})$. Additionally, $r + \tilde{r}$ is the truncation error of the interface algorithm. In a computation with $\lambda_I = const.$, apparently, we have the following statement according to the requirement on the truncation error.

Proposition 3.4.3 *In a computation using (3.26), let the two subdomains have the same grid spacings but different algorithms. Then, (1) the interface algorithm has the consistency with the PDE if and only if*

$$r + \tilde{r} \to 0 \tag{3.38}$$

as $\Delta x, \Delta t \to 0$. *(2) When* r *and* $\tilde{r}$ *do not affect each other, the interface algorithm is consistent if and only if*

$$\overline{f}^2_{I-1/2} - \overline{f}^1_{I-1/2} = \Delta t \tilde{r} \tag{3.39}$$

where $\tilde{r} \to 0$ *as* $\Delta x, \Delta t \to 0$.

In the aforementioned, r and $\tilde{r}$ do not affect each other in the sense that they do not cancel each other and do not generate a singularity so the order of $(r + \tilde{r})$ depends on the orders of both r and $\tilde{r}$, which is true in general. It is easy to utilize condition (3.39) to check the consistency of an interface algorithm. The consistency of an interface algorithm ensures it approximates the underlying PDE, and the inconsistency should be avoided. Note that it is equivalent to apply (3.39) at $(I + 1/2)$ when examining the consistency.

To facilitate the discussion of inconsistency at a grid interface, let us use the Lax–Wendroff and the Lax–Friedrichs scheme in the subdomains 1 and 2, respectively, and interface algorithm (3.26) to compute initial value problem (3.1). In this case, one has

$$\begin{aligned}
&\overline{f}^{LF}_{I+1/2} - \overline{f}^{LW}_{I+1/2}\\
=&\{(f_I + f_{I+1})/2 - (u_{I+1} - u_I)/(2\lambda)\}\\
&- \{(f_I + f_{I+1})/2 - \lambda a_{I+1/2}(f_{I+1} - f_I)/2\}\\
=&\{(f_I + f_{I+1})/2 - (u_{I+1} - u_I)/(2\lambda)\}\\
&- \{(f_I + f_{I+1})/2 - \lambda a^2_{I+1/2}(u_{I+1} - u_I)/2 + \mathcal{O}(\Delta^2)\}\\
=&(1/(2\lambda) - \lambda a^2_{I+1/2}/2)(u_{I+1} - u_I) + \mathcal{O}(\Delta^2)\\
=&\mathcal{O}(\Delta)\\
=&\Delta t \mathcal{O}(\Delta/\Delta t)
\end{aligned}$$

where $\Delta = max(\Delta x, \Delta t)$, $a_{I+1/2}$ is evaluated as in Eq. (3.16), and $1/(2\lambda) - \lambda a^2_{I+1/2}/2 \neq 0$. As long as $1/\lambda - \lambda a^2_{I+1/2} \neq 0$, which implies that the local CFL number does not equal 1, and Δx and Δt are of the same order of magnitude, the aforementioned becomes

$$\overline{f}^{LF}_{I+1/2} - \overline{f}^{LW}_{I+1/2} = \Delta t \mathcal{O}(1) \tag{3.40}$$

that is $\tilde{r} = \mathcal{O}(1)$, which, according to proposition (3.4.3), indicates that interface algorithm (3.26) becomes inconsistent.

Although the aforementioned inconsistency problem due to the difference in discretization schemes in the two subdomains happens only at the interface (note that the consistency holds inside the two subdomains), it could lead to significant numerical errors in computation, as illustrated in the following examples.

Example 3.4.3 Numerical examples of errors due to the inconsistency of the interface algorithm in the computation of two flows are shown in Fig. 3.5. The first flow is a quasi-1D nozzle inviscid flow, whose governing equations are as follows, slightly modified from the Euler equations [14]:

$$Q_t + F(Q)_x = -S(Q, x)$$

in which Q and F are defined as in (3.3). $S(Q, x) = (0, pdA(x)/dx, 0)^T$, with $A(x)$ being the cross-section area, and $A(x) = 1.398 + 0.347tanh(0.8x - 4)$, $0 < x < 10$. Let us compute the quasi-1D nozzle inviscid flow using two patched grids in the zonal method. In the computation, conservative algorithm (3.26) is adopted. The Lax–Wendroff scheme and Lax–Friedrichs scheme are used in the left and right grids, respectively. The computed flow is presented in Figs. 3.5a and b. The flow in Fig. 3.5a is supersonic, with continuous, smooth flow variables. The figure shows that the numerical solution for the flow is somewhat disconnected at the interface, which, compared to the analytical solution, is an artifact. In Fig. 3.5b, the flow is supersonic at the entrance but becomes subsonic at the exit, containing a stationary shock right at the grid interface. It is seen that the numerical solution for the shock is erroneous in both its location and strength against the analytical solution.

The second flow is about a 1D Riemann problem, whose governing equations are the Euler equations. The computation uses two patched grids, and it adopts the Lax–Friedrichs scheme and Lax–Wendroff scheme in the left and right grids, respectively, with conservative interface algorithm (3.26). As seen in Fig. 3.5c, two shocks are initially generated at the interface, and then one travels toward the left, and the other propagates to the right. The right shock passes a grid interface. The figure indicates that the numerical solution becomes artificially discontinuous at the interface. Moreover, the Riemann problem is computed again, but with two overlapping grids and conservative algorithm (3.27) and (3.28) for the interfaces. The computed flow is shown in Fig. 3.5d. As seen in the figure, the artifact still takes place at the interfaces. Additionally, in the grid overlapping region, the solutions in the two subdomains are different, or there are multiple solutions.

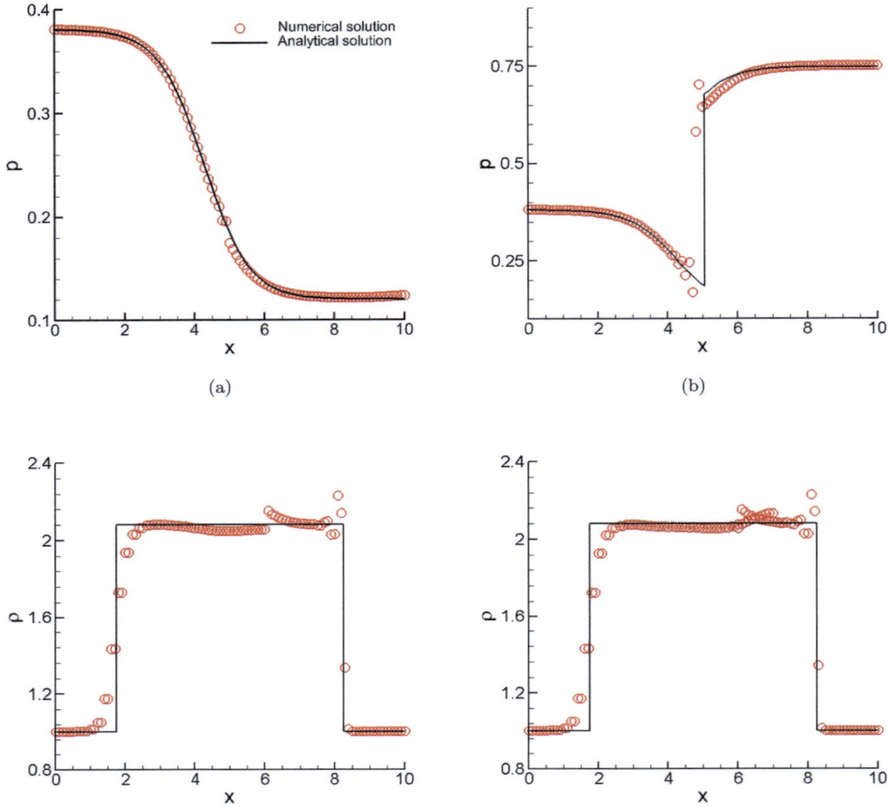

(a) (b) (c) (d)

Fig. 3.5 Computation of inviscid flows using different schemes in subdomains. The Lax-Wendroff and Lax-Friedrichs schemes are employed in subdomains 1 and 2, respectively, in (**a**) and (**b**). The Lax-Friedrichs and the Lax-Wendroff schemes are adopted in subdomains 1 and 2, respectively, in (**c**) and (**d**). $\Delta x_1, \Delta x_2 = 0.1$, and $CFL = 0.95$. (**a**) Supersonic flow in the divergent nozzle. $(\rho, u, p) = (0.502, 1.299, 0.381)$ at entrance. $x_I = 5$. (**b**) Subsonic flow (at the exit) in the divergent nozzle. $(\rho, u, p) = (0.502, 1.299, 0.381)$ at entrance, $(\rho, u, p) = (0.776, 0.5, 0.75)$ at exit. $x_I = 5$. (**c**) Riemann problem. $(\rho_l, u_l, p_l) = (1, 1, 1)$, $(\rho_r, u_r, p_r) = (1, -1, 1)$, and $t = 3.5$. $x_I = 6$. (**d**) Same as (**c**) but with overlapping grids, $x_1 = 6$, $x_I = 7$

If subdomain schemes are different but in the same order of accuracy, the aforementioned inconsistency may not occur. To illustrate this, let us use the Lax–Wendroff and MacCormack schemes in the left and right subdomains, respectively. Note that details for the MacCormack are available in [8]. Then, it is obtained that

$$
\begin{aligned}
&\overline{f}^{MacC}_{I+1/2} - \overline{f}^{LW}_{I+1/2} \\
=&\{f(u^n_I - \lambda(f_{I+1} - f_I)/2) + f_{I+1}/2\} \\
&- \{(f_I + f_{I+1})/2 - \lambda a_{I+1/2}(f_{I+1} - f_I)/2\}
\end{aligned}
$$

$$
\begin{aligned}
=&\{(f_I/2 - \lambda a_I(f_{I+1} - f_I)/2 + \lambda\mathcal{O}(\Delta^2) + f_{I+1}/2\} \\
&- \{(f_I + f_{I+1})/2 - \lambda a_{I+1/2}(f_{I+1} - f_I)/2 + \lambda\mathcal{O}(\Delta^2)\} \\
=&\lambda(a_{I+1/2} - a_I)(f_{I+1} - f_I)/2 + \mathcal{O}(\Delta^2) \\
=&\Delta t\mathcal{O}(\Delta)
\end{aligned}
$$

which leads to

$$
\overline{f}^{MacC}_{I+1/2} - \overline{f}^{LW}_{I+1/2} = \Delta t\mathcal{O}(\Delta) \tag{3.41}
$$

According to Proposition 3.4.3, it is known that the interface algorithm is consistent. This is also true if we use the Lax–Wendroff and a second-order total variation diminishing (TVD) scheme of Harten in subdomains 1 and 2, respectively. The TVD scheme's numerical flux is constructed in such a way that [14]

$$
\overline{f}^{TVD}_{I+1/2} - \overline{f}^{LW}_{I+1/2} = \mathcal{O}(\Delta^2)
$$

it is readily shown that still (3.39) holds, and thus, the interface algorithm is consistent. As a result, interface algorithm (3.26), (3.27), and (3.28) are consistent, and these conservative interface algorithms exhibit no inconsistency problem when different but same-order accurate schemes are used in subdomains. In this situation, the artifacts and errors in the aforementioned simulations will be removed or reduced, as demonstrated in the following example.

Example 3.4.4 Let us compute the inviscid flows of Figs. 3.5a and d, keeping all the settings the same except that the MacCormack scheme is used to replace the Lax-Friedrichs scheme in subdomains 2 and 1, respectively. As shown in Eq. (3.41), a conservative interface algorithm, e.g., (3.26), or (3.27) and (3.28), will be consistent. The errors of numerical solutions in the two figures basically disappear, see Fig. 3.6.

Computation with Different Grid Spacings Now, we examine the second situation, that is, the same scheme but different grid spacings are used in subdomains 1 and 2. Again, here the discussion is made using algorithm (3.26) for the patched grids (Fig. 3.2a). In this situation, still, it is expected that its conclusion holds for algorithm (3.27) and (3.28) for overlapping grids (Fig. 3.2b). Now, let us consider the adoption of the Lax–Friedrichs scheme in both subdomains but with different grid spacings in them. From (3.11) and (3.26) and by Taylor expansion with a smooth function, $u(x, t)$, it is readily derived that

$$
u_I^{n+1} - q_I^{n+1}(u_{I-1}^n, u_I^n, u_{I+1}^n) = \Delta t\left(\frac{\Delta x_2 - \Delta x_1}{\Delta t}u_{xI}^n + \mathcal{O}(\Delta)\right)
$$

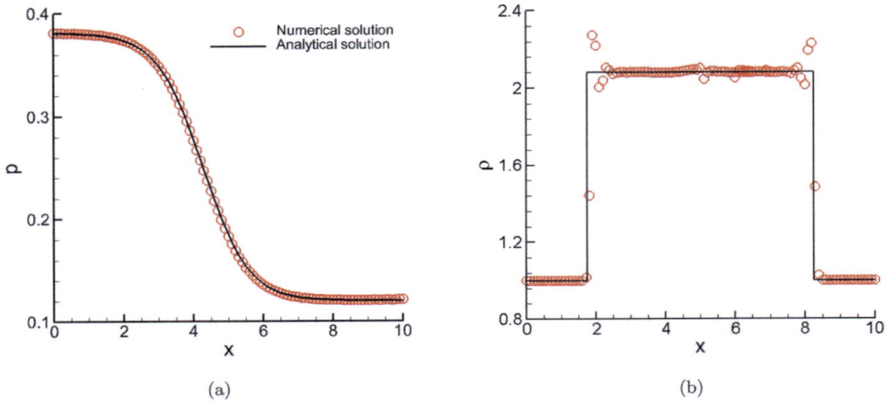

Fig. 3.6 Computation of inviscid flows using different but same accuracy-order schemes in subdomains. In (**a**) and (**b**), all are the same as those in Figs. 3.5a and d, respectively, except that the MacCormack scheme replaces the Lax-Friedrichs scheme

It is seen that the terms in the parenthesis will not tend to zero as long as Δx_1 and Δx_2 are not equal to each other. When $(\Delta x_2 - \Delta x_1)/\Delta t \sim \mathcal{O}(1)$, the aforementioned becomes

$$u_I^{n+1} - q_I^{n+1}(u_{I-1}^n, u_I^n, u_{I+1}^n) = \Delta t \mathcal{O}(1) \tag{3.42}$$

Then, according to (3.12), it is known that the interface algorithm becomes inconsistent, or it lacks consistency with the PDE. In fact, the aforementioned derivation indicates that, instead of the conservation law in (3.1), the interface algorithm approximates a different equation:

$$u_t + f(u)_x = \frac{\Delta x_2 - \Delta x_1}{\Delta t} u_x + \mathcal{O}(\Delta)$$

As the grid spacings of the two grids are more different, the magnitude of $(\Delta x_2 - \Delta x_1)/\Delta t$ becomes bigger, and such inconsistency is anticipated to be more severe. Similarly, when the Lax–Wendroff scheme is used in the two subdomains,

$$u_I^{n+1} - q_I^{n+1}(u_{I-1}^n, u_I^n, u_{I+1}^n) = \Delta t \left(\frac{\Delta t(\Delta x_2 - \Delta x_1)}{2(\Delta x_1 + \Delta x_2)^2} a(u_I^n)^2 u_{x}{}_I^n + \mathcal{O}(\Delta) \right)$$

which, when $\Delta t(\Delta x_2 - \Delta x_1)/[2(\Delta x_1 + \Delta x_2)^2] \sim \mathcal{O}(1)$, also leads to (3.42) and inconsistency at the interface. In this situation, the interface algorithm approximates

$$u_t + f(u)_x = \frac{\Delta t(\Delta x_2 - \Delta x_1)}{2(\Delta x_1 + \Delta x_2)^2} a(u)^2 u_x + \mathcal{O}(\Delta)$$

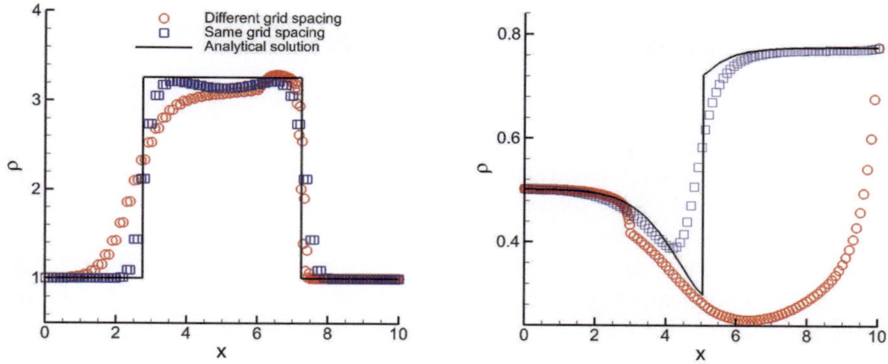

(a) (b)

Fig. 3.7 Computation of inviscid flows using different grid spacings in subdomains. The Lax-Friedrichs scheme is used. $CFL = 0.95$. (**a**) Riemann problem. $(\rho_l, u_l, p_l) = (1, 2, 1)$, $(\rho_r, u_r, p_r) = (1, -2, 1)$, and $t = 2.5$. $x_I = 5$. Same grid spacings: $\Delta x_1, \Delta x_2 = 0.1$. Different grid spacings: $\Delta x_1 = 0.1$, $\Delta x_2 = 0.03333$. (**b**) Subsonic flow (at the exit) in the divergent nozzle. $(\rho, u, p) = (0.502, 1.299, 0.381)$ at entrance, $(\rho, u, p) = (0.776, 0.5, 0.75)$ at exit. $x_I = 3$. Same grid spacing: $\Delta x_1, \Delta x_2 = 0.1$, different grid spacings: $\Delta x_1 = 0.025$, $\Delta x_2 = 0.1$

It is expected that the inconsistency problem of the conservative interface algorithm due to the difference in grid spacing of the two grids may lead to problems in numerical solutions. Relevantly, analysis also indicates that conservative interface algorithm (3.26) is only weakly stable, and it may lead to nonunique solutions in the overlapping regions of subdomains. For these, the reader is referred to [18, 19].

Example 3.4.5 This example is about flows for a Riemann problem and in a divergent nozzle. Still, the Lax-Friedrichs scheme is applied, and conservative interface algorithm (3.26) is adopted. For the Riemann problem, first, the same grid spacing is used in the two subdomains (i.e., the two grids comprise a uniform grid), and the numerical solution performs well, as seen in Fig. 3.7a. Second, the grid is refined in the right subdomain. Then, the solution becomes problematic (Fig. 3.7a); while it gets better in terms of shock strength in the right subdomain after the mesh refinement, its accuracy in the left subdomain degrades in comparison to that with the same grid spacing. For the divergent nozzle flow, the problem becomes even worse (Fig. 3.7b); the solution with different grid spacings is wrong and non-physical in the right subdomain after the mesh refinement. Actually, it cannot settle down to a steady state but keeps changing over time.

This example tells an interesting fact: in the computation by the multiblock method, mesh refinement only within one subdomain may decrease a solution's accuracy or may lead to a wrong solution instead of improving it. Such a problem is attributed to the inconsistency in the interface algorithm, which is formulated in Eq. (3.42).

Remedy for Inconsistency Problem The inconsistency problem of a conservative interface algorithm constructed by enforcing numerical flux is related to how its numerical fluxes are constructed. In the previously discussed two scenarios, particularly the difference in discretization accuracy and in grid spacing of the two subdoamins, the inconsistency problem comes from the fact that, contingently, the interface algorithms take numerical fluxes constructed with grid spacing and time step, see (3.14) and (3.17) for instance. Then, it is expected that a conservative interface algorithm may exhibit no inconsistency problem if it adopts numerical fluxes independent of grid spacing and time step. For example, in the Godunov scheme, the numerical flux comes from the analytical solution of the local solution of the Riemann problem, and it is not related to grid spacing and time step, as seen in (3.18). This is demonstrated in the following example.

Example 3.4.6 Let us compute the aforementioned divergent nozzle flow and keep all settings the same. However, the Godunov scheme is adopted in both subdomains. In this situation, the numerical fluxes, particularly at nodes $I - 1/2$ and $I + 1/2$, of the interface algorithm (3.26) are computed using the fluxes of the Godunov scheme (see Eq. (3.18)):

$$\overline{f}^1_{I-1/2} = \overline{f}^G_{I-1/2} = f(u(x_{I-1/2}/t, v^n_{I-1}, v^n_I))$$
$$\overline{f}^2_{I+1/2} = \overline{f}^G_{I+1/2} = f(u(x_{I+1/2}/t, v^n_I, v^n_{I+1}))$$

which are independent of grid spacing and time step. In the Godunov scheme, a local Riemann problem for the Euler equations, particularly the corresponding local solution for the Euler equations, between every two adjacent grid nodes, is solved to evaluate the numerical flux. The result is shown in Fig. 3.8, which shows that no wrong or non-physical solution is present, and the whole computation works well.

Example 3.4.7 This computation concerns a shock reflection on a wall, an initial-boundary value problem. The governing equations are the 2D Euler equations, which are an extension of 1D Euler equations (3.2) and (3.3). The computational domain is L in the x-direction and H in the y-direction, with $L = 0.00762$ m and

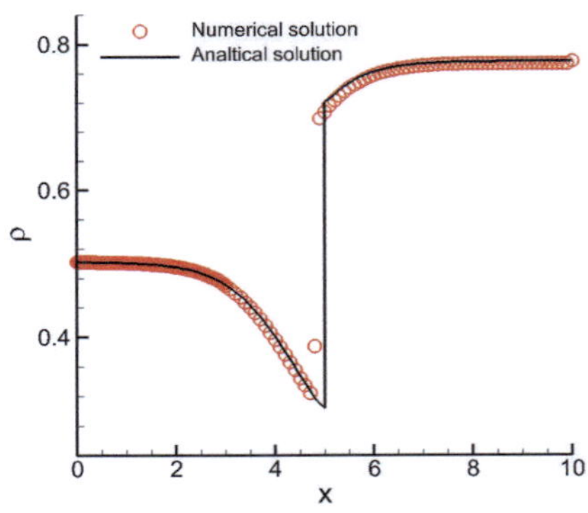

Fig. 3.8 Subsonic flow (at the exit) in the divergent nozzle. All parameters remain the same as in Fig. 3.7b, except that the Godunov scheme is used in the two subdomains and the interface algorithm

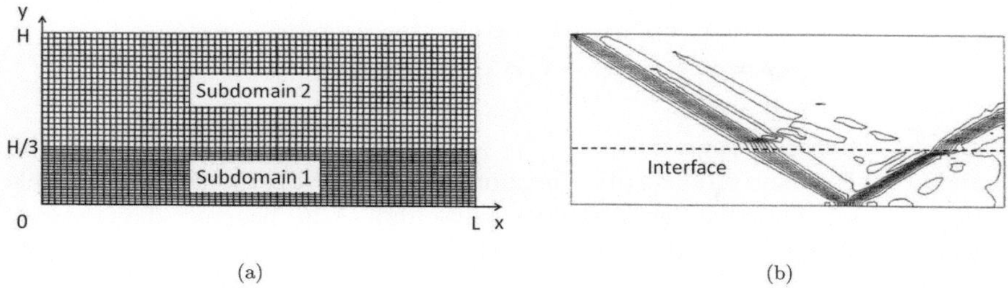

Fig. 3.9 Computation of shock reflection. (**a**) Subdomains and meshes. (**b**) Computed contours of pressure

$H = 0.00313$ m, see Fig. 3.9a. The flow entrance is on the left side ($x = 0$), and the wall is at the bottom ($y = 0$). The conditions at the entrance: $Ma = 2$ (Ma - Mach number), $p = 1.65 \times 10^5$ Pa, $T = 293$°K. The shock's angle of incidence is 32.6°. The initial condition in the entire computational domain is specified as a uniform flow identical to that at the entrance.

The domain is divided into Subdomain 1, $0 < x < L$ and $0 < y < H/3$, Subdomain 2, $0 < x < L$ and $H/3 < y < H$, and they patch at $y = H/3$, see Fig. 3.9a. Both subdomains 1 and 2 use a grid 71×21, with 71 and 21 being in x- and y-directions, respectively. The two grids patch as a bigger grid $i \times j$ =71×41, with i and j being in x- and y-directions, respectively, for the whole computational domain. The interface of the two grids is along $y = y_J$ ($1 \leq i \leq 71$, $J = 21$) in the bigger grid. The MacCormack scheme is used in both subdomains.

The simulation presents an unsteady flow process: Shortly after the computation starts, a shock forms near the top boundary, and it gradually swings downward, passing the interface, touching the wall, and then reflecting at it. When conservative interface algorithm (3.26) is used in the vertical direction, or along j-grid lines), the computation is unstable; it blows up in a short time after the shock reflects at the wall, leading to no stable solution.

As a remedy, instead of the MacCormack scheme, the numerical flux of the Godunov scheme, which is evaluated by the local solutions of Riemann problems as in Eq. (3.18), is adopted in interface algorithm (3.26). Particularly, let $V^n_{i,j}$ be the numerical solution vector to approximate $Q(x_i, y_j, t_n)$, the solution vector of the 2D Euler equations. The numerical flux in the y-direction at $j = J - 1/2$, $J + 1/2$ is computed by

$$\overline{F}^G_{i,J-1/2} = F(Q(y_{J-1/2}/t, V^n_{i,J-1}, V^n_{i,J}))$$
$$\overline{F}^G_{i,J+1/2} = F(Q(y_{J+1/2}/t, V^n_{i,J}, V^n_{i,J+1}))$$

where F is the y-flux of the 2D Euler equations, and $\overline{F}$ is the corresponding numerical flux. $Q(y_{J-1/2}/t, V^n_{i,J-1}, V^n_{i,J})$ and $Q(y_{J+1/2}/t, V^n_{i,J}, V^n_{i,J+1})$ are the analytical solution of the Riemann problem at $j = J - 1$, J as follows:

$$\begin{cases} Q_t + F(Q)_y = 0, \ t > 0 \\ Q = V^n_{i,j}, \qquad x = x_i, \ y < y_{j+1/2}, \ t = 0 \\ Q = V^n_{i,j+1}, \qquad x = x_i, \ y > y_{j+1/2}, \ t = 0 \end{cases}$$

Here is a 1D Riemann problem in y-direction, and its analytical solution is available [7, 8, 15]. As a result, the flux of interface algorithm (3.26) becomes the Godunov scheme. Additionally, the aforementioned Godunov–scheme flux is also used in the computation of inner nodes at nodes along lines $J - 1$ and $J + 1$, and thus, a 2D version of conservation condition (3.22) is maintained. With the remedy in the numerical flux, the computation becomes stable, and it captures the flow process; after the shock reflects at the wall, it gradually settles down at a steady solution (Fig. 3.9b).

3.5 Nonconservative Interface Algorithm

3.5.1 Analysis and Property

Nonconservative interface algorithms are widely used, especially in the simulation of real-world problems with complex geometries. Although pursuing conservative interface algorithms has been mainstream in research, frequently, it is not straightforward but challenging to construct them in practical computation, especially in situations of high dimensions and overset grids. In these situations, however, a nonconservative algorithm or treatment, e.g., interpolation, is possible, and it is relatively straightforward and easy to realize. This section discusses the nonconservative interface treatment and explores its properties, with the content developed from [16, 17, 20, 21].

Let us start the discussion with the total variation of a numerical solution for initial value problem (3.1), which is defined as

$$TV(v_i) \equiv \sum_{i=-\infty}^{\infty} \mid v_i - v_{i-1} \mid$$

Here, it is assumed that this solution has finite support. Suppose the problem is computed by an explicit scheme, particularly (3.24) and (3.29). Notice that linear interpolation (3.29) implies that $|v^n_I - v^n_{I-1}| + |v^n_{I+1} - v^n_I| = |v^n_{I+1} - v^n_{I-1}|$, one has

$$\begin{aligned} TV(v^n_i) &= \textstyle\sum_{i \le I-1, i \ge I+2} \mid v^n_i - v^n_{i-1} \mid + \mid v^n_{I+1} - v^n_I \mid + \mid v^n_I - v^n_{I-1} \mid \\ &\ge \textstyle\sum_{i \le I-1, i \ge I+2} \mid v^n_i - v^n_{i-1} \mid + \mid v^n_{I+1} - v^n_{I-1} \mid \\ &= TV^I(v^n_i) \end{aligned}$$

where $TV^I(v^n_i)$ stands for the total variation of a numerical solution obtained with interpolation (3.29). Therefore, we have the following statement.

Proposition 3.5.1 *In the computation of (3.24), interface algorithm (3.29) produces the least amount of total variation.*

Reduction of the total variation has been a target in research to suppress numerical oscillations, and it is the basic idea for TVD schemes [14]. The aforementioned proposition indicates that a nonconservative interface algorithm based on interpolation helps reduce numerical oscillations, and numerical evidence is available in the next subsection.

As discussed previously, the solution for problem (3.1) satisfies conservation condition (3.6). When a nonconservative interface algorithm is adopted, such a condition, or, its numerical version (3.22), is not guaranteed. Let conservation error between time t_m and t_n be defined as

$$S(t_m, t_n) = \int_{-\infty}^{+\infty} (v_\Delta(x, t_n) - v_\Delta(x, t_m))dx \tag{3.43}$$

The conservation error is directly related to the error of the numerical solution. Consider a step function from the numerical solution as defined in (3.31). Utilizing initial value evaluation (3.23), one has

$$\begin{aligned}
&\int_{-\infty}^{+\infty} v_\Delta(x, 0)dx \\
&= \int_{-\infty}^{x_{I-1/2}} v_\Delta(x, 0)dx + \int_{x_{I-1/2}}^{x_{I+1/2}} v_\Delta(x, 0)dx + \int_{x_{I+1/2}}^{+\infty} v_\Delta(x, 0)dx \\
&= \int_{-\infty}^{x_{I-1/2}} u_0(x)dx + \int_{x_{I-1/2}}^{x_{I+1/2}} u_0(x)dx + \int_{x_{I+1/2}}^{+\infty} u_0(x)dx \\
&= \int_{-\infty}^{+\infty} u(x, 0)dx \\
&= \int_{-\infty}^{+\infty} u(x, t)dx
\end{aligned}$$

Here, conservation condition (3.6) for the analytical solution to initial value problem (3.1) is used. The aforementioned leads to

$$\begin{aligned}
S(0, t) &= \int_{-\infty}^{+\infty} (v_\Delta(x, t) - v_\Delta(x, 0))dx \\
&= \int_{-\infty}^{+\infty} (v_\Delta(x, t) - u(x, t))dx
\end{aligned}$$

which leads to the following conclusion.

Proposition 3.5.2 *The absolute value of the conservation error between the initial time and a later time has the error of the numerical solution as its upper bound:*

$$| S(0, t) | \leq \int_{-\infty}^{+\infty} | v_\Delta(x, t) - u(x, t) | \, dx = ||v_\Delta(x, t) - u(x, t)||_1 \quad (3.44)$$

Now, consider computation of (3.24) via nonconservative interface algorithm (3.29). By the definition of conservation error (3.43), one can readily derive the conservation error between two time steps as

$$\begin{aligned} S(t_n, t_{n+1}) = & \frac{\Delta t \Delta x_2}{\Delta x_1}(\overline{f}^1_{I-3/2} - \overline{f}^1_{I-1/2}) \\ & + \Delta t(\overline{f}^1_{I-1/2} - \overline{f}^2_{I+1/2}) \\ & + \frac{\Delta t \Delta x_1}{\Delta x_2}(\overline{f}^2_{I+1/2} - \overline{f}^2_{I+3/2}) \end{aligned}$$

The magnitude of conservation error depends on the numerical solutions near the interface. When there is no discontinuity near x_I and $v_i - v_{i+1} = \mathcal{O}(\Delta)$, it is readily shown from the above that

$$S(t_n, t_{n+1}) = \mathcal{O}(\Delta^2) \quad (3.45)$$

Here, $\Delta = max(\Delta t, \Delta x_1, \Delta x_2)$. When there is a discontinuity near the interface, (3.45) is usually no longer true and $S(t_n, t_{n+1})$ will become large in magnitude. These are further concluded as follows.

Proposition 3.5.3 *If the solution to (3.24) is bounded within a neighborhood of* x_I, *then* $S(0, T)$ *is also bounded for a computation from* 0 *to* T *in time.*

Proof It is known from (3.43) that

$$\begin{aligned} S(0, T) &= \textstyle\sum_{n=0}^{N-1} S_\Delta(t_n, t_{n+1}) \\ &= \textstyle\sum_{n=0}^{N-1} \Delta t(-\overline{f}^1_{I-1/2} + \overline{f}^2_{I+1/2}) + (v_I^N - v_I^0)(\Delta x_1 + \Delta x_2)/2 \end{aligned}$$

where $\Delta t = T/N$. This equation yields

$$|S(0, T)| \leq \Delta t \textstyle\sum_{n=0}^{N-1} | - \overline{f}^1_{I-1/2} + \overline{f}^2_{I+1/2}| + |v_I^N - v_I^0|(\Delta x_1 + \Delta x_2)/2$$

In view that the numerical solution is bounded at the interface, the numerical fluxes in subdomains 1 and 2 are also bounded there. As a result,

$$|v_i^n| < c_1, \quad |\overline{f}^1_{I-1/2}|, \quad |\overline{f}^2_{I+1/2}| < c_2$$

where $c_1, c_2 = const$. The aforementioned leads to

$$|S(0, T)| \leq C$$

in which $C = c_1(\Delta x_A + \Delta x_B) + 2c_2 T$. This completes the proof of the proposition. □

The aforementioned two propositions help in understanding the conservation error in practical computation. Actually, it can further be shown that, as grid spacing goes to zero, the conservation error will tend to zero regardless of the smoothness of the solution [22].

In the following, we present a conclusion on whether a converged solution is a weak solution. Still, let us define a step function of the solution by (3.31) and build a step function of numerical flux by (3.32). Then, algorithm (3.24) can be expressed as

$$\frac{v_\Delta(x,t+\Delta t)-v_\Delta(x,t)}{\Delta t}-\frac{f_\Delta(x+\Delta x_1/2,t)-f_\Delta(x-\Delta x_1/2,t)}{\Delta x_1}=0,$$

$$x\in(-\infty,x_{I-1/2})$$

$$\frac{v_\Delta(x,t+\Delta t)-v_\Delta(x,t)}{\Delta t}-\frac{f_\Delta(x+\Delta x_2/2,t)-f_\Delta(x-\Delta x_2/2,t)}{\Delta x_2}=0,$$

$$x\in[x_{I+1/2},+\infty)$$

$$\frac{v_\Delta(x,t+\Delta t)-q_I^{n+1}}{\Delta t}=0,\ x\in[x_{I-1/2},x_{I+1/2}) \tag{3.46}$$

Theorem 3.5.1 *Suppose $v_\Delta(x,t)$ in (3.46) is uniformly bounded and converges to a function $u(x,t)$ as $\Delta\to 0$, where Δ=max(Δx_1, Δx_2, Δt). Then, either of the following conditions is sufficient for $u(x,t)$ to become a weak solution of initial value problem (3.1):*

Condition A: $u(x,t)$ is continuous across the interface x_I, except at the moment t_m, where m is at most countable.

Condition B: $u(x,t)$ is always discontinuous across x_I but

$$\lim_{\Delta\to 0}(f_\Delta(x_{I-1/2},t)-f_\Delta(x_{I+1/2},t))=0,\ \ a.e. \tag{3.47}$$

Proof Multiplying the three equations of (3.46) with a scalar function $\phi(x,t)\in C_0^\infty(R\times R^+)$. Then, by integrating the resulting first, second, and third equations over $(-\infty,x_I-\Delta x_1/2]\times(0,+\infty)$, $(x_I+\Delta x_2/2,+\infty)\times(0,+\infty)$, $(x_I-\Delta x_1/2,x_I+\Delta x_2/2]\times(0,+\infty)$, respectively, and introducing changes of variables, one has

$$\int_{\Delta t}^{+\infty}\int_{-\infty}^{\infty}\frac{\phi(x,t-\Delta t)-\phi(x,t)}{\Delta t}v_\Delta(x,t)dxdt$$

$$-\int_0^{\Delta t}\int_{-\infty}^{\infty}\frac{\phi(x,t)v_\Delta(x,t)}{\Delta t}dxdt+\int_0^{\infty}\int_{x_{I-1/2}}^{x_{I+1/2}}\phi(x,t)\frac{v_\Delta(x,t)-q_I^{n+1}}{\Delta t}dxdt$$

$$+\int_{0}^{+\infty}\int_{-\infty}^{x_{I-1}}\frac{\phi(x-\Delta x_1/2,t)-\phi(x+\Delta x_1/2,t)}{\Delta x_1}f_\Delta(x,t)dxdt$$

$$+\int_{0}^{+\infty}\int_{x_{I+1}}^{+\infty}\frac{\phi(x-\Delta x_2/2,t)-\phi(x+\Delta x_2/2,t)}{\Delta x_2}f_\Delta(x,t)dxdt$$

$$+\int_{0}^{+\infty}\big(\phi(\xi-\Delta x_1/2,t)f_\Delta(x_{I-1/2},t)-\phi(\eta+\Delta x_2/2,t)f_\Delta(x_{I+1/2},t)\big)\,dt=0$$

where $\xi\in(x_{I-1},x_I)$ and $\eta\in(x_I,x_{I+1})$. In the above, the first three integrals come from the terms of temporal derivatives, and the last three integrals result from the terms of spatial derivatives. Details of their derivation may refer to those of proof for Theorems 3.2.1 and 3.4.1. Let $\Delta\to 0$. Then, the third integral becomes zero (note that its integrand tends to the product of the test function and the time derivative of the solution, and the derivative is bounded, a.e.). Moreover, because of the Lebesgue bounded convergence theorem, Proposition 3.2.1, and (3.23), the aforementioned equation becomes

$$\int_{0}^{+\infty}\int_{-\infty}^{+\infty}\big(\phi_t(x,t)u(x,t)+\phi_x(x,t)f(u(x,t))\big)dxdt+\int_{-\infty}^{+\infty}\phi(x,0)u_0(x)dx$$

$$+\int_{0}^{+\infty}\phi(x_I,t)\lim_{\Delta\to 0}\big(f_\Delta(x_{I-1/2},t)-f_\Delta(x_{I+1/2},t)\big)\,dt=0$$

Since $\phi(x,t)$ is an arbitrary function belonging to $C_0^\infty(R\times R^+)$, the aforementioned suggests that $u(x,t)$ is a weak solution of (3.1) if and only if

$$\lim_{\Delta\to 0}(f_\Delta(x_{I-1/2},t)-f_\Delta(x_{I+1/2},t))=0,\qquad a.e.$$

Because either condition A or B leads to the above equation at the limit, $u(x,t)$ is a weak solution of (3.1). This completes the proof of the theorem. □

Conditions A and B in Theorem 3.5.1 correspond to two typical situations in practical calculations. In condition A, a discontinuity, such as a shock or a contact discontinuity, propagates across a grid interface (for at most countable times). In condition B, a stationary discontinuity stays right at an interface. It is easy to verify that condition B includes this special case: solution $u(x,t)$ has a stationary discontinuity at x_I, and

$$\lim_{\Delta\to 0}f_\Delta(x_{I\pm 1/2},t)=f(u(x_I^{\pm},t))$$

Actually, in this situation, the solution satisfies the Rankine–Hugoniot condition (3.5), with $s=0$, along $x=x_I$. Moreover, as seen in its proof, the theorem imposes no particular restrictions on the interface algorithm for q_I^{n+1}, such as whether it is conservative or how it is constructed, as long as the corresponding integral tends to zero. Finally, it is expected that, when a nonconservative interface is used and either condition holds, the converged solution may be a piecewise smooth solution.

According to Theorem 3.5.1, one can detect whether the limit of a numerical solution will be a weak solution. Particularly, one may conduct mesh refinement tests and then detect if condition A or B holds at $\Delta \to 0$. The detection can be made on whether (3.47) will hold since both conditions A and B lead to it. In general, it is not a straightforward task to show how to ensure that conditions A and B are satisfied. But we may resort to numerical evidence of whether these conditions can be satisfied (e.g., [23]).

Generally, an explicit nonconservative interface algorithm can be written as a perturbation of a conservative algorithm. For instance, when two grids patch with each other (Fig. 3.2a), as a perturbation of conservative interface algorithm (3.26), a nonconservative interface algorithm is

$$q_I^{n+1} = v_I^n - \lambda_I(\overline{f}^2_{I+1/2} - \overline{f}^1_{I-1/2}) + \Delta t \overline{q}_I^n. \tag{3.48}$$

Therefore, this neoconservative interface algorithm consists of a conservative algorithm and a perturbation term, and $\overline{q}_I^n$ is the perturbation term. An example of $\overline{q}_I^n$ will be given in the later content.

Theorem 3.5.2 *In the computation of (3.46), let $\overline{q}_I^{n+1}$ be determined by (3.48). Then, if $u_\Delta(x,t)$ is uniformly bounded and converges to a function $u(x,t)$ as $\Delta \to 0$, where $\Delta = \max(\Delta x_1, \Delta x_2, \Delta t)$, $u(x,t)$ is a weak solution of the problem if*

$$\lim_{\Delta \to 0} \overline{q}_I^n = 0. \tag{3.49}$$

Proof Multiplying all the equations of (3.46) with a scalar function $\phi(x,t) \in C_0^\infty(R \times R^+)$, integrating the resulting three equations as before, introducing changes in variables, and then letting $\Delta \to 0$, one has

$$\int_0^{+\infty}\int_{-\infty}^{+\infty} (\phi_t(x,t)u(x,t) + \phi_x(x,t)f(u(x,t)))\,dxdt + \int_{-\infty}^{+\infty} \phi(x,0)u_0(x)dx$$
$$+ \lim_{\Delta \to 0}\int_0^{+\infty}\int_{x_{I-1/2}}^{x_{I+1/2}} \phi(x_I,t)\overline{q}^n dxdt = 0$$

The derivation is the same as that for Theorem 3.4.1, which yields the first two integrals on the left-hand side, while the third integral results from the perturbation term in (3.48). In view that $\phi(x,t)$ has finite support, one has

$$\lim_{\Delta \to 0}\int_0^{+\infty}\int_{x_{I-1/2}}^{x_{I+1/2}} \phi(x_I,t)\overline{q}^n dxdt = \int_0^{+\infty}\int_{x_{I-1/2}}^{x_{I+1/2}} \phi(x_I,t)\lim_{\Delta \to 0}\overline{q}^n dxdt = 0$$

Therefore, $u(x,t)$ is a weak solution. This completes the proof. □

Theorem 3.5.2 indicates that a nonconservative interface algorithm can provide a weak solution if it tends to a conservative treatment as mesh spacing and time step shrink. Condition (3.49) serves as a guide for its construction; a nonconservative interface algorithm is "almost" a conservative treatment or a "small" perturbation of a conservative treatment, and it tends to be conservative when the grid spacing and time step go to zero. Additionally, as discussed previously, see (3.45), conservation error is minor when the solution is smooth at the interface. The error becomes pronounced when it is discontinuous there. With such an understanding, an adaptive interface scheme may be constructed (e.g., [16, 17, 23]); it adopts a conservative interface algorithm when there is a discontinuity near an interface. Otherwise, it uses a nonconservative interface algorithm.

The analyses in this subsection are about 1D space and patched grids. A further study shows that in more general situations, such as in the 2D space and on overset grids, still these analyses and conclusions, such as Theorems 3.5.1 and 3.5.2, are extendable [20, 21]. In the following, numerical examples are presented to illustrate the analyses and conclusions of this subsection.

3.5.2 Numerical Experiment

As analyzed in the previous subsection, a nonconservative interface algorithm works for practical computation under certain conditions, although it does not strictly preserve a conservation property for the associated numerical solution. In the following, numerical examples illustrate the analysis, showing the evidence and the performance of nonconservative interface algorithms. The reader is referred to [17, 20] for details of these experiments.

Example 3.5.1 Consider the following initial value problem:

$$\begin{aligned} u_t(x,t) + u_x(x,t) &= 0 \\ u(x,0) &= \begin{cases} -1, & x < 0 \\ 1, & x > 0 \end{cases} \end{aligned}$$

Two subdomains patch at $x_I = 0$, and $\Delta x_1 = \Delta x_2$. Also, the computation is made using the Lax-Friedrichs scheme, with $CFL = 1$, which yields $\Delta x = \Delta t$, and $v_I^0 = 0$. Then, according to (3.13), one has ($i \neq I$)

$$\begin{aligned} v_i^{n+1} &= \frac{v_{i-1}^n + v_{i+1}^n}{2} - \frac{\Delta t}{2\Delta x}\left(f(v_{i+1}^n) - f(v_{i-1}^n)\right) \\ &= \frac{v_{i-1}^n + v_{i+1}^n}{2} - \frac{1}{2}\left(v_{i+1}^n - v_{i-1}^n\right) \\ &= v_{i-1}^n \end{aligned}$$

Nonconservative interface algorithm (3.29) yields

$$v_I^{n+1} = \frac{v_{I-1}^{n+1} + v_{I+1}^{n+1}}{2}$$

By this computation, it is easy to check that the numerical solution increases monotonically in magnitude, from –1 to 1 and with the total variation of $|1 - (-1)| = 2$, with i as the grid spacing goes to zero. As a result, the numerical solution exhibits the least total variation possible for the initial value problem that is also 2, thereby, verifying Proposition 3.5.1. Additionally, it shows direct evidence for condition A. Consequently, according to Theorem 3.5.1, the limit of the numerical solution is a weak solution.

Example 3.5.2 Now, consider the initial value problem of an equation slightly modified from the Burgers' equation:

$$u_t + \left(u^2\right)_x = 0, \quad t > 0$$
$$u = \begin{cases} 1, & x < 0,\ t = 0 \\ -1, & x > 0,\ t = 0 \end{cases}$$

The problem has an analytical solution that is a stationary discontinuity at $x = 0$. In the computation, set $x_I = 0$, $\Delta x_1 = \Delta x_2$, and adopt the Lax-Friedrichs scheme in the two subdomains. Let $v_I^0 = 0$ and use interface algorithm (3.29). By straightforward calculations, it is known that the computation leads to

$$v_I^{n+1} = 0, \quad v_{I-1}^{n+1} = -v_{I+1}^{n+1}$$

which, according to (3.14), gives rise to

$$\begin{aligned}
f_\Delta(x_{I-1/2}, t) &= \frac{1}{2}\left(f(v_{I-1}^n) + f(v_I^n)\right) - \frac{\Delta x}{2\Delta t}(v_I^n - v_{I-1}^n) \\
&= \frac{1}{2}\left({v_{I-1}^n}^2 + 0\right) - \frac{\Delta x}{2\Delta t}(0 - v_{I-1}^n) \\
&= \frac{1}{2}\left({v_{I+1}^n}^2 + 0\right) - \frac{\Delta x}{2\Delta t}(0 + v_{I+1}^n) \\
&= \frac{1}{2}\left(v_I^{n2} + {v_{I+1}^n}^2\right) - \frac{\Delta x}{2\Delta t}(v_{I+1}^n - v_I^n) \\
&= \frac{1}{2}\left(f(v_I^n) + f(v_{I+1}^n)\right) - \frac{\Delta x}{2\Delta t}(v_{I+1}^n - v_I^n) \\
&= f_\Delta(x_{I+1/2}, t)
\end{aligned}$$

that is, (3.47) in condition B is satisfied. Therefore, the numerical solution converges to a weak solution according to Theorem 3.5.1. Actually, when $CFL = 1$, it can be directly verified that the numerical solution is stationary and converges to the problem's analytical solution. Moreover, it can be checked that the conservation error of the numerical solution is zero in this example.

Example 3.5.3 Consider the computation of the initial value problem for the Burgers' equation:

$$u_t + \left(\frac{1}{2}u^2\right)_x = 0, \quad t > 0$$

$$u = \begin{cases} 1, & x < 0,\ t = 0 \\ 1 - x, & 0 \leq x \leq 1,\ t = 0 \\ 0, & x > 1,\ t = 0 \end{cases}$$

In this problem, a shock forms gradually as the solution evolves in time, and the solution is continuous at a subdomain interface except at the moment when the shock passes through it.

The Lax–Friedrichs and Lax–Wendroff schemes are used in subdomains 1 and 2, respectively. The grid interface is at $x_I = 2$, and interface algorithm (3.29) is adopted, with $T = 5$, $\Delta x_1 = \Delta x_2$, and $CFL = 0.95$. Figure 3.10 and Table 3.1 show the results. The computed shock moves a little slower in comparison to the analytical solution, see Figs. 3.10a and b. As analyzed previously, conservation error takes place mainly when the shock moves across the interface, see Figs. 3.10c and d.

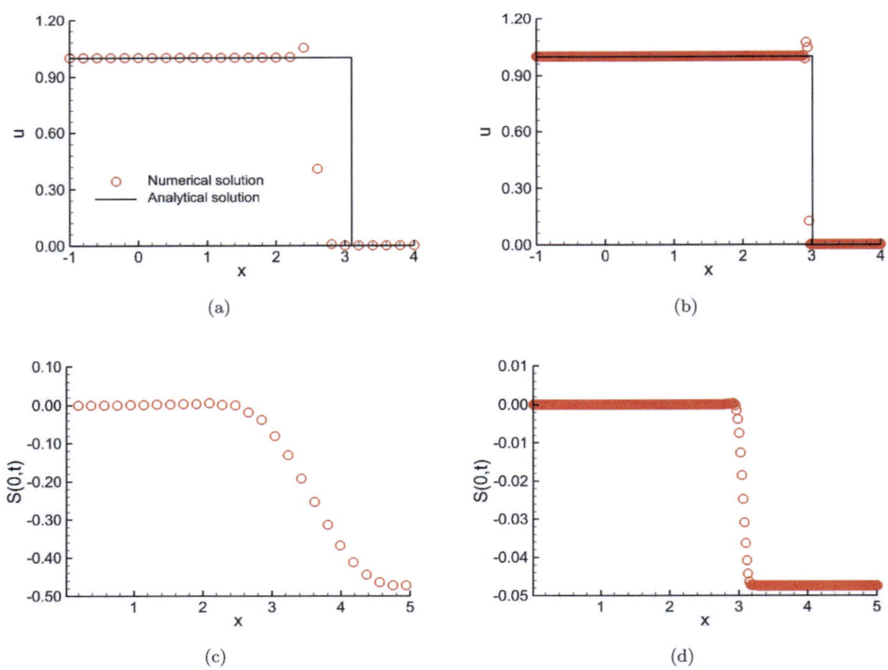

Fig. 3.10 Solution for initial value problem for the Burgers' equation. (**a**) Numerical solution for $\Delta x_1 = 0.2, t = T$. (**b**) Numerical solution for $\Delta x_1 = 0.02, t = T$. (**c**) History of conservation error $S(0, t)$ for (**a**). (**d**) History of conservation error $S(0, t)$ for (**b**)

Table 3.1 Convergence test of the initial value problem for the Burgers' equation

$\Delta x_1, \Delta x_2$	0.2	0.02	0.002	0.0002
$S_\Delta(0, T)$	-4.720×10^{-1}	-4.754×10^{-2}	-4.644×10^{-3}	-8.523×10^{-5}
$\|v_\Delta(x, T) - u(x, T)\|_1$	5.276×10^{-1}	6.008×10^{-2}	5.838×10^{-3}	7.553×10^{-4}

Table 3.2 Convergence test of the Riemann problem for the Euler equations. $V_\Delta^m(x, t)$ is the step function constructed from V_i^n. Superscript $m(= 1, 2, 3)$ refers to components of $V_\Delta^m(x, t)$ and $Q^m(x, t)$

$\Delta x_1, \Delta x_2$	$0.1, 0.05$	$0.01, 0.005$	$0.001, 0.0005$
$\|\|S_\Delta(0, T)\|\|_\infty$	1.657×10^{-1}	5.655×10^{-3}	5.045×10^{-4}
$\left(\sum_m \|\|V_\Delta^m(x, T) - Q^m(x, 0)\|\|_1{}^2\right)^{1/2}$	2.412×10^0	3.032×10^{-1}	3.244×10^{-2}

Particularly, these two figures show that the conservation error starts as zero. Then, it suddenly deviates from zero at about $t = 3$ when the shock passes the interface. After this moment, its value remains about the same. The calculation also shows the evidence that, as the grid spacing shrinks, condition A holds true, conservation error decreases, and accuracy for shock location improves, see Figs. 3.10b, d and Table 3.1. As the grid spacing tends to zero, the numerical solution will converge to the analytical one, as evidenced in the figures and the table. The results in the table show a linear correlation between grid spacing and solution error, suggesting a convergence rate of approximately order 1 in L_1-norm.

Example 3.5.4 Consider the computation of a Riemann problem of the Euler equations with the following initial conditions:

$$
\begin{aligned}
(\rho_l, u_l, p_l) &= (1, 1, 1), \quad\;\; x < 0\\
(\rho_r, u_r, p_r) &= (1, -1, 1), \quad x > 0
\end{aligned}
$$

In the computation, two grids patch with each other, $x_I = 6$, and $T = 3.5$. The Lax-Friedrichs and the Lax-Wendroff schemes are used in subdomains 1 and 2, respectively, $\Delta x_1 = 2\Delta x_2$, and $CFL = 0.95$. Let V_i^n be a numerical solution for $Q(x_i, t_n)$, the analytical solution to the Euler equations (3.2) and (3.3). Theorem 3.5.2 indicates that a nonconservative interface algorithm can be constructed as a perturbation of a conservative interface algorithm. In the vector version of interface algorithm (3.48), the perturbation term, $\overline{Q}_I^n$, is constructed as

$$
\overline{Q}_I^n = (\Delta x_1 + \Delta x_2)^2 \left(\frac{V_{I+1}^n - V_I^n}{\Delta x_2} - \frac{V_I^n - V_{I-1}^n}{\Delta x_1} \right)
$$

which satisfies a vector version of condition (3.49) as long as the numerical solution is bounded. By the Taylor expansion, it is known that $\overline{Q}_I^n$ is a 2nd-order derivative, and thus, it may be considered an artificial viscosity term. Since V_i^n is a vector, the conservation error, $S_\Delta(0, T)$, from 0 to T in time, is also a vector, and it is evaluated as $||S_\Delta(0, T)||_\infty$. As seen in Table 3.2, the interface algorithm works, and the numerical solution converges at a rate of order 1.

Various interpolation methods are widely used as nonconservative interface schemes in the overset-grid method to compute practical flow problems, even when discontinuities propagate across grid interfaces. The following 2D example illustrates the performance of such a scheme in overset grids. In this example, the conservation error is defined similarly as in 1D situation:

$$S_\Delta(0, T) = \int_{\Omega_1 \cup \Omega_2} (V_\Delta(x, y, T) - V_\Delta(x, y, 0))dxdy$$

Again, $S_\Delta(0, T)$ is a vector, and it is quantified as $\|S_\Delta(0, T)\|_\infty$. Besides, grids G_1 and G_2 overlap arbitrarily, without restriction on their orientation and position, and they are called Chimera grids [24]. Chimera grids are popular in applications to compressible flows, such as the flow around a space shuttle [25].

Example 3.5.5 Consider an inviscid flow with a shock past interfaces of overset square grids G_1 and G_2, with side-lengths of 1 and 0.5, respectively, see Fig. 3.11a. As shown, the interface nodes of G_2 are on its four sides, while those of G_1 are the nodes marked by small squares. The region enclosed by the small squares is called a blanked zone of the subdomain covered by G_1. The solution in this zone is not computed on its own grid but instead on G_2. The flow is simulated as an initial-boundary value problem.

In the computation, 2D Euler equations are solved with boundary conditions

$$(\rho_l, u_l, v_l, p_l) = (3.948, 4.359, 0, 5.005)$$
$$(\rho_r, u_r, v_r, p_r) = (1.4, 3, 0, 1)$$

where subscripts l and r stand for the left edge (behind shock) and right edge (before shock), respectively, of G_1, and v is the velocity in y-direction. Initially, (ρ_l, u_l, v_l, p_l) is assigned at grid nodes on the left edge of G_1, and (ρ_r, u_r, v_r, p_r) is assigned at all other nodes of G_1 and G_2. On the upper and lower boundaries of G_1, the analytical solution of the problem is prescribed. In the two subdomains, a scheme extended from the 1D scheme TVD by Harten is used [14], and $CFL = 0.95$. The solution at an interface node of each grid is determined by the bilinear interpolation from the four nodes of its host cell in the other grid.

The solution in Fig. 3.11b shows that the shock is a little distorted, and the speed of the numerical shock is a little faster than that of the analytical solution for the shock, whose location can be seen on the upper and lower boundaries of grid G_1. But, there is evidence that condition A holds (precisely, its 2D version, see [20]). Accordingly, by increasing mesh resolution (Fig. 3.11c), the solution for the shock gets better, and conservation error is also reduced (Fig. 3.11d). The conservation error is computed and listed in the caption of Fig. 3.11.

All the aforementioned experiments indicate that nonconservative interface algorithms work as predicted by the analyses. However, the following is a counterexample, and it shows that discretion is needed, especially as a discontinuity occurs at a grid interface.

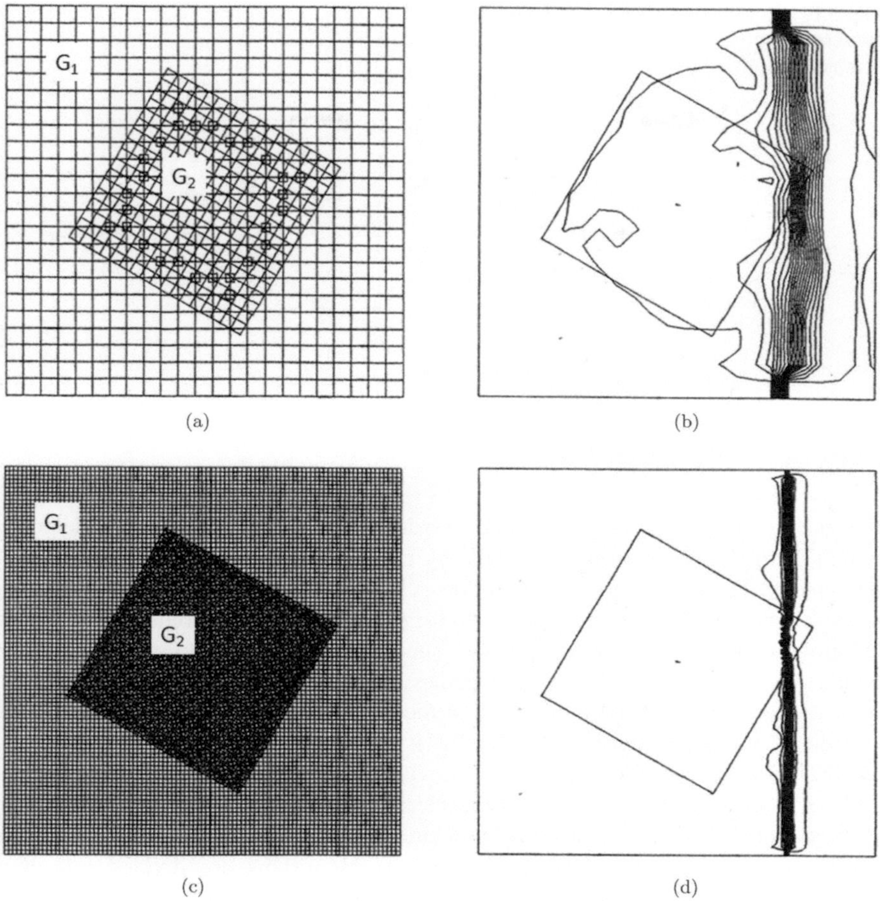

Fig. 3.11 Shock traveling on Chimera grids. $T = 0.13$. (**a**) Meshes. G_1: 18×18 , G_2: 24×24. (**b**) Pressure contours computed on the grids in (**a**), $\|S_\Delta(0, T)\|_\infty = 2.4 \times 10^{-3}$. (**c**) Meshes. G_1: 69×69, G_2: 93×93. (**d**) Pressure contours computed on the grids in (**c**), $\|S_\Delta(0, T)\|_\infty = 7.1 \times 10^{-5}$

Example 3.5.6 Again, consider the computation of the stationary-shock problem in Example 3.4.2. As already shown in Fig. 3.4, nonconservative algorithm (3.29) leads to a wrong solution. Actually, the computation shows that the conservation error accumulates as time goes on, see Fig. 3.12. As predicted by Eq. (3.44), such accumulation leads to that the difference between the calculated and analytical solutions gets larger as time progresses, and this is visualized in Fig. 3.4. Additionally, the computation shows that a stationary artificial discontinuity is always located at the interface, and no improvement is present with the mesh refinement. Actually, its value shows that the numerical solution violates (3.47) in condition B.

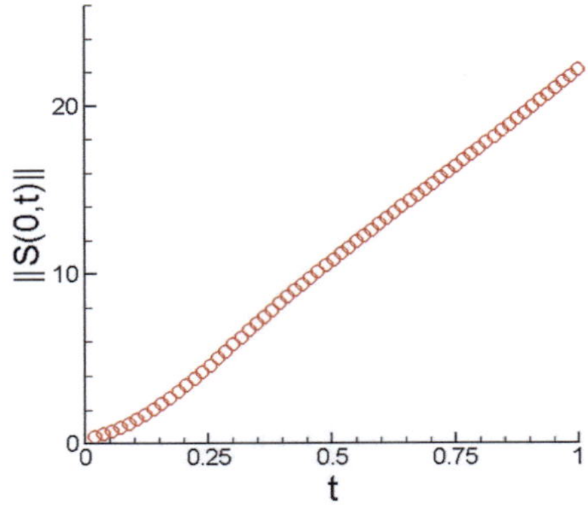

Fig. 3.12 Time history of conservation error in the computation of Fig. 3.4

Summary and Question

Summary This chapter begins with conservation laws and their numerical methods, which form the foundation for simulating compressible flows. It introduces fundamental concepts, including hyperbolic systems, weak solutions, and piecewise smooth solutions. The theory states that a piecewise smooth solution is a weak solution under certain conditions, and such a solution satisfies a conservation condition. Conservative numerical schemes are then introduced, accompanied by typical examples. A theorem shows that the converged solution of a conservative scheme is a weak solution. Additionally, it is demonstrated that such a solution satisfies a discrete form of the conservation condition.

The chapter then proceeds to numerical computations using the multiblock method, with a particular focus on computations involving patched and overlapping grids, especially at their interfaces. Interface algorithms are divided into two categories: conservative and nonconservative algorithms. For both patched and overlapping grids, conservative interface algorithms based on enforcing numerical fluxes are presented. Nonconservative interface algorithms built on interpolation are also discussed. The setup of initial values for the computations is specified.

When a conservative interface algorithm is applied, it is shown that the numerical solution satisfies the conservation condition and the converged solution is a weak solution. These theoretical results are supported by numerical experiments, which demonstrate the advantages of conservative interface algorithms over nonconservative ones. In particular, while a nonconservative treatment may introduce errors in the strength and location of discontinuities or even yield a non-physical or incorrect solution, conservative treatments avoid such issues.

However, conservative interface algorithms, particularly those based on enforcing numerical fluxes, are not without challenges. First, analysis shows that when two subdomains have the same grid spacing but use different numerical schemes, such a interface algorithm may become inconsistent with the PDE, effectively computing a different equation at the interface. Second, when two subdomains use the same numerical scheme but differ in grid spacing, a similar inconsistency may arise. Numerical experiments reveal that such inconsistencies can lead to artifacts at grid interfaces, erroneous solutions, and even the breakdown of the computational process. As a remedy, numerical fluxes that are independent of grid spacing and time step, such as those used in the Godunov scheme, can mitigate the issue.

Finally, the chapter turns to nonconservative interface treatments. It introduces the concept of conservation error and analyzes its magnitude in numerical computations. A theoretical foundation is provided for nonconservative treatments, proving that a converged solution can still be a weak solution under two specific conditions. These conditions may be verified numerically. Corresponding experiments are presented to validate the performance of nonconservative interface algorithms under these conditions. Additionally, counterexamples are provided to demonstrate that when these conditions are not met, the resulting solution becomes invalid.

This chapter deals exclusively with explicit schemes and does not include Schwarz iteration. However, implicit schemes are also widely used. It is expected that the analyses, methods, and conclusions presented in this chapter may still hold to some extent when implicit schemes are employed. Nonetheless, this remains an interesting topic for further investigation. Additionally, it would be worthwhile to explore whether the work from the last chapter can be extended through techniques such as linearizing the conservation laws into linear advection equations. Moreover, since nonconservative interface algorithms are widely used in practice, constructing effective algorithms of this type, guided by Theorem 3.5.2, would be a meaningful topic for further research.

Questions The following questions are provided to aid in understanding, discussing, and investigating the content. Their answers and clues are embedded within the chapter.

1. What are the classical and weak solutions for conservation laws? Why is the weak solution, rather than the classical solution, discussed in the computation of the conservation laws?
2. What is a piecewise smooth solution, and what is its relation to a weak solution of a conservation law?
3. What is a conservative scheme, and what is its numerical flux? Discuss the relationship between the consistency condition for the numerical flux and that for the scheme.
4. In the computation of conservation laws, different schemes, including non-conservative schemes, are available for discretization. However, a conservative scheme is preferred. Discuss the reasons why it is preferred.
5. The Lax-Friedrichs scheme, the Lax-Wendroff scheme, and the Godunov scheme are widely used in theoretical analysis and practical computations. Derive their accuracy orders using Taylor series expansions.
6. Interface algorithms can be classified into two categories based on whether the numerical solution satisfies the conservation condition or not. What are these categories, and how can algorithms in each category be designed?

7. Algorithms (3.26) and (3.29) are typical examples of conservative and nonconservative interface algorithms, respectively. Derive their orders of accuracy.
8. This chapter demonstrates that a conservative interface algorithm may exhibit inconsistency with the PDE. Identify the consistency conditions for the following situations in the two subdomains: (1) different schemes with the same grid spacing, and (2) the same scheme with different grid spacings.
9. In the proof of Theorem 3.5.1, the integral term involving the interface algorithm tends to zero under given conditions (e.g., the solution is uniformly bounded). Discuss whether such conditions, especially those related to the integral term, can be relaxed or changed.
10. Continuing from the above question, construct requirements on a nonconservative interface algorithm that ensure the integral tends to zero, thereby yielding a weak solution.
11. Discuss conditions A and B in Theorem 3.5.1. Construct your examples where these conditions hold and the solutions converge to weak solutions.
12. Examples 3.4.2 and 3.5.6 illustrate cases that violate the conditions of Theorem 3.5.1 and result in incorrect solutions. Consider whether you can construct another such case.
13. The chapter suggests that a nonconservative interface algorithm may yield a weak solution when it is a perturbation of a conservative interface algorithm. Example 3.5.4 provides an example of such a perturbation term. Construct another perturbation term for the example according to Theorem 3.5.2.

References

1. L. Cambier, W. Ghazzi, J.P. Veuillot, H. et Viviand, Une approche par domaines pour le calcul d'ecoulements compressibles. Compressibles Cinquierae Colloque International sur les Methodes de Calcul Sclentifique et Technique. INRIA, Versailles, France, 14–18 Decembre, 1981
2. R. Glowinski, J. Periaux, Numerical solution of partial differential equations using interfacing and/or overlapping grids of sub-domains. Application to Navier-Stokes equations and transonic potential flows for inviscid fluids. Symposium on Multigrid Methods, NASA-Ames Research Center, October 21–22, 1981
3. K.A. Hessenius, T.H. Pulliam, A zonal approach to solution of the Euler equations. AIAA paper 82-0969, 1982
4. P.D. Lax, Weak solutions of nonlinear hyperbolic equations and their numerical computation. Comm. Pure Appl. Math. **7**, 159–193 (1954)
5. P.D. Lax, Hyperbolic systems of conservation laws II. Comm. Pure Appl. Math. **10**, 537–566 (1957)
6. P.K. Kundu, *Fluid Mechanics* (Academic Press, 1990)
7. E.F. Toro, *Riemann Solvers and Numerical Methods for Fluid Dynamics - A Practical Introduction, 3rd edition* (Springer, 2009)
8. G.A. Sod, *Numerical Methods in Fluid Dynamics – Initial and Initial Boundary-Value Problems* (Cambridge Universiy Press, 1985)
9. P.D. Lax, *Hyperbolic Systems of Conservation Laws and the Mathematical Theory of Shock Waves* (Society for Industrial and Applied Mathematics, Philadelphia, 1973)

10. L.A. Ying, Z.H. Teng, *Hyperbolic System of Conservation Laws and Their Finite Difference Method* (Science Press, Beijing, 1991. In Chinese)
11. L.C. Evans, *Partial Differential Equations, Graduate Studies in Mathematics* (American Math Society, 1991)
12. M.M. Rai, A conservative treatment of zonal boundaries for Euler equation calculations. J. Comput. Phys. **62**, 472–503 (1986)
13. M.J. Berger, On conservation at grid interfaces. SIAM J. Numer. Anal. **24**, 967–984 (1987)
14. A. Harten, High resolution schemes for hyperbolic conservation laws. J. Comput. Phys. **49**, 393–393 (1983)
15. H.S. Tang, D. Huang, A second-order accurate capturing scheme for 1D inviscid flows of gas and water with vacuum zones. J. Comput. Phys. **128**, 301–318 (1996)
16. H.S. Tang, C.H. Lee, On the consistency of conservative algorithms for internal interface in simulation of inviscid flow by zonal method. Chinese J. Comput. Phys. **13**(4), 445–453 (1996)
17. H.S. Tang, D.L. Zhang, C.H. Lee, Comments on algorithms for grid interfaces in simulating Euler flows. Commun. Nonlinear Sci. Numer. Simul. **1**, 50–54 (1996)
18. E. Part-Enander, B. Sjogreen, Conservative and non-conservative interpolation between overlapping grids for finite volume solutions of hyperbolic problems. Comput. Fluids **23**, 551–574 (1994)
19. Z.L. Wu, On uniqueness of steady state solutions for difference equations on overlapping grids. SIAM J. Numer. Anal. **33**(4), 1336–1357 (1996)
20. H.S. Tang, T. Zhou, On nonconservative algorithms for grid interfaces. SIAM J. Numer. Anal. **37**, 173–193 (1999)
21. Tang, H. S. and Zhou, T. (1997). Why nonconservative interface algorithms may be applicable: analysis for Chimera grids. Proc. 7th Int. Symp on Computational Fluid Dynamics. Int. Academic Pub. Eds. F. G. Zhuang, 336–341
22. H.Z. Tang, G. Warnecke, On convergence of a domain decomposition method for a scalar conservation law. SIAM J. Numer. Anal. **45**(4), 1453–1471 (2007)
23. S. Karni, Hybrid multifluid algorithms. SIAM J. Sci. Comput. **17**, 1019–1039 (1996)
24. J.L. Steger, F.C. Dougherly, J.A. Benek, A Chimera grid scheme. ASME Mini-Symposium on Advance in Grid Generation, Houston, Texas, June 1983
25. P.G. Buning, R.J. Gomez, 20+ years of Chimera grid development for the space shuttle. 10th Symposium on Overset Composite Grid and Solution Technology, Moffett Field, CA, September 20–23, 2010

Chapter 4
Incompressible Flow

4.1 Governing Equation and Solution

4.1.1 Governing Equation

Since the late 1980s, research on the multiblock method for incompressible flows has quickly become a popular topic in the computational fluid dynamics community [1, 2]. Similar to that for compressible flows, a primary driving force of the method for incompressible flows at the early time is to overcome the difficulty of mesh generation caused by complex flow boundaries. A comprehensive review of the multiblock method for incompressible flows is available in [3].

The governing equations describing the motion of an incompressible fluid flow consist of the continuity equation, the momentum equations, i.e., the Navier–Stokes (NS) equations, and the energy equation, which are respectively as follows [4, 5]:

$$\nabla \cdot \mathbf{u} = 0 \tag{4.1}$$

$$\begin{aligned} \mathbf{u}_t + \nabla \cdot \mathbf{u}\mathbf{u} = & -\frac{1}{\rho_0}\nabla p + \nabla \cdot ((\nu + \nu_t)\nabla \mathbf{u}) \\ & -g\left(1 - \alpha(T - T_0)\right)\mathbf{k} \end{aligned} \tag{4.2}$$

$$T_t + \nabla \cdot (\mathbf{u}T) = \nabla \cdot \left((\frac{\nu}{Pr} + \frac{\nu_t}{Pr_t})\nabla T \right) \tag{4.3}$$

In the aforementioned equations, $\mathbf{u}$ is the velocity vector, with components u, v, and w in the Cartesian coordinates x-, y- and z-directions, respectively, p the pressure, and T the temperature. ρ_0 is a reference density, ν the kinematic viscosity, ν_t the kinematic viscosity of turbulence, and T_0 a reference temperature. g is the

H. Tang, *Multiblock Method for Fluid Flow*, Nonlinear Systems and Complexity 40,
https://doi.org/10.1007/978-3-031-78568-9_4

acceleration of gravity, α the thermal expansion coefficient, and **k** the unit vector in the vertical direction. Pr is the molecular Prandtl number, which takes the value of 7.1 for thermal stratification in water, and Pr_t is the turbulence Prandtl number, which is set to 0.85 for shear flows [6]. For a turbulence flow, a turbulence closure, such as the mixing length model or the $k-\epsilon$ model, is needed to evaluate ν_t and Pr_t [7, 8]. Note that, by convention, the term 'NS equations' refers collectively to Eqs. (4.1) and (4.2). However, strictly speaking, the NS equations correspond to Eq. (4.2) alone, and this distinction will be maintained hereafter. Moreover, satisfactorily enforcing Eq. (4.1) — often referred to as the divergence-free condition — is essential in simulating an incompressible flow.

Different coordinates are available to express the governing equations. In computation, adopting the generalized curvilinear coordinates is a natural approach to simulate flows contained by a boundary with complex geometry. In such coordinates, continuity equation (4.1) and momentum equations (4.2) are transformed as ($k, l, m = 1, 2, 3$, and repeated indices imply summation)

$$\frac{\Gamma}{J}\frac{\partial Q}{\partial t} + \frac{\partial}{\partial \xi^l}(F^l - F_\nu^l) + S = 0 \tag{4.4}$$

where

$$\begin{aligned}
\Gamma &= diag(0, 1, 1, 1)\\
Q &= (P, u^1, u^2, u^3)^T\\
F^l &= \frac{1}{J}(U^l, u^1U^l + p\xi^l_{x^1}, u^2U^l + p\xi^l_{x^2}, u^3U^l + p\xi^l_{x^3})^T\\
F_\nu^l &= \frac{1}{J}(0, g^{ml}\frac{\partial u^1}{\partial \xi^m} + R_{m1}\xi^l_{x^m}, g^{ml}\frac{\partial u^2}{\partial \xi^m} + R_{m2}\xi^l_{x^m}, g^{ml}\frac{\partial u^3}{\partial \xi^m} + R_{m3}\xi^l_{x^m})^T\\
S &= \frac{g}{J}(0, 0, 0, -(1 - \alpha(T - T_0)))^T
\end{aligned}$$

Energy equation (4.3) becomes

$$\frac{1}{J}\frac{\partial T}{\partial t} + \frac{\partial}{\partial \xi^l}\left(\frac{1}{J}\left(U^lT - \left(\frac{\nu}{Pr} + \frac{\nu_t}{Pr_t}\right)g^{ml}\frac{\partial T}{\partial \xi^m}\right)\right) = 0 \tag{4.5}$$

In the above equations, x^1, x^2, and x^3 are the Cartesian coordinates x, y, and z, respectively, and ξ^1, ξ^2, and ξ^3 are the curvilinear coordinates ξ, η, and ζ, respectively. $P = p/\rho_0$, and u^1, u^2, and u^3 are u, v, and w, respectively. U^l is the contravariant velocity component in ξ^l-direction. The relations between U^l and u^m are

$$U^l = u^m\xi^m_{x^l}, \quad u^m = x^m_{\xi^l}U^l \tag{4.6}$$

Here, $\xi^m_{x^l}$ and $x^m_{\xi^l}$ are the metric of the geometric transformation $\partial\xi^m/\partial x^l$ and $\partial x^m/\partial \xi^l$, respectively, and g^{ml} is the contravariant metric tensor ($= \xi^m_{x^k}\xi^l_{x^k}$).

R_{ml} is the velocity gradient tensor ($= (\partial u^m/\partial \xi^k)\xi^k_{x^l}$), and J is the Jacobian of the geometric transformation ($= 1/det(\partial x^m/\partial \xi^l)$), see Appendix A.

4.1.2 *Discretization and Solution*

The governing equations in (4.4) are discretized using a second-order accurate, implicit finite difference method. Particularly, the method adopts three-point, backward finite difference for the temporal derivative and a second-order accurate central finite difference for the spatial derivatives:

$$\frac{\Gamma}{J_{i,j,k}}\frac{3Q^{n+1}_{i,j,k} - 4Q^{n}_{i,j,k} + Q^{n-1}_{i,j,k}}{2\Delta t} + \left(\delta_{\xi^l}\overline{F}^l + \delta_{\xi^l}\overline{F}^l_v\right)^{n+1}_{i,j,k} = 0 \tag{4.7}$$

In the discretization, Δt is the time step, and the center difference operator is defined as

$$\delta_{\xi^1}(\,)_{i,j,k} = \frac{(\,)_{i+1/2,j,k} - (\,)_{i-1/2,j,k}}{\Delta\xi^1}$$
$$\delta_{\xi^2}(\,)_{i,j,k} = \frac{(\,)_{i,j+1/2,k} - (\,)_{i,j-1/2,k}}{\Delta\xi^2}$$
$$\delta_{\xi^3}(\,)_{i,j,k} = \frac{(\,)_{i,j,k+1/2} - (\,)_{i,j,k-1/2}}{\Delta\xi^3}$$

$\overline{F}^l$ ($l = 1, 2, 3$) is the numerical flux calculated as

$$\begin{aligned}
\overline{F}^1{}_{i+1/2,j,k} &= \frac{1}{2}\left(F^1{}_{i,j,k} + F^1{}_{i+1,j,k}\right) + D^1{}_{i+1/2,j,k} \\
\overline{F}^2{}_{i,j+1/2,k} &= \frac{1}{2}\left(F^2{}_{i,j,k} + F^2{}_{i,j+1,k}\right) + D^2{}_{i,j+1/2,k} \\
\overline{F}^3{}_{i,j,k+1/2} &= \frac{1}{2}\left(F^3{}_{i,j,k} + F^3{}_{i,j,k+1}\right) + D^3{}_{i,j,k+1/2}
\end{aligned} \tag{4.8}$$

where $\Delta\xi^1$, $\Delta\xi^2$, and $\Delta\xi^3$ are grid spacings in the curvilinear coordinates, and they may be set to 1 in the computation. $\overline{F}^l$ has two parts in the above. The first part is the average of F^l, which is the convection term across the cell interfaces at $i + 1/2$, $j + 1/2$, and $k + 1/2$. The second part is D^l, which is artificial dissipation. $\overline{F}^l_v$ is the approximation to the viscous term F^l_v, which contains derivatives of velocities, and its calculation requires evaluation of these derivatives at the cell interfaces. These derivatives are evaluated using a second-order accurate central difference. For example, in the calculation of $\overline{F}^1_{v\,i+1/2,j,k}$, the following is adopted:

$$\left(\frac{\partial u^1}{\partial \xi^1}\right)_{i+1/2,j,k} \approx \delta_{\xi^1}(u^1)_{i+1/2,j,k}$$
$$\left(\frac{\partial u^1}{\partial \xi^2}\right)_{i+1/2,j,k} \approx \big(\delta_{\xi^2}(u^1)_{i,j-1/2,k} + \delta_{\xi^2}(u^1)_{i,j+1/2,k} + \delta_{\xi^2}(u^1)_{i+1,j-1/2,k} + \delta_{\xi^2}(u^1)_{i+1,j+1/2,k}\big)/4$$
$$\left(\frac{\partial u^1}{\partial \xi^3}\right)_{i+1/2,j,k} \approx \big(\delta_{\xi^3}(u^1)_{i,j,k-1/2} + \delta_{\xi^3}(u^1)_{i,j,k+1/2} + \delta_{\xi^3}(u^1)_{i+1,j,k-1/2} + \delta_{\xi^3}(u^1)_{i+1,j,k+1/2}\big)/4$$

In the discretization, the Jacobian and metrics of the geometric transformation at a cell interface are calculated using the arithmetic average.

Several schemes for D^l, including scalar and matrix-valued artificial dissipation and flux-difference splitting unwinding, have been evaluated [9]. With the same mesh resolution, the matrix-valued artificial dissipation scheme introduces the least overall amount of artificial dissipation, thus leading to superior accuracy. The matrix-valued artificial dissipation D^l is calculated as follows:

$$D^1_{i+1/2,j,k} = \epsilon\delta_{\xi^1}(|A^1|\delta_{\xi^1}\delta_{\xi^1})Q_{i+1/2,j,k}$$
$$D^2_{i,j+1/2,k} = \epsilon\delta_{\xi^2}(|A^2|\delta_{\xi^2}\delta_{\xi^2})Q_{i,j+1/2,k}$$
$$D^3_{i,j,k+1/2} = \epsilon\delta_{\xi^3}(|A^3|\delta_{\xi^3}\delta_{\xi^3})Q_{i,j,k+1/2}$$

where ϵ is a constant ($0 < \epsilon < 1$) controlling the amount of artificial dissipation, and $|A^l|$ is the absolute value of the Jacobian matrix $A^l = \partial F^l/\partial Q$. The exact form of $|A^l|$ is given in Appendix B.

Since Γ is a singular matrix, discretization algorithm (4.7) cannot march directly in time to advance the solution to the next time step. Moreover, the algorithm is fully implicit, and its solution requires an iterative scheme. Such a scheme can be designed using the dual time-stepping artificial compressibility (AC) method. With this method, a dual or pseudo-time derivative is introduced into Eq. (4.7) as follows:

$$\frac{dQ^{n+1}_{i,j,k}}{d\tau} + R^{n+1}_{i,j,k} = 0 \tag{4.9}$$

where τ is the pseudo-time, and

$$R^{n+1}_{i,j,k} = \Gamma\frac{3Q^{n+1}_{i,j,k} - 4Q^n_{i,j,k} + Q^{n-1}_{i,j,k}}{2\Delta t} + J_{i,j,k}(\delta_{\xi^l}\overline{F}^l + \delta_{\xi^l}\overline{F}^l_v)^{n+1}_{i,j,k} \tag{4.10}$$

(4.9) is a system of ordinary differential equations evolving in the pseudo-time. It is expected that the pseudo-time derivative term in the system tends to zero as the pseudo-time becomes sufficiently large. Then, the system can be integrated in the pseudo-time until the pseudo-time derivative reduces to zero or a prescribed small tolerance. When the derivative reaches zero, the system (4.9) reduces to discretization (4.7), and the solution at the next physical time step $n + 1$ is obtained.

An approach to integrate Eq. (4.9) in the pseudo-time from pseudo-time τ to $\tau + \Delta\tau$, $\Delta\tau$ being the pseudo-time step, is via a four-stage, pointwise implicit, Runge–Kutta algorithm as follows ($s = 1, 2, 3, 4$):

$$Q_{i,j,k}^{n+1,s} = Q_{i,j,k}^{n+1,0} - \alpha_s \Delta\tau [\Gamma \frac{3Q_{i,j,k}^{n+1,s} - 4Q_{i,j,k}^{n} + Q_{i,j,k}^{n-1}}{2\Delta t} + J(\delta_{\xi^l}\overline{F}^l + \delta_{\xi^l}\overline{F}_v^l)_{i,j,k}^{n+1,s-1}] \tag{4.11}$$

where $Q^{n+1,s}$ is the sth stage value, and α_s are the Runge–Kutta coefficients ($=$ 1/4, 1/3, 1/2, and 1 for $s = 1, 2, 3, 4$, respectively). Moreover, $Q^{n+1,0} = \{Q^{n+1}\}^{\tau}$, and $Q^{n+1,4} = \{Q^{n+1}\}^{\tau+\Delta\tau}$. Here, $\{Q^{n+1}\}^{\tau}$ is the estimate of Q^{n+1} at pseudo-time level τ. With the Runge–Kutta algorithm, Eq. (4.9) is integrated in the pseudo-time step by step, from $\tau = 0$, to $\Delta\tau$, $2\Delta\tau$, ..., $M\Delta\tau$, M being the total number of the pseudo time step. At $\tau = M\Delta\tau$, the computation converges, or the pseudo-time derivative term in Eq. (4.9) becomes zero.

In algorithm (4.11), the implicit treatment of Q^{n+1}, in terms of s, in the approximation of the physical time derivative is essential for enhancing the iterative algorithm's robustness, resulting in a point-wise implicit iterative scheme. The efficiency and robustness of the scheme are further enhanced by implicit residual smoothing, local dual-time stepping, and 2- or 3-level V-cycle multigrid acceleration [10–12]. The details of implementing the local time stepping can be found in Appendix C. Here, on the basis of (4.11), we present an algorithm with implicit residual smoothing, which, after some algebraic manipulations and rearrangements, can be formulated as

$$Q_{i,j,k}^{n+1,s} = Q_{i,j,k}^{n+1,s-1} + \left(I + \alpha_s \frac{3\Delta\tau}{2\Delta t}\Gamma\right)^{-1} \Im^{-1}[Q_{i,j,k}^{n+1,0} - Q_{i,j,k}^{n+1,s-1} - \alpha_s \Delta\tau R_{i,j,k}^{n+1,s-1}] \tag{4.12}$$

where I is the identity matrix, and $\Im$ is the implicit residual smoothing operator [12]:

$$\Im = (1 - \epsilon_1 \delta_{\xi^1}\delta_{\xi^1})(1 - \epsilon_2 \delta_{\xi^2}\delta_{\xi^2})(1 - \epsilon_3 \delta_{\xi^3}\delta_{\xi^3})$$

where $\epsilon_l > 0$. In the calculations of this chapter, scheme (4.12) will be used, together with $\epsilon_1 = \epsilon_2 = \epsilon_3 = 0.5$.

In solving Eq. (4.12), the converged solution for $\{Q^{n+1}\}^{\tau}$ is considered to be Q^{n+1} when the following residual criterion has been met:

$$log \frac{||\{f^{n+1}\}^{\tau+\Delta\tau} - \{f^{n+1}\}^{\tau}||_{\infty}}{||\{f^{n+1}\}^{\tau_0+\Delta\tau} - \{f^{n+1}\}^{\tau_0}||_{\infty}} < \epsilon_4 \tag{4.13}$$

f being p or u or v or w. $\tau_0 = 0$, the initial value for the pseudo-time, and ϵ_4 being a prescribed tolerance.

For the computation of a steady state, the iterative algorithm is considered convergent when the difference between two successive physical time steps has reduced to a prescribed value. The steadiness criterion, in this case, reads as follows:

$$log\frac{||f^{n+1}-f^n||_\infty}{||f^2-f^1||_\infty} < \epsilon_5$$

with ϵ_5 being a prescribed tolerance.

In the computations of the following sections in this chapter, the methods described earlier will be applied to all subdomains of the overset-grid method. Additionally, the conventional Schwarz method will be used to solve them as manifested in Sect. 1.2. Since the discretization, i.e., (4.9), is implicit, Schwarz iteration is necessary when the interface algorithm is also implicit.

4.2 Interface Algorithm and Computation Procedure

4.2.1 *Standard Interpolation*

Consider the multiblock computation on two overset grids, G_1 and G_2, as shown in Fig. 4.1a. The two grids are also called composite meshes [13], Chimera grids [14], etc. In the figure, Γ, or grid line $i = 1$, is an interface of G_1. To carry out the computation in G_1, a boundary condition along Γ is necessary. Such a condition will be obtained from the solution on G_2. Particularly, each grid node of G_1 on the interface, referred to as an interface node of the grid, falls into a grid cell of G_2, and

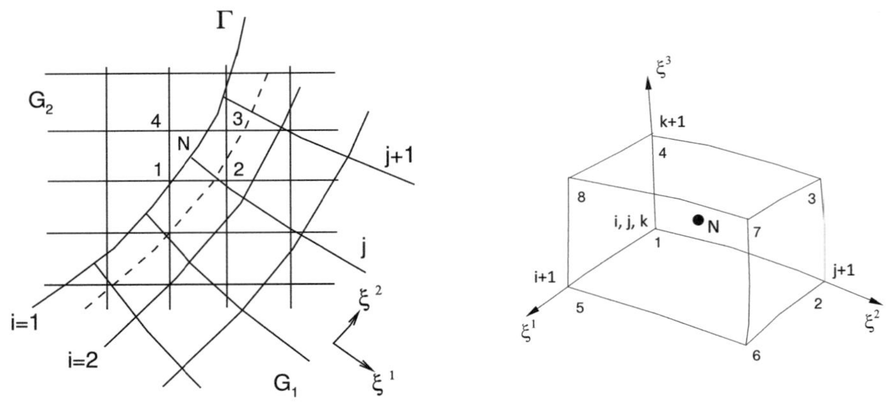

Fig. 4.1 Connectivity of overset grids. (**a**) Overset grids G_1 and G_2. (**b**) An interface node and its host cell

such a cell is called a host cell of the node. For instance, in Fig. 4.1a, node N is an interface node of G_1, and its host cell is cell $1-2-3-4$ in G_2.

For the multiblock computation, it is necessary to determine grid connectivity, the relative position between the two grids. Specifically, it is needed to identify the host cells of interface nodes and then determine the relative locations of these nodes in the cells. Let us use a 3D situation, as shown in Fig. 4.1b, to illustrate how the connectivity is determined. In the figure, N is an interface node, and its host cell is the cubic enclosed by $[\xi_i^1, \xi_{i+1}^1] \times [\xi_j^2, \xi_{j+1}^2] \times [\xi_k^3, \xi_{k+1}^3]$. This host cell is identified by a search in grid G_2 until a cell, i.e., the host cell, is found to contain N, particularly satisfy the following:

$$\mathbf{r}_{42} \times \mathbf{r}_{13} \cdot \mathbf{r}_{aN} \geq 0, \ \mathbf{r}_{75} \times \mathbf{r}_{68} \cdot \mathbf{r}_{bN} \geq 0$$
$$\mathbf{r}_{81} \times \mathbf{r}_{54} \cdot \mathbf{r}_{cN} \geq 0, \ \mathbf{r}_{63} \times \mathbf{r}_{72} \cdot \mathbf{r}_{dN} \geq 0$$
$$\mathbf{r}_{16} \times \mathbf{r}_{52} \cdot \mathbf{r}_{eN} \geq 0, \ \mathbf{r}_{38} \times \mathbf{r}_{74} \cdot \mathbf{r}_{fN} \geq 0$$

Here, $\mathbf{r}$ is a vector between two points. For instance, $\mathbf{r}_{aN}$ is the vector from point a to N. Subscripts a, b, c, d, e, and f represent the centers on the surfaces of the host cell at $\xi^1 = \xi_i^1$, $\xi^1 = \xi_{i+1}^1$, $\xi^2 = \xi_j^2$, $\xi^2 = \xi_{j+1}^2$, $\xi^3 = \xi_k^3$, and $\xi^3 = \xi_{k+1}^3$, respectively. The coordinates of the interface node are the averages of those for its eight vertices, and the coordinates of these surface centers are averages of their four vertices. For instance, for the interface node N $(l = 1, 2, 3)$,

$$x_N^l = \frac{1}{4}\,(x_{i,j,k}^l + x_{i,j+1,k}^l + x_{i,j,k+1}^l + x_{i,j+1,k+1}^l$$
$$+ x_{i+1,j,k}^l + x_{i+1,j+1,k}^l + x_{i+1,j,k+1}^l + x_{i+1,j+1,k+1}^l)$$

The location of the interface node in the host is described by its coordinates in a local coordinate system (r_1, r_2, r_3). The coordinates are defined as

$$r_1 = \frac{\xi_N^1 - \xi_i^1}{\xi_{i+1}^1 - \xi_i^1}, \quad r_2 = \frac{\xi_N^2 - \xi_j^2}{\xi_{j+1}^2 - \xi_j^2}, \quad r_3 = \frac{\xi_N^3 - \xi_k^3}{\xi_{k+1}^3 - \xi_k^3} \tag{4.14}$$

Consider the situation that the local coordinate system is orthogonal (or ξ^1, ξ^2, and ξ^3 are orthogonal to each other). Let the location of the interface point in the Cartesian coordinates be x, y, and z, which are denoted as f. Such a location in the local coordinate system can be determined by linear interpolation, particularly trilinear interpolation:

$$f_N = a_1 r_1 r_2 r_3 + a_2 r_1 r_2 + a_3 r_1 r_3 + a_4 r_2 r_3 + a_5 r_1 + a_6 r_2 + a_7 r_3 + a_8 \tag{4.15}$$

where $r_1, r_2, r_3 \in [0, 1]$, and

$$
\begin{aligned}
a_1 &= f_{i+1,j+1,k+1} - a_2 - a_3 - a_4 - a_5 - a_6 - a_7 - a_8 \\
a_2 &= f_{i+1,j+1,k} - a_5 - a_6 - a_8 \\
a_3 &= f_{i+1,j,k+1} - a_5 - a_7 - a_8 \\
a_4 &= f_{i,j+1,k+1} - a_6 - a_7 - a_8 \\
a_5 &= f_{i+1,j,k} - a_8 \\
a_6 &= f_{i,j+1,k} - a_8 \\
a_7 &= f_{i,j,k+1} - a_8 \\
a_8 &= f_{i,j,k}
\end{aligned}
\tag{4.16}
$$

Assigning f to be x, y, and z, Eq. (4.15) leads to a 3×3 nonlinear system of r_1, r_2, and r_3. The nonlinear system is solved iteratively using the Newton–Raphson method. The iteration method exhibits a quadratic convergence rate as it converges, and software packages for the method are available [15]. For the iteration, the initial values are computed as

$$
r_1 = \frac{\mathbf{r}_{aN}}{\mathbf{r}_{ab}}, \quad r_2 = \frac{\mathbf{r}_{cN}}{\mathbf{r}_{cd}}, \quad r_3 = \frac{\mathbf{r}_{eN}}{\mathbf{r}_{ef}}
\tag{4.17}
$$

It has been shown that (4.17) is actually a first-order accurate solution to Eq. (4.15), and one or two iterations lead to convergence [16].

Once the grid connectivity is determined, interpolation of the solution from G_2 to G_1 can be carried out. A direct method is to interpolate all flow variables in the curvilinear coordinates. For the convenience of discussion, let contravariant velocity components U^1, U^2, and U^3 be expressed as U, V, and W, respectively. Consider interpolation at interface nodes along grid line $i = 1$ in Fig. 4.1a. The interpolation for the solution at interface nodes $(1,j,k)$ (k is not shown in the figure) becomes

$$
\begin{aligned}
U_{1,j,k} &= U^I_{1,j,k} \\
V_{1,j,k} &= V^I_{1,j,k} \\
W_{1,j,k} &= W^I_{1,j,k} \\
p_{1,j,k} &= p^I_{1,j,k}
\end{aligned}
\tag{4.18}
$$

in which superscript I depicts the solution interpolated from grid G_2. In the Cartesian coordinate system, the direct method can be expressed as

$$
\begin{aligned}
u_{1,j,k} &= u^I_{1,j,k} \\
v_{1,j,k} &= v^I_{1,j,k} \\
w_{1,j,k} &= w^I_{1,j,k} \\
p_{1,j,k} &= p^I_{1,j,k}
\end{aligned}
\tag{4.19}
$$

A difference between algorithms (4.18) and (4.19) is that the latter does not involve any geometric parameters, i.e., transformation between the Cartesian and curvilinear coordinates. Hereafter, (4.18) and (4.19) are referred to as the standard interpolation (SI). The interpolation in (4.18) and (4.19) can be performed simply by letting f

be flow variables, such as u, v, w, and p, in (4.15) and (4.16). Note that the SI is a nonconservative interface algorithm.

Second-order accuracy is achieved by interpolation (4.15) and (4.16). This can be shown by plugging (4.14) and (4.16) into (4.15), which leads to

$$
\begin{aligned}
f_N = f_{i,j,k} &+ \left((\xi^1 - \xi^1_i)\frac{\partial}{\partial \xi^1} + (\xi^2 - \xi^2_j)\frac{\partial}{\partial \xi^2} + (\xi^3 - \xi^3_k)\frac{\partial}{\partial \xi^3} \right)_{i,j,k} f \\
&+ \left((\xi^1 - \xi^1_i)(\xi^1_{i+1} - \xi^1_i)\frac{\partial^2}{\partial {\xi^1}^2} + (\xi^2 - \xi^2_j)(\xi^2_{j+1} - \xi^2_j)\frac{\partial^2}{\partial {\xi^2}^2} \right. \\
&\left. + (\xi^3 - \xi^3_k)(\xi^3_{k+1} - \xi^3_k)\frac{\partial^2}{\partial {\xi^3}^2} \right)_{i,j,k} f \\
&+ \left((\xi^1 - \xi^1_i)(\xi^2 - \xi^2_i)\frac{\partial^2}{\partial \xi^1 \partial \xi^2} + (\xi^1 - \xi^1_j)(\xi^3 - \xi^3_j)\frac{\partial^2}{\partial \xi^1 \partial \xi^3} \right. \\
&\left. + (\xi^2 - \xi^2_k)(\xi^3 - \xi^3_k)\frac{\partial^2}{\partial \xi^2 \partial \xi^3} \right)_{i,j,k} f \\
&+ \mathcal{O}(\Delta^3)
\end{aligned}
$$

Let its exact value be f^{exact}, Taylor expansion leads to

$$
\begin{aligned}
f^{exact} = f_{i,j,k} &+ \left((\xi^1 - \xi^1_i)\frac{\partial}{\partial \xi^1} + (\xi^2 - \xi^2_j)\frac{\partial}{\partial \xi^2} + (\xi^3 - \xi^3_k)\frac{\partial}{\partial \xi^3} \right)_{i,j,k} f \\
&+ \left((\xi^1 - \xi^1_i)^2\frac{\partial^2}{\partial {\xi^1}^2} + (\xi^2 - \xi^2_j)^2\frac{\partial^2}{\partial {\xi^2}^2} + (\xi^3 - \xi^3_k)^2\frac{\partial^2}{\partial {\xi^3}^2} \right)_{i,j,k} f \\
&+ \left((\xi^1 - \xi^1_i)(\xi^2 - \xi^2_i)\frac{\partial^2}{\partial \xi^1 \partial \xi^2} + (\xi^1 - \xi^1_j)(\xi^3 - \xi^3_j)\frac{\partial^2}{\partial \xi^1 \partial \xi^3} \right. \\
&\left. + (\xi^2 - \xi^2_k)(\xi^3 - \xi^3_k)\frac{\partial^2}{\partial \xi^2 \partial \xi^3} \right)_{i,j,k} f \\
&+ \mathcal{O}(\Delta^3)
\end{aligned}
$$

Then, one has

$$
f^{exact} - f_N = \mathcal{O}(\Delta^2) \tag{4.20}
$$

It is expected that the second-order accuracy leads to an approximation of flow variables represented by spatial derivatives such as vorticity, enabling them to pass grid interfaces continuously.

4.2.2 Mass Flux Balance Interpolation

It is straightforward to directly interpolate and determine all flow variables, i.e., U, V, W, p (or u, v, w, and p), at grid interfaces, and the SI interface algorithm in the last subsection is an example. However, in general, the adoption of a complete set of interpolated values, such as U, V, W, and p, at the interfaces leads to over-imposition of the boundary conditions, or it results in redundancy in the boundary conditions at the interfaces. Such over-imposition tends to deteriorate the properties of the governing equations, such as the divergence-free condition, at interfaces, and thus could cause problems in computation. A remedy is discussed in the following content.

It is well recognized that a crucial issue in the simulation of an incompressible flow is to ensure that the divergence-free condition is well satisfied. Therefore, it is important to enforce such a condition along grid interfaces. For this purpose, a mass-flux-balance interpolation (MFBI) algorithm is developed [16, 17]. The basic idea of the algorithm is that, along a grid interface, numerical fluxes computed on two adjacent grids should have the same value. For the Chimera grids in Fig. 4.1a, let ξ, η, and ζ be along the directions of i, j, and k, respectively, with the direction of k and ζ into the paper (consider a 3D situation and follow the right-hand rule). Along the middle line between $i = 1$ and $i = 2$, i.e., $i = 3/2$, the dashed line in the figure, such an idea is expressed as

$$\overline{F}^1{}_{3/2,j,k}\left(U_{1,j,k}, U_{2,j,k}, \ldots\right)\Big|_{G_1} = \overline{F}^1{}_{3/2,j,k}\left(U_{1,j,k}, U_{2,j,k}, \ldots\right)\Big|_{G_2} \tag{4.21}$$

where $\overline{F}^1{}_{3/2,j,k}$ stands for the ξ^1-direction numerical mass flux across the middle line. Equation (4.21) requires that the numerical mass flux computed from the solution on G_1 be the same as that from the solution on G_2, and it enforces balance or conservation of the numerical mass flux between the two grids. Recall (3.10), the consistency condition of a numerical flux requires that, as the grid spacing and the time step go to zero and all its arguments become the same (consider solution continuity), the numerical flux tends to a corresponding physical flux, such as the mass or momentum flux. Therefore, (4.21) is physics-based: it enforces the conservation of mass flux across the interface.

As shown in (4.20), when the solution on grid G_2 is interpolated by (4.15) onto grid G_1, one has $U^I = U + \mathcal{O}(\Delta^2)$. As a result of the Taylor expansion, it is known that

$$\overline{F}^1{}_{3/2,j,k}\left(U^I_{1,j,k}, U^I_{2,j,k}, \ldots\right)\Big|_{G_1} = \overline{F}^1{}_{3/2,j,k}\left(U_{1,j,k}, U_{2,j,k}, \ldots\right)\Big|_{G_2} + \mathcal{O}(\Delta^2)$$

which, combined with (4.21), leads a second-order accurate formula

$$\overline{F}^1{}_{3/2,j,k}\left(U_{1,j,k}, U_{2,j,k}, \ldots\right)\Big|_{G_1} = \overline{F}^1{}_{3/2,j,k}\left(U^I_{1,j,k}, U^I_{2,j,k}, \ldots\right)\Big|_{G_1} \tag{4.22}$$

The aforementioned equation means that the numerical mass flux along a grid interface computed using the solution on G_1 should be the same as the flux there computed using the solution interpolated from G_2. According to (4.8), the numerical mass flux (considering no artificial dissipation) for the finite difference in ξ^1-direction is

$$
\begin{aligned}
\overline{F}^1_{3/2,j,k}(U_{1,j,k}, U_{2,j,k})\Big|_{G_1} &= \frac{1}{2}\left[\left(\frac{U}{J}\right)_{1,j,k} + \left(\frac{U}{J}\right)_{2,j,k}\right] \\
\overline{F}^1_{3/2,j,k}(U^I_{1,j,k}, U^I_{2,j,k})\Big|_{G_2} &= \frac{1}{2}\left[\left(\frac{U}{J}\right)^I_{1,j,k} + \left(\frac{U}{J}\right)^I_{2,j,k}\right]
\end{aligned}
\tag{4.23}
$$

The right hand sides are evaluated on G_1. Applying the above to (4.22), one has

$$
\frac{1}{2}\left[\left(\frac{U}{J}\right)_{1,j,k} + \left(\frac{U}{J}\right)_{2,j,k}\right] = \frac{1}{2}\left[\left(\frac{U}{J}\right)^I_{1,j,k} + \left(\frac{U}{J}\right)^I_{2,j,k}\right]
$$

This sets up a condition for velocity in ξ^1-direction, i.e., $U_{1,j,k}$. This condition, together with interpolation of other variables, leads to an interface algorithm

$$
\begin{aligned}
U_{1,j,k} &= J_{1,j,k}\left[\left(\frac{U}{J}\right)^I_{1,j,k} + \left(\frac{U}{J}\right)^I_{2,j,k} - \left(\frac{U}{J}\right)_{2,j,k}\right] \\
V_{1,j,k} &= V^I_{1,j,k} \\
W_{1,j,k} &= W^I_{1,j,k} \\
p_{1,j,k} &= p^I_{1,j,k}
\end{aligned}
\tag{4.24}
$$

which provides boundary conditions at interface nodes. Equation (4.24) is an MFBI algorithm in the curvilinear coordinates, and it is referred to as the MFBI1 algorithm. It is seen that the MFBI algorithm is just a slight modification of SI algorithm (4.18); instead of directly imposing the interpolated value as in SI, its condition for U is derived from mass flux balanced condition (4.21), which is based on physics, essentially the mass conservation or the divergence-free condition. The MFBI is a "relaxation" of SI and a remedy for the over-imposing problem.

The MFBI can also be implemented on variables in Cartesian grids. Consider the stencil of 2 for the numerical mass flux:

$$
\begin{aligned}
&\overline{F}^1_{3/2,j,k}(U_{1,j,k}, U_{2,j,k}) \\
&= \overline{F}^1_{3/2,j,k}(U_{3/2,j,k}, U_{3/2,j,k}) \\
&\quad + \frac{\partial \overline{F}_{3/2,j,k}(\cdot, U_{3/2,j,k})}{\partial (U/J)_{1,j,k}}\Big|_{3/2,j,k}\left(\left(\frac{U}{J}\right)_{1,j,k} - \left(\frac{U}{J}\right)_{3/2,j,k}\right) \\
&\quad + \frac{\partial \overline{F}_{3/2,j,k}(U_{3/2,j,k}, \cdot)}{\partial (U/J)_{2,j,k}}\Big|_{3/2,j,k}\left(\left(\frac{U}{J}\right)_{2,j,k} - \left(\frac{U}{J}\right)_{3/2,j,k}\right) + \mathcal{O}(\Delta^2) \\
&= \left(\frac{U}{J}\right)_{3/2,j,k} + \frac{1}{2}\left(\left(\frac{U}{J}\right)_{1/2,j,k} - 2\left(\frac{U}{J}\right)_{3/2,j,k} + \left(\frac{U}{J}\right)_{2,j,k}\right) + \mathcal{O}(\Delta^2) \\
&= \frac{U_{1,j,k} + U_{2,j,k}}{2J_{3/2,j,k}} + \frac{1}{2}\left(\left(\frac{U}{J}\right)_{1/2,j,k} - 2\left(\frac{U}{J}\right)_{3/2,j,k} + \left(\frac{U}{J}\right)_{2,j,k}\right) + \mathcal{O}(\Delta^2) \\
&= \frac{U_{1,j,k} + U_{2,j,k}}{2J_{3/2,j,k}} + \mathcal{O}(\Delta^2)
\end{aligned}
$$

where $\Delta = max(\Delta\xi^1, \Delta\xi^2, \Delta\xi^3)$. In the aforementioned, consistency condition $\overline{F}(U, U) = F(U)$ (see (3.10) for reference) and Eq. (4.23) are utilized. Plugging the above into (4.22), one has

$$U_{1,j,k} - U^I_{1,j,k} = U^I_{2,j,k} - U_{2,j,k} + \mathcal{O}(\Delta^2)$$

which, in view of the relation between contravariant and Cartesian velocities in (4.6), leads to

$$\begin{aligned}
&(u_{1,j,k} - u^I_{1,j,k})(\xi^1_x)_{1,j,k} + (v_{1,j,k} - v^I_{1,j,k})(\xi^2_x)_{1,j,k}\\
&+(w_{1,j,k} - w^I_{1,j,k})(\xi^3_x)_{1,j,k}\\
=\ &(u^I_{2,j,k} - u_{2,j,k})(\xi^1_x)_{2,j,k} + (v^I_{2,j,k} - v_{2,j,k})(\xi^2_x)_{2,j,k}\\
&+(w^I_{2,j,k} - w_{2,j,k})(\xi^3_x)_{2,j,k} + \mathcal{O}(\Delta^2)
\end{aligned}$$

Note that $(\xi^m_x)_{2,j,k} = (\xi^m_x)_{1,j,k} + \mathcal{O}(\Delta)$ $(m = 1, 2, 3)$, it is readily to check that the aforementioned is satisfied if

$$\begin{aligned}
u_{1,j,k} - u^I_{1,j,k} &= u^I_{2,j,k} - u_{2,j,k}\\
v_{1,j,k} - v^I_{1,j,k} &= v^I_{2,j,k} - v_{2,j,k}\\
w_{1,j,k} - w^I_{1,j,k} &= w^I_{2,j,k} - w_{2,j,k}
\end{aligned}$$

The aforementioned formulas, together with pressure interpolation, lead to the following interface algorithm:

$$\begin{aligned}
u_{1,j,k} &= u^I_{1,j,k} + u^I_{2,j,k} - u_{2,j,k}\\
v_{1,j,k} &= v^I_{1,j,k} + v^I_{2,j,k} - v_{2,j,k}\\
w_{1,j,k} &= w^I_{1,j,k} + w^I_{2,j,k} - w_{2,j,k}\\
p_{1,j,k} &= p^I_{1,j,k}
\end{aligned} \tag{4.25}$$

which is the MFBI algorithm in the Cartesian coordinates, and it is referred to as the MFBI2 algorithm.

4.2.3 Procedure of Multiblock Computation

In the computation by the multiblock method, the conventional Schwarz approach introduced in Sects. 1.2 and 2.2, particularly that illustrated in algorithms (1.12) and (2.18), will be adopted when marching in time. Particularly, consider the computation over subdomains Ω_i $(i = 1, 2, ...)$ and in a time interval $[0, T]$. The computation marches in time step by step: $[t_0, t_1]$, $[t_1, t_2]$, ..., $[t_{N-1}, t_N]$, $t_n = n\Delta t$, $n = 0, 1, 2, ..., N$. Δt is the time step, and $\Delta t = T/N$. To make it simple, the time step is the same in all subdomains. Since discretization (4.7) is an implicit scheme, and if an implicit interface algorithm, e.g., via (4.18), is adopted, the Schwarz iteration among solutions of subdomains becomes necessary when marching in time. Then, the computation by the conventional Schwarz method proceeds with the following procedure:

$$Q_1^0 = Q_0, \quad \mathbf{x} \in \Omega_1$$
$$Q_2^0 = Q_0, \quad \mathbf{x} \in \Omega_2$$

......

Do $n = 0, 1, ..., N-1$

$$Q_1^{n+1,0} = Q_1^n, \quad \mathbf{x} \in \Omega_1$$
$$Q_2^{n+1,0} = Q_2^n, \quad \mathbf{x} \in \Omega_2$$

......

Do $m = 0, 1, 2, ...$

$$\begin{cases} \mathcal{S}_1(Q_1^{n+1,m+1}) = 0, & \mathbf{x} \in \Omega_1 \\ Q_1^{n+1,m+1} = g(\mathbf{x}, t_{n+1}), & \mathbf{x} \in \partial\Omega_1 \backslash \cup_j \Gamma_{1j}, \ j \neq 1 \\ Q_1^{n+1,m+1} = \mathcal{I}_1(Q_2^{n+1,m}, Q_3^{n+1,m}, ...), & \mathbf{x} \in \cup_j \Gamma_{1j}, \ j \neq 1 \end{cases}$$

$$\begin{cases} \mathcal{S}_2(Q_2^{n+1,m+1}) = 0, & \mathbf{x} \in \Omega_2 \\ Q_2^{n+1,m+1} = g(\mathbf{x}, t_{n+1}), & \mathbf{x} \in \partial\Omega_2 \backslash \cup_j \Gamma_{2j}, \ j \neq 2 \\ Q_2^{n+1,m+1} = \mathcal{I}_2(Q_1^{n+1,m}, Q_3^{n+1,m}, ...), & \mathbf{x} \in \cup_j \Gamma_{2j}, \ j \neq 2 \end{cases}$$

......

End Do

End Do

(4.26)

in which Q is solution vector, $Q = (P, u, v, w, ...)$, and Q_0 is its initial value. $\mathcal{S}$ is the flow solver as discussed in Sect. 4.1.2, i.e., (4.12). $\mathcal{I}$ represents an interface algorithm, i.e., (4.18) or (4.19) in Sect. 4.2.1, or (4.24) or (4.25) in Sect. 4.2.2. Γ_{ij} $(i, j = 1, 2, ...)$ is the interface between subdomain Ω_i and subdomain Ω_j. g is the boundary condition of the flow problem. m is the index of the Schwarz iteration among the subdomain solutions.

In (4.26), when the solver adopts AC algorithm (4.12), the solution will march in the pseudo time by $\Delta\tau$ for every Schwarz iteration, m. At the convergence, $m = M$, and it marches M steps in the pseudo time, reaching $\tau = M\Delta\tau$. Also, at the convergence, (4.13) holds, $Q_k^{m,n+1} \to Q_k^{n+1}$, and the computation between time steps n and $n+1$ is finished. Note that when no iteration is adopted (i.e., when $M = 1$), issues arise, including a non-zero pseudo-time derivative in the AC method, as well as accuracy degeneration and numerical instability at the grid interfaces (e.g., [18]).

4.3 Illustrative Simulation

4.3.1 Impulsive Cavity Flow

To demonstrate the performance of the multiblock method for incompressible flows, numerical examples for several typical flow problems are presented in this and the following few subsections. A comparison of MFBI1 and MFBI2 will be made in terms of their performance. Additionally, the advantages of MFBI over the SI algorithms will be illustrated. More relevant examples are available in the literature [16, 17, 19–22].

An important aspect of an interface algorithm, especially in incompressible flow simulations, is the satisfaction of mass conservation. In a well-posed computational problem, velocity boundary conditions for a subdomain should satisfy the divergence-free condition or the continuity equation. In discretization (4.7), the continuity equation at node (i,j,k) is

$$\frac{\overline{F}^1_{i+1/2,j,k} - \overline{F}^1_{i-1/2,j,k}}{\Delta\xi^1} + \frac{\overline{F}^2_{i,j+1/2,k} - \overline{F}^2_{i,j-1/2,k}}{\Delta\xi^2} + \frac{\overline{F}^3_{i,j,k+1/2} - \overline{F}^3_{i,j,k-1/2}}{\Delta\xi^3} = 0$$

This is a conservative scheme. Let I, J, and K be the number of grid nodes in the directions of i, j, and k, respectively. A summation of the aforementioned equation at all inner nodes ($i = 2, ..., I-1, j = 2, ..., J-1$, and $k = 2, ..., K-1$) of the subdomain leads to

$$\begin{aligned}
&\Delta\xi^2\Delta\xi^3 \textstyle\sum_{j,k}(\overline{F}^1_{I-1/2,j,k} - \overline{F}^1_{3/2,j,k}) \\
&+ \Delta\xi^1\Delta\xi^3 \textstyle\sum_{i,k}(\overline{F}^2_{i,J-1/2,k} - \overline{F}^2_{i,3/2,k}) \\
&+ \Delta\xi^1\Delta\xi^2 \textstyle\sum_{i,j}(\overline{F}^3_{i,j,K-1/2} - \overline{F}^3_{i,j,3/2}) \\
&= 0
\end{aligned}$$

Here $\Delta\xi^1$, $\Delta\xi^2$, $\Delta\xi^3 = const$. Note that, in the summation, the fluxes for inner nodes cancel with each other, and only those next to the boundaries remain. In the computation, it is desired that the aforementioned equation is satisfied. However, frequently, the equation is not exactly satisfied, or the summation of the left-hand side is not precisely zero, and there is an error in conservation. To quantify the error, conservation error is defined as

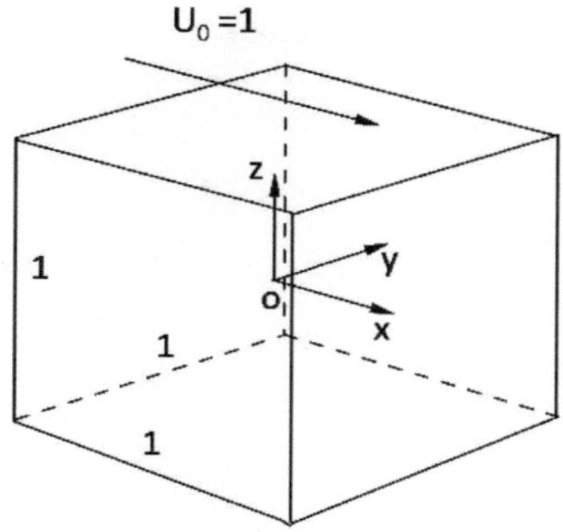

Fig. 4.2 A schematic representation of impulsively started cavity flow

$$\begin{aligned} Conservation\ error = |\ & \Delta\xi^2\Delta\xi^3 \textstyle\sum_{j,k}(\overline{F}^1_{I-1/2,j,k} - \overline{F}^1_{3/2,j,k}) \\ + \ & \Delta\xi^1\Delta\xi^3 \textstyle\sum_{i,k}(\overline{F}^2_{i,J-1/2,k} - \overline{F}^2_{i,3/2,k}) \\ + \ & \Delta\xi^1\Delta\xi^2 \textstyle\sum_{i,j}(\overline{F}^3_{i,j,K-1/2} - \overline{F}^3_{i,j,3/2})\ | \end{aligned}$$

Example 4.3.1 This is a cavity flow, particularly an impulsive cavity flow. The fluid inside a cavity, or a cubic box, is at rest initially, at $t = 0$. Then, it suddenly begins to flow due to the constant - speed motion of the lid at the top, $z = 1/2$, in the x-direction. The Reynolds number of the flow is $Re = 400$, based on the lid's side length and moving speed. In dimensionless form, the cavity has side lengths of $1\times1\times1$, and the lid speed is $U_0 = 1$, as shown in Fig. 4.2. AC method (4.12) is adopted for the simulation. Two overset grids are used, G_1 of $33\times33\ \times33$ and G_2 of $21\times21\ \times21$, see Fig. 4.3a. The background subdomain, or its grid G_1, has a hole called as a blanked zone, which is covered by the other subdomain, or its grid G_2, see Fig. 4.3b. The solution is not computed at the nodes inside the blanked zone of G_1, but is instead obtained from the nodes of G_2 lying within the zone. The time step is $\Delta t = 0.05$, and the dissipation is $\epsilon = 0.005$, together with 3-level multigrids. For each physical time step, the governing equations are integrated for 2,500 pseudo time steps, i.e., $M = 2{,}500$, to ensure that a full convergence is reached and no long-term numerical instabilities develop. We also calculate the flow using a single grid, G_1, for validation purposes.

Figure 4.4 presents the overset-grid simulations associated with interface algorithms (4.24) and (4.25), the MFBI1 and MFBI2, respectively, together with that from the single-grid simulation. This figure shows instant pressure and velocity contours on the symmetric planes of $x = 0$ and $y = 0$. At this instant, fluid motion is primarily confined in the vicinity of the moving lid, and the velocities at the lower part of the cavity are rather weak. The simulations indicate that the flow is symmetric about the line $y = 0$ in the $x = 0$ plane. The overset-grid results agree with those of the single grid; their pressure and u- and w-velocity fields are identical, except for the small discrepancies in the v-velocity contours near the hole. At this time, the flow is weak in this vicinity, and v-velocity there is small in magnitude.

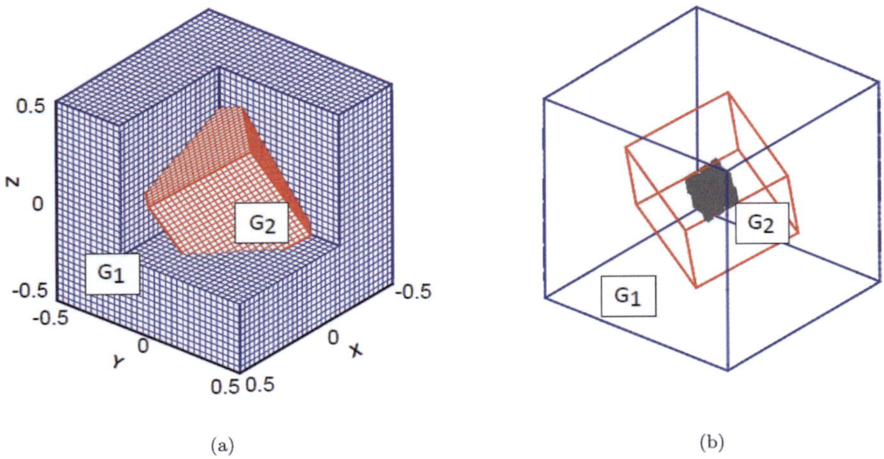

Fig. 4.3 Grids for the impulsive cavity flow. (**a**) Grids G_1 and G_2 overset with each other. (**b**) Grid G_1 has a hole (shadow), or a blanked zone

As time passes, the simulation shows that the motion diffuses downward, and the fluid in the lower part of the cavity starts moving at higher velocities, finally reaching a steady state [23]. At the steady state, the pressure and all three velocity components obtained using the overset-grid methods are practically identical, at least to the plotting scale of the figure, to the results of the single-grid computation, including the v-velocity contours in the vicinity of the hole. For instance, the aforementioned discrepancy in v-velocity disappears. The calculated time histories for the pressure and the velocity components at the center of the cavity are shown in Fig. 4.5. As it is seen, the steady-state solution is achieved after approximately 15 time units. Moreover, both interface algorithms yield identical time histories for all flow quantities at the scales of the plots. Figures 4.4 and 4.5 show that interface algorithms MFBI1 and MFBI2 lead to solutions that are almost the same to each other.

To compare the efficiency of interface algorithms MFBI1 and MFBI2, convergence histories over 10 physical time steps are shown in Fig. 4.6, in which the residual is computed as the term of the left-hand side in Eq. (4.13). A full convergence is typically achieved within 200 pseudo-time iterations for both algorithms. Beyond 200 iterations, all residuals oscillate at their respective final residual levels. Note that a single-precision computer arithmetic has been used to carry out the calculations. The entire integration interval exhibits no signs of long-term numerical instabilities. Even though both interface algorithms appear to have similar convergence characteristics, it is worth noting that they lead to a little difference. For instance, algorithm MFBI1 does yield somewhat faster convergence for v-velocity in the first few time steps. On the other hand, algorithm MFBI2 appears to consistently yield slightly lower residual levels for pressure and u-velocity.

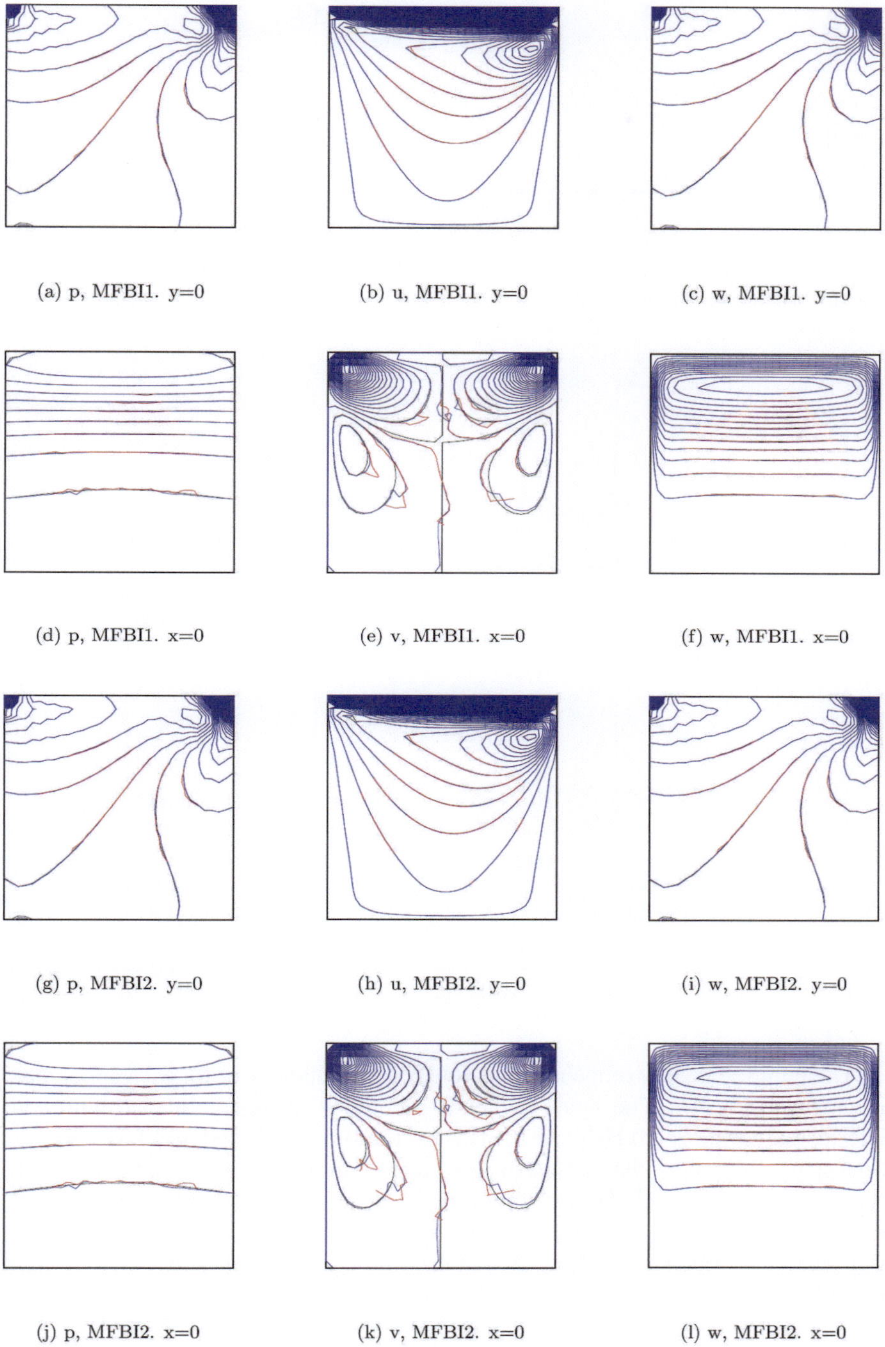

(a) p, MFBI1. y=0 (b) u, MFBI1. y=0 (c) w, MFBI1. y=0

(d) p, MFBI1. x=0 (e) v, MFBI1. x=0 (f) w, MFBI1. x=0

(g) p, MFBI2. y=0 (h) u, MFBI2. y=0 (i) w, MFBI2. y=0

(j) p, MFBI2. x=0 (k) v, MFBI2. x=0 (l) w, MFBI2. x=0

Fig. 4.4 Transient solution of the impulsive cavity flow obtained with algorithms MFBI1 and MFBI2, (4.24) and (4.25), respectively, and with the single grid. $t = 1$. Blue—solution on G_1, red—solution on G_2, black—solution of single grid

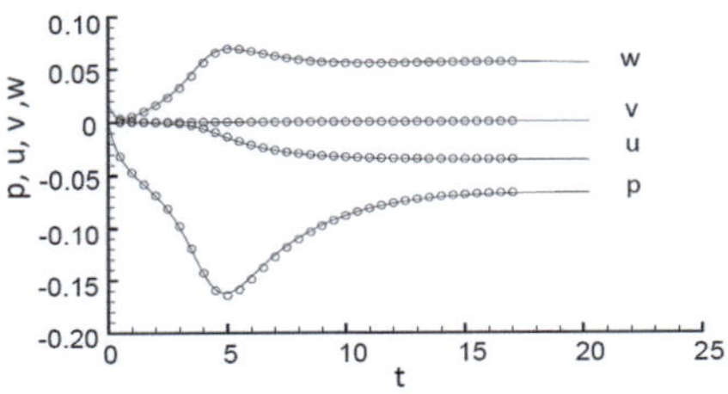

(a) MFBI1 (b) MFBI2

Fig. 4.5 Time history of pressure and velocity of the impulsive cavity flow at $x, y, z = 0$. Solid lines—multiblock, circles—single grid

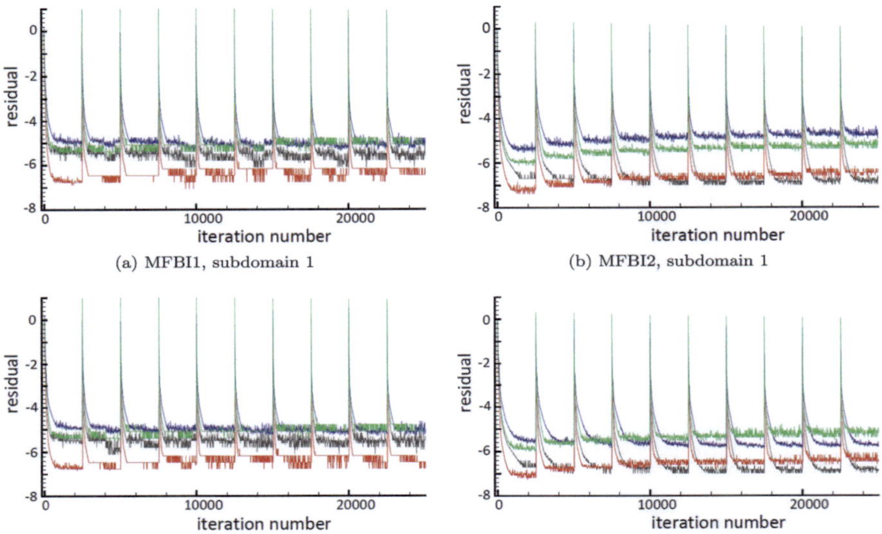

(a) MFBI1, subdomain 1 (b) MFBI2, subdomain 1

(c) MFBI1, subdomain 2 (d) MFBI2, subdomain 2

Fig. 4.6 Convergence history for the impulsive cavity flow associated with algorithms MFBI1 and MFBI2, (4.24) and (4.25), respectively. Black—p, red—u, blue—v, green—w

The iteration history of the above-defined conservation error for each subdomain is shown in Fig. 4.7. The figure depicts the iteration histories obtained with both interface algorithms MFBI1 and MFBI2 over a simulated interval of 10 physical time steps. Both algorithms result in small conservation errors for each subdomain, typically about 10^{-5}. It is also important to point out that, for both algorithms, the conservation error decreases rapidly during the iteration at each physical time step and then remains constant thereafter without exhibiting signs of long-term instabilities.

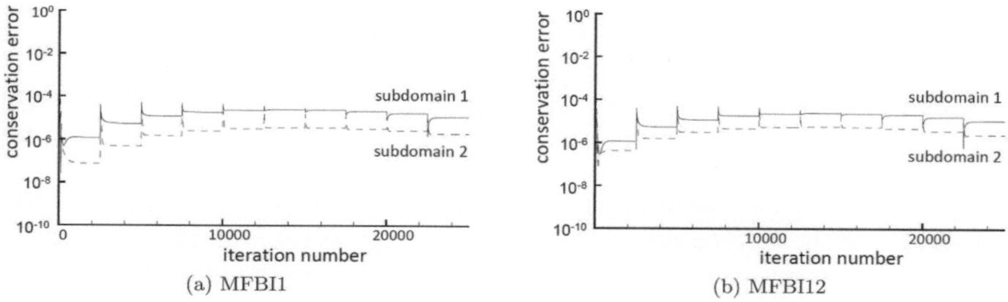

Fig. 4.7 Conservation errors for the impulsive cavity flow. (**a**) Algorithm MFBI1 (4.24). (**b**) Algorithm MFBI2 (4.25)

Now, we consider the solution accuracy associated with the overset-grid method, which includes the NS solver, the grid connectivity, and the MFBI algorithm. Let $\Delta = max(\Delta x, \Delta y, \Delta z, \Delta t)$, and f^{Δ} be the numerical solution corresponding to Δ and being a γth-order approximation to its exact value f^{exact}:

$$f^{\Delta} \approx f^{exact} + C\Delta^{\gamma} \tag{4.27}$$

It is assumed that the solution changes gradually in time and space and contains no singular point, and approximately C is a constant over the whole flow field and time interval. With such an assumption, it is derived that (e.g., [23])

$$\gamma \approx \frac{log(||f^{4\Delta} - f^{2\Delta}||_2/||f^{2\Delta} - f^{\Delta}||_2)}{log2} \tag{4.28}$$

With this formula, a numerical test is made as follows on the accuracy of the overset-grid method.

Example 4.3.2 To estimate the accuracy of the solution associated with the overset-grid method, this example computes impulsive cavity flow as described in the last example. The computation is carried out on three sets of grid spacings and time steps:

$$\begin{array}{llll} \text{I}: & G_1 : 13 \times 13 \times 13, & G_2 : 7 \times 7 \times 7, & \Delta t = 0.05 \\ \text{II}: & G_1 : 25 \times 25 \times 25, & G_2 : 13 \times 13 \times 13, & \Delta t = 0.025 \\ \text{III}: & G_1 : 49 \times 49 \times 49, & G_2 : 25 \times 25 \times 25, & \Delta t = 0.0125 \end{array}$$

To exclude its effect, artificial viscosity is turned off, $\epsilon = 0$. The 2-level multigrids are employed. The accuracy estimate formula (4.28) is applied to the solutions at $t = 0.5$, obtained by the three sets of grid spacings and time steps. The formula is only applied to the solutions on G_2 but not on G_1. This is because G_1 has a hole with an irregular boundary (Fig. 4.3b), and it cannot be applied directly. The result is shown in Table 4.1. The table shows that the accuracy orders associated with

Table 4.1 Estimated accuracy order γ associated with MFBI

MFBI	p	u	u	w
MFBI1	2.53	1.90	2.02	1.81
MFBI2	2.48	1.93	2.01	1.86

MFBI1 and MFBI2 are almost the same, about 2, for both pressure and each of the velocity components. The orders in the table are consistent with the fact that the NS solver is second-order (globally) and provides second-order (globally) accurate numerical solutions. Such orders have also been verified numerically on a single grid computation [24]. It is noted that, however, MFBI algorithms (4.24) and (4.25) are second-order, but locally, and their accuracy is only first-order globally. This is to say that MFBI algorithms are one order lower than the solver. In general, the deduction of accuracy at the interface may affect the overall accuracy of the solutions. However, the table shows that the impact of the lower accuracy at the interface is not significant on the overall solution accuracy in this test.

The two examples in this subsection show that the multiblock method proposed in the chapter works as intended. Additionally, the MFBI1 and the MFBI2 have similar performances. In the following subsections, only MFBI1 will be used, and, hereafter, it is just referred to as MFBI.

4.3.2 *Oscillatory Cavity Flow*

Before starting the example, consider the numerical mass flux across a grid interface. Particularly, let us consider it in the computation of a cavity flow on two overlapping subdomains, shown in Fig. 4.8. The grid of subdomain 1 is $k = 1, 2, \ldots$, and its interface is at $k = 1$. The grid of subdomain 2 is $k = 1, 2, \ldots, K$, and its interface is at $k = K$. Here $k = const.$ is a grid plane in the ξ^3-direction, the z-direction in the figure. The numerical mass fluxes at the two interfaces of the two subdomains are computed as

$$IntF_{3/2} = \Delta\xi^1 \Delta\xi^2 \sum_{i,j} \overline{F}^3_{i,j,3/2}$$
$$IntF_{K-1/2} = \Delta\xi^1 \Delta\xi^2 \sum_{i,j} \overline{F}^3_{i,j,K-1/2}$$

The numerical mass fluxes approximates mass fluxes in physics, particularly mass flow rates across the interfaces.

Example 4.3.3 Similar to the cavity flow in the previous subsection, the oscillatory cavity flow starts from the stationary state at $t = 0$. But its lid velocity is time-dependent, particularly $U_0 = cos(2\pi t)$. The Reynolds number of the flow is $Re = 200$, based on the lid's side length and its maximum velocity. As a benchmark, first, the simulation is made on a single grid, $25 \times 25 \times 25$. The time step is $\Delta t = 0.05$, and

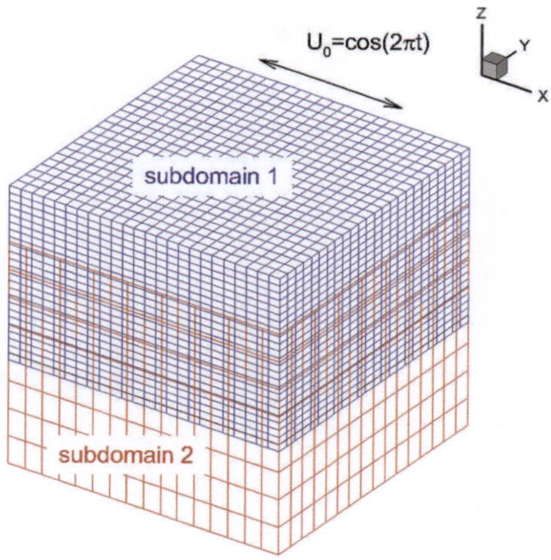

Fig. 4.8 The configuration and meshes of oscillatory cavity flow

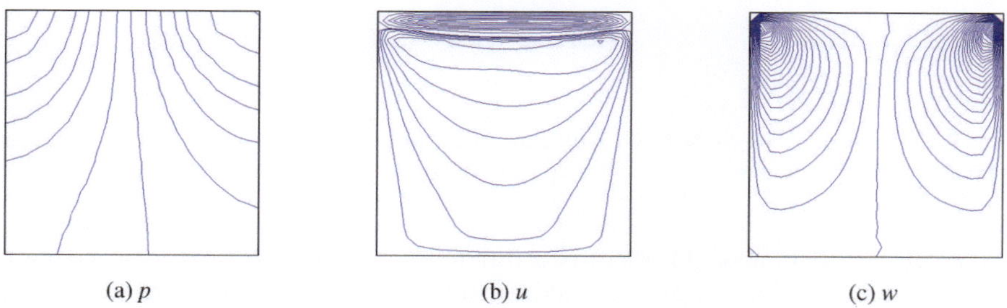

Fig. 4.9 Simulation of oscillatory cavity flow, single grid, $t = 0.25$, $y = 0$

no artificial viscosity is used, i.e., $\epsilon = 0$. The computation adopts 3-level multigrids. To ensure a full convergence, the Schwarz iteration number is set as $M = 2{,}000$ between two adjacent time physical steps.

Subdomain 1 resides in $-0.5 \le x, y \le 0.5$, $-0.125 \le z \le 0.5$, and subdomain 2 occupies $-0.5 \le x, y \le 0.5$, $-0.5 \le z \le 0.3$ (the origin $x, y, z = 0$ is at the center of the cubic). The grids for subdomains 1 and 2 are uniform, with $25 \times 25 \times 25$ and $17 \times 17 \times 9$, in x-, y-, and z-directions, respectively, see Fig. 4.8. The grid spacing of the former to that of the latter is 1.47 in the x and y directions and 3.56 in the z-direction.

An instantaneous solution obtained with the single grid is presented in Fig. 4.9. The solution is on the symmetric plane, $y = 0$. For the two-grid computation, the solution obtained with SI algorithm (4.19) is shown in Fig. 4.10. In comparison to the solution on the single grid, it is seen that the computed solution on the two grids largely captures the features of the flow. There is a little dissimilarity between the single-grid and two-grid solutions, which is attributed to the difference in their mesh resolution.

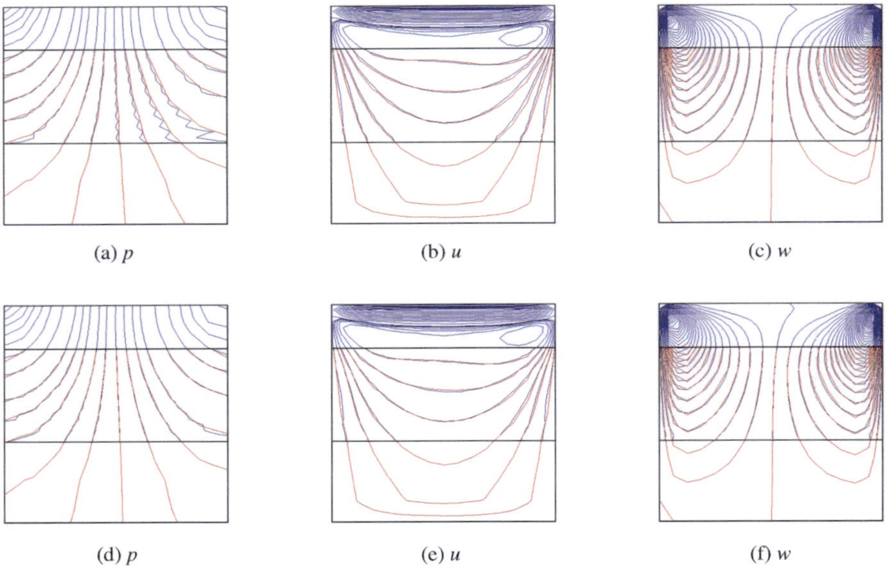

Fig. 4.10 Simulation of oscillatory cavity flow, two grids. $t = 0.25$, $y = 0$. Top—solution obtained with SI. Bottom—solution obtained with MFBI

When the SI is adopted, there is a difference between the solutions obtained on the two grids, or there are multiple solutions, in their overlapping region (Figs. 4.10a, b, and c). Moreover, the solution for p in subdomain 1 presents numerical oscillation there, which is an artifact, or a non-physical solution (Fig. 4.10a). In general, such difference and oscillation become more pronounced as the disparity increases in the two subdomains' grid spacings. Not only is such oscillation non-physical, but it also pollutes the accuracy of the solution. Therefore, it is desired to prevent or reduce such oscillation. The MFBI algorithm, particularly (4.24), effectively overcomes the difficulties encountered in the SI algorithm, as seen in Figs. 4.10d, e, and f. The figures show that, besides less difference in solutions of the two subdomains, i.e., for p and u, in the overlapping region, the solutions associated with the MFBI algorithm present much less oscillation, especially in p.

In addition, the MFBI algorithm leads to better convergence than SI. Their convergence processes in the computation of the first five time steps are shown in Fig. 4.11. The figure shows that the convergence associated with MFBI is better than that with SI in terms of lower residuals and faster speeds. Moreover, the conservation errors are computed, see Fig. 4.12a. It is seen that the errors fluctuate substantially before the convergence is achieved. The error fluctuates less with respect to magnitude when MFBI is adopted. At the convergence, the errors settle to values close to zero in both subdomains for both the SI and MFBI algorithms.

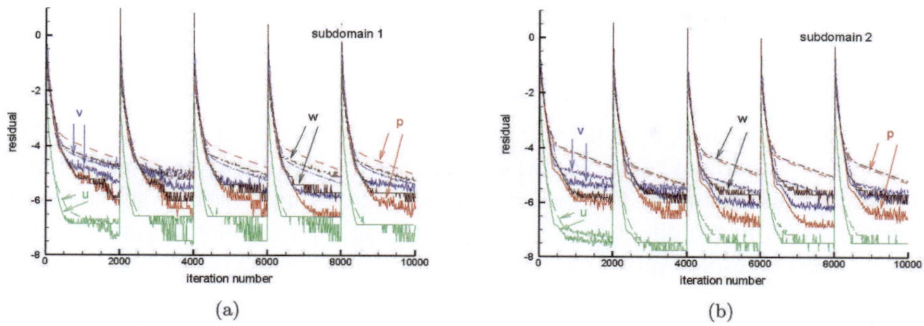

Fig. 4.11 Convergence in the first five time steps for the oscillatory cavity flow. Each variable, e.g., p, points to two lines. The higher one is for SI, and the lower one is for MFBI

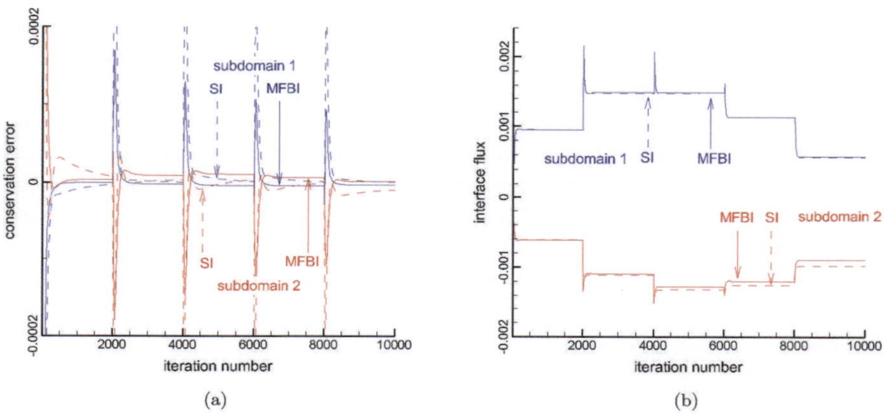

Fig. 4.12 Conservation error and interface numerical mass flux for the oscillatory cavity flow. (**a**) Conservation error. (**b**) Interface numerical mass flux

Now consider the numerical mass fluxes at subdomains' interfaces $k = 3/2$, $k = K - 1/2$ $(K = 9)$. In this situation, it is expected that the two fluxes are close in values due to mass conservation; the mass leaving one subdomain should equal that entering the other. The computed interface fluxes fluctuate during the iteration at each physical time step, and they settle down at certain values as the iteration finishes for the step, as shown in Fig. 4.12b. Note that the fluxes are in the normal direction of a surface, and a flux is positive and negative in value as it is outward and inward, respectively. The interface flux of subdomain 1 is positive, and that of subdomain 2 is negative, implying that the flux leaves the former and enters the latter (Fig. 4.12b).

4.3.3 Pipe-Bend Flow

Example 4.3.4 A steady flow in a 90-degree pipe bend is simulated. Figure 4.13 shows the flow's configurations and meshes. The Reynolds number is $Re = 500$, based on the maximum velocity at the entrance and the pipe diameter. A set of coarse meshes and another set of fine meshes are used. The coarse mesh for subdomain 1 is $61 \times 17 \times 49$, in the directions of the axial (along pipe's centerline), radius, and circumference, respectively. The coarse mesh for subdomain 2 is $61 \times 13 \times 13$, in the direction of the axial, width, and height, respectively. The fine meshes are $85 \times 25 \times 53$ and $85 \times 21 \times 21$ for subdomains 1 and 2, respectively. A fully developed (parabolic) velocity profile is specified at the entrance. The other parameters are set as $\epsilon = 0.005$, 3-level multigrids, and $M = 10{,}000$. Both SI algorithm (4.18) and MFBI algorithm (4.24) are used for the computation. Note that, as a practical approach to simulating a steady flow, only one time step is computed, with a very large time step, effectively rendering the time derivative term in Eq. (4.7) negligible.

Figure 4.14 shows the simulated flow at a horizontal plane within the bend. It is seen that the two interface algorithms present a certain dissimilarity in pressure, while they lead to little difference in velocity. Figure 4.15 depicts the pressure and axial velocity, further illustrating the discrepancy between the solutions associated with the SI and the MFBI. Notably, as the SI is adopted, both the pressure and velocity of the two subdomains exhibit discrepancy in the overlapping region of the two grids. Moreover, pressure presents oscillation there, which is an unmistakable artifact. When the MFBI is employed, such discrepancy and oscillation are reduced effectively.

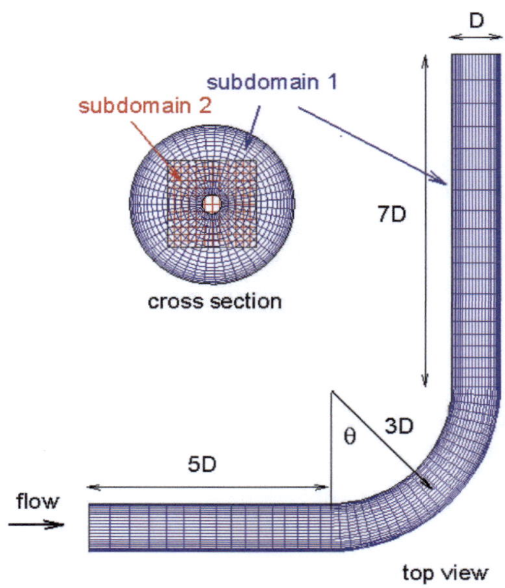

Fig. 4.13 Configurations and meshes for the flow in the pipe bend. D is the pipe diameter

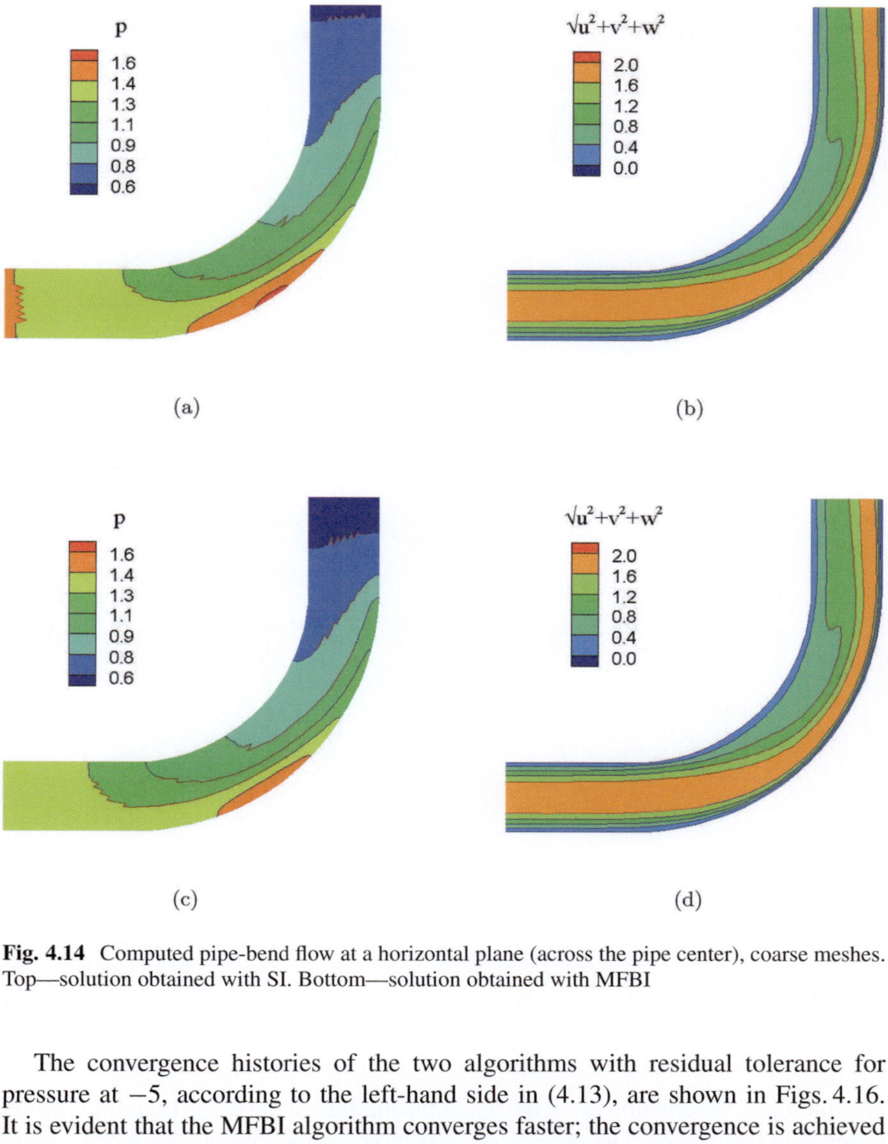

Fig. 4.14 Computed pipe-bend flow at a horizontal plane (across the pipe center), coarse meshes. Top—solution obtained with SI. Bottom—solution obtained with MFBI

The convergence histories of the two algorithms with residual tolerance for pressure at -5, according to the left-hand side in (4.13), are shown in Figs. 4.16. It is evident that the MFBI algorithm converges faster; the convergence is achieved after 5,010 and 3,560 iterations for the SI algorithm and the MFBI algorithm, respectively. The conservation errors are presented in Fig. 4.17a. It is seen that before convergence, the errors fluctuate less in value when the MFBI is adopted. Moreover, at convergence, i.e., at the iteration number of 10,000, the errors produced by the MFBI are smaller in magnitude. Since the errors approximate the flow rates across the boundary of a subdomain, ideally, the summation should be zero because of the divergence-free condition. Therefore, the smaller the magnitude

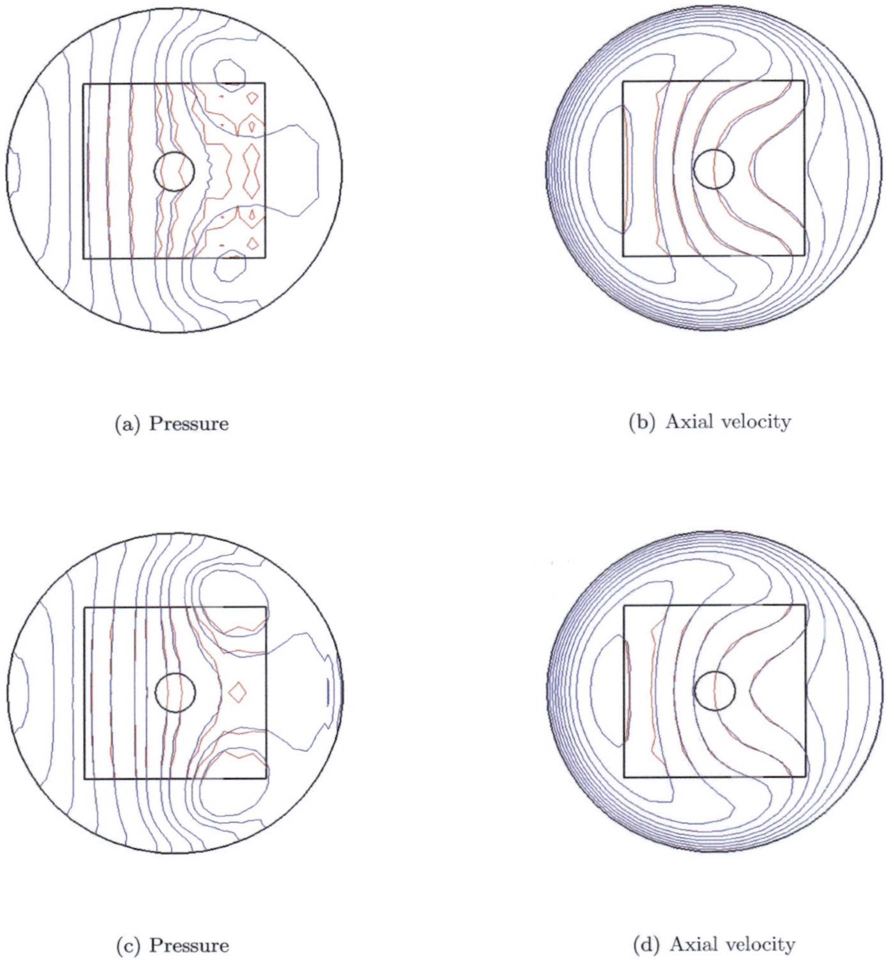

Fig. 4.15 Computed pipe-bend flow, at cross-section $\theta = 45°$, coarse meshes. Top—The SI algorithm. Bottom—MFBI algorithm

of a interface conservation error, the better the interface algorithm. Therefore, MFBI is better in this regard. The numerical mass fluxes at the interfaces are depicted in Fig. 4.17b. They behave similarly to the conservation errors.

Besides the aforementioned aspects, the SI and MFBI algorithms produce velocity profiles that agree with the measurements. This is shown in Fig. 4.18, which compares the calculated axial velocity profiles with the measurements at the plane of symmetry reported in [25]. Results obtained on both the fine and coarse grids are shown in this figure. In the figure, U_0 is the maximum of axial velocity at the entrance, and U_θ is the axial velocity. r_i and r_θ are the radii of the bend's inner

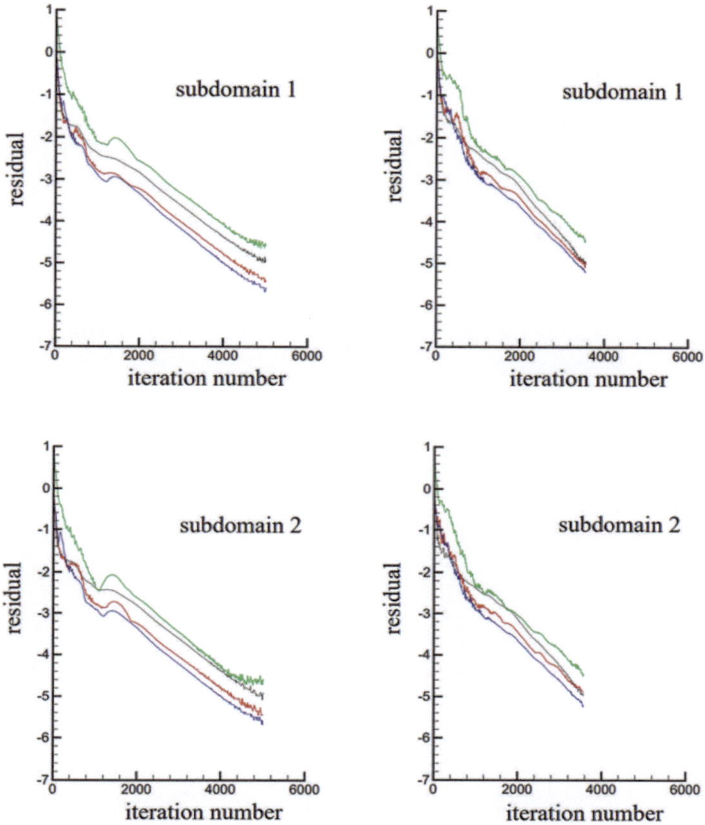

(a) SI (b) MFBI

Fig. 4.16 Convergence history of pipe-bend flow, with coarse grids. Black—p, red—u, blue—v, green—w

and outer curvatures, respectively. As seen, the two algorithms yield practically identical velocity profiles that are in overall agreement with the measurements. The only notable difference between the two algorithms is that the MFBI appears less sensitive to grid refinement. Notice, for instance, on the coarse and fine meshes, the velocity profiles obtained with the MFBI are indistinguishable, while small, albeit clearly visible, differences are observed between those obtained with the SI.

This example shows that, as in the previous cavity flows, the MFBI leads to a better convergence than the SI in terms of its speed and residual in the pipe-bend flow. Additionally, more information on the performance of the SI and the MFBI and the results of their comparison to experimental data is available in [23].

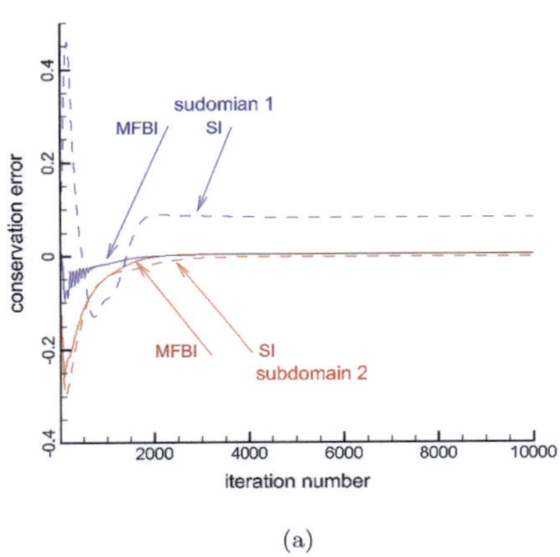

Fig. 4.17 Conservation error and numerical interface flux. Coarse grids. (**a**) Conservation error. (**b**) Interface numerical flux

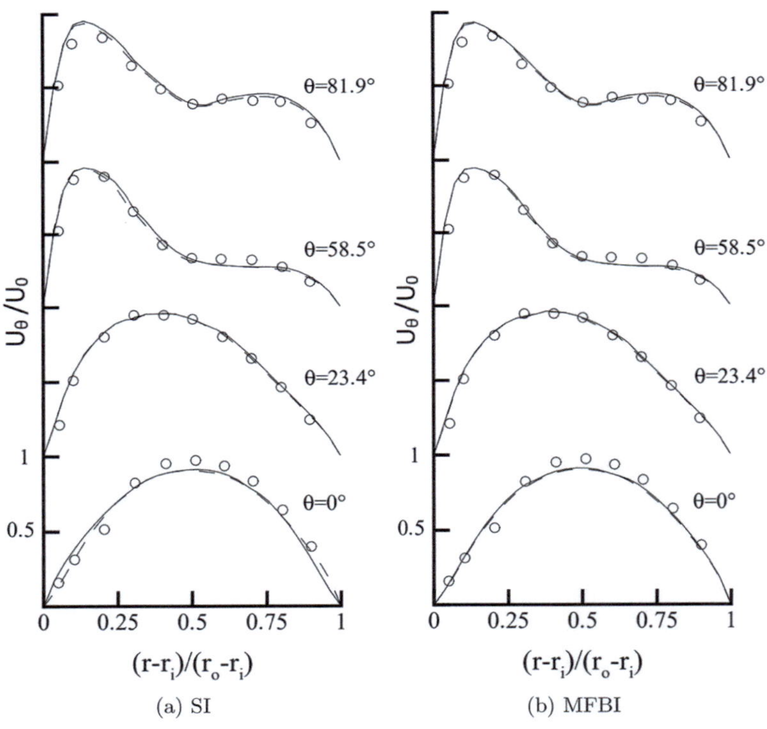

Fig. 4.18 Computed and experimental axial velocity at different locations in the horizontal plane passing the bend centerline. Circle—experiment, solid line—coarse grids, and dash line—fine grids

4.3.4 *Flow Past Cylinder*

The flow past a circular cylinder is complex, and it has been a benchmark problem in fluid simulation [26]. The simulation of such a cylinder flow serves as a more comprehensive test of the multiblock method presented in this chapter.

Example 4.3.5 A simulation is made for the 3D, unsteady, laminar flow past a cylinder mounted between two plates. The incoming flow speed is U_0, the diameter of the cylinder is D, and the length of the cylinder is L, which equals $16D$. The Reynolds number of the flow (based on the incoming flow speed and cylinder diameter) is 100. The computation is made on half of the overall flow domain, particularly the domain between a plate and the symmetry/center plane between the two plates. In the dimensionless form, the computational domain is depicted in Fig. 4.19. In this figure, $z = 0$ is the endplate, and $z = 8$ is the center plane. The flow and the computational domain parameters are identical to those adopted in the computation of [27].

A uniform velocity profile is specified at the entrance, particularly, $u = 1$, $v = 0$, and $w = 0$. At the exit, extrapolation is used for the velocity field. The no-slip condition is specified on the endplate and cylinder surfaces. Symmetry boundary conditions are specified on the symmetry planes: the bottom plane before the endplate, the center plane, and the lateral boundaries. For the pressure field, $p = 0$ is set at the center point of the exit plane. On all six surface planes of the computational domain and the surface of the cylinder, the pressure is calculated via linear extrapolation from the interior. In the whole computational domain, the initial velocity is set as $u = 1$, $v = 0$, and $w = 0$, and the initial pressure is set to $p = 0$.

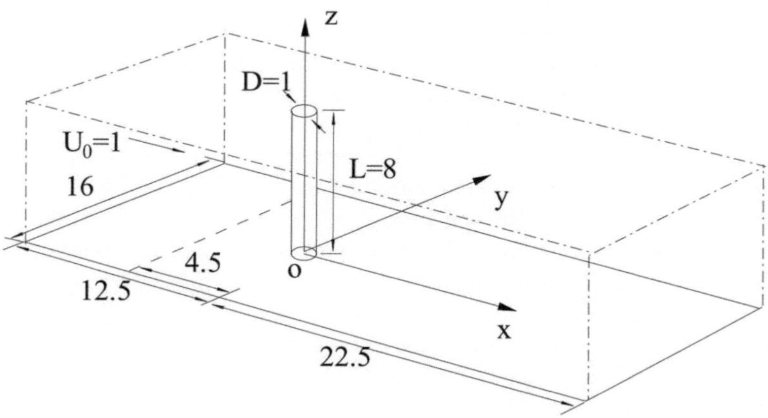

Fig. 4.19 Computational domain for cylinder flow. Inlet: $x = -12.5$; Exit: $x = 22.5$; Symmetry planes: $y = -8, 8$ (lateral sides), $z = 8$ (center plane between endplates), and $x < -4.5$ with $z = 0$ (bottom before cylinder); Wall: $x > -4.5$ with $z = 0$ (endplate surface), and $x^2 + y^2 = 1$ with $0 < z < 8$ (cylinder surface)

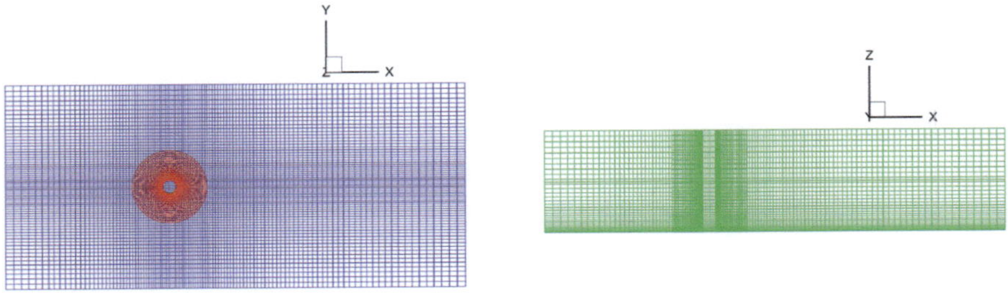

(a) Top view (b) Side view

Fig. 4.20 Grids for cylinder flow

An H-type grid and an O-type grid are used, and they overlap with each other, see Fig. 4.20. More particularly, the H-type grid, or its subdomain, covers the whole region, except a blanked zone, i.e., a hole that is next to the cylinder. The O-type grid, or its subdomain, wraps the cylinder and also encloses the blanked zone. MFBI algorithm (4.24) is applied to the interface of the O-type grid. The interface of the H-type grid is the edge of the blanked zone, which is not along a grid line, and MFBI is not applied there. Then, SI algorithm (4.18) is used at the interface of the H-type grid. The H-type grid has nodes of $129 \times 81 \times 49$ in the longitudinal, lateral, and vertical directions, respectively. The O-type grid has nodes of $37 \times 121 \times 53$ in radius, circumferential, and vertical directions, respectively. The two grids have 749,000 nodes in total. For both grids, hyperbolic tangent stretching is used to cluster the grid nodes near the walls (Fig. 4.20). The minimum grid spacing in the O-type grid at the walls is $0.0061D$. The computational adopts the AC method.

Relevant parameters are time step $\Delta t = 0.05$, $M = 20$, and $\epsilon = 0.004$, together with 3-level multigrids. It is noted that after approximately 20 iterations between the subdomains, i.e., $M = 20$, the pressure residual decreases by about one order of magnitude, while the velocity residual decreases by about three orders of magnitude. Continuing iteration is found to have little effect on the residual levels and the evolution of the flow field. For this reason, also for the sake of computational efficiency, the number of iterations at each physical time step is limited to 20 only.

The time history of the lift and drag coefficients computed from the numerical solutions is plotted in Fig. 4.21. The figure shows that the two coefficients experience an initial transient stage and then become periodic. The figure's stable mean values of the lift and drag coefficients are approximately 0.075 and 1.315, respectively. The periodic nature of the coefficients is clearly established by the calculated power spectra for the lift and drag coefficients, which suggest the presence of a single frequency, see Fig. 4.22. The Strouhal numbers, defined as $St = fU_0/D$, where f is the dominant frequency, for C_l and C_d are $St_l = 0.151$ and $St_d = 0.303$, respectively. Here, the calculated dominant frequency for C_d is almost exactly twice that for C_l, and this trend is in agreement with the 2D theory developed by Wu [28]. Note that three small spikes in the drag spectrum are related rationally to the main frequency—$St_d/6$, $2St_d$, and $3St_d$, respectively.

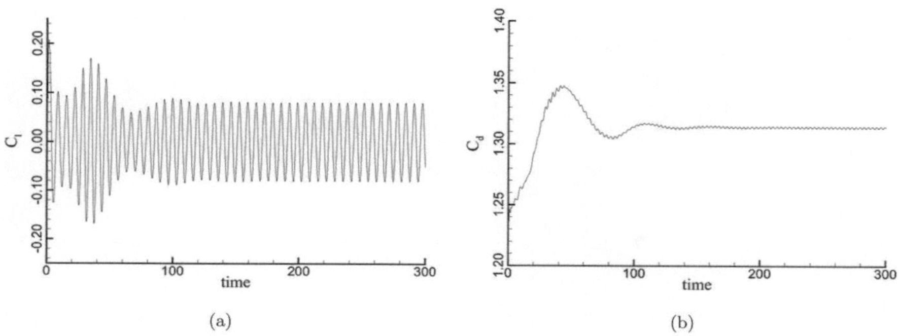

Fig. 4.21 Time history of lift and drag coefficients

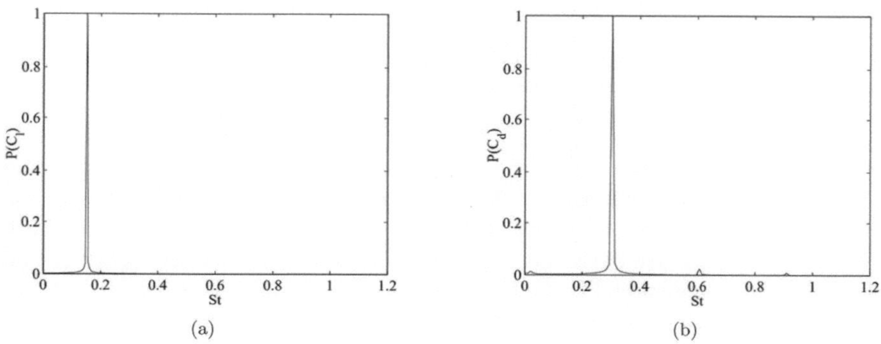

Fig. 4.22 Power spectra of the time history of lift and drag coefficients

The simulated solutions are in the range of data obtained from other sources. The computed average value for the drag coefficient is within 2% of the value reported by Mittal [27]. However, the computed amplitude of the lift oscillation, on the other hand, is considerably higher than Mittal's simulations—0.075 vs. 0.047, respectively (the discrepancy could be attributed to the numerical algorithm, grid layout, grid spacing, and time step). Additionally, $St_l = 0.145$ is reported in [27]. An experiment for $Re = 101$ suggests a step-function-like variation of St_l with L/D. St_l is approximately constant and equal to 0.141 for aspect ratios less than 30 [29].

Instantaneous limiting streamlines (or friction lines) on the solid plate are shown in Fig. 4.23. The calculated topology of the flow appears to be rather stable in time and is broadly consistent with the results of previously reported calculations for cylinder-plate juncture flows at low Reynolds numbers, e.g., Visbal [30]. The topology consists of a line of convergence that wraps around the cylinder, its associated saddle point S_A, a line of divergence, and its associated node N_A. The line of convergence is an indication of 3D separation of the approach boundary layer and is the footprint on the wall of the necklace (or horseshoe) vortex that wraps

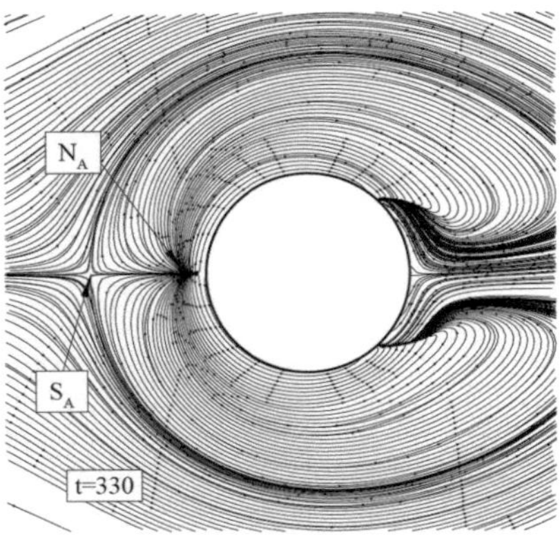

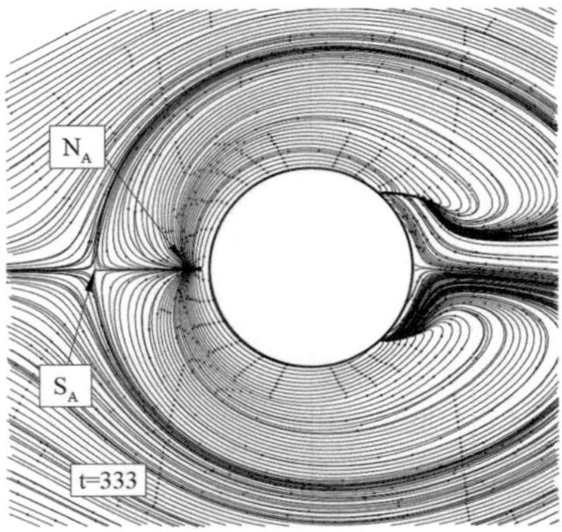

Fig. 4.23 Instantaneous streamline patterns next to the endplate

around the cylinder. Note that there is a significant portion of upstream originating streamlines that are attracted by the two, symmetrically placed concerning the horizontal direction, spiral-in foci at the downstream end of the cylinder. These two singular points also attract all streamlines between the line of convergence and the surface of the cylinder. For the fluid continuity to be satisfied, this flow topology suggests that near-wall trajectories attracted by the two foci at the rear of the cylinder must be ejected upward in the spanwise direction.

The ascending secondary motion in the wake of the cylinder is illustrated in Fig. 4.24, which shows the calculated trajectories of fluid particles. Particularly, fluid particles are released continuously from fixed locations next to the cylinder as well as the endplate, and their trajectories are integrated in a time-accurate fashion using the particle-tracking algorithm described in Appendix D. In Fig. 4.24a, particles are released just above the endplate along two concentric circles around the cylinder. As seen, right after their release, all particles remain very close to the endplate until they approach the downstream side of the cylinder, where they are ejected upward along the span. Then, the particle trajectories braid together to form tornado-like Lagrangian structures that are clearly distorted in the spanwise direction. Finally, at approximately a quarter span, the particles' ascending motion appears to stop as all trajectories turn sharply and begin to move in the downstream direction. It is interesting to note that very near the endplate, the flow appears to be only weakly unsteady, as all trajectories stay close together at all times. Moreover, the stirring of blue and red particles that originated from very near the endplate appears relatively weak in the wake of the cylinder. As seen in Fig. 4.24b, trajectories frequently cross the vertical plane of symmetry.

Figure 4.25 presents an instantaneous solution for the computed flow field in terms of the 2D contour of pressure and vorticity components, and it shows the pattern of the vortex shedding in the wake of the cylinder. It is seen in the figure

Fig. 4.24 Trajectories of particles. Blue and red colors indicate the particles released from two sides of the cylinder. (**a**) 3D view. (**b**) Top view

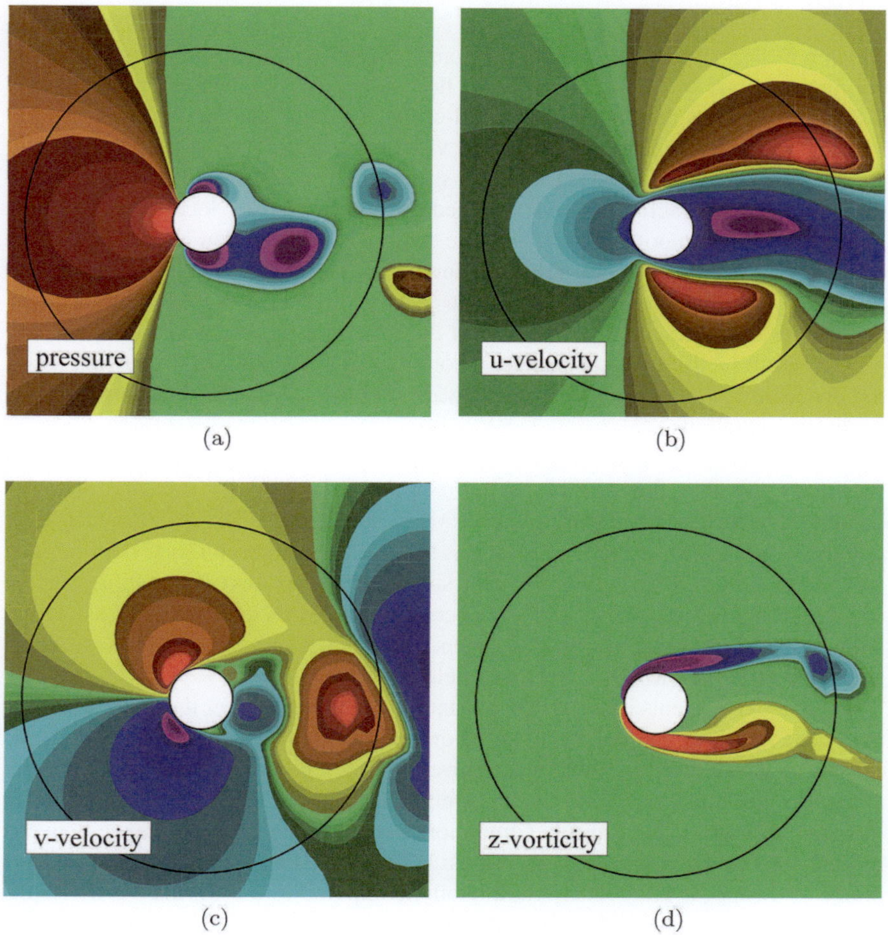

Fig. 4.25 Instantaneous contours of pressure, velocity, and vorticity at the symmetrical plane $z = 8$. The solid line is the grid interface

that vortical structures generated at the surface of the cylinder continuously cross the interface of the cylinder subdomain into the background subdomain without distortion or artifact. More particularly, the computed vorticity is continuous across the interface, which is the derivative of the velocity, and this could imply that the MFBI algorithm has certain approximation for derivatives, although it only exchanges velocity, see MFBI algorithm (4.24).

The aforementioned example indicates that the multiblock method, including the MFBI algorithm, works as intended for a complex flow. More information of the simulation and solution is available in [17].

Summary and Question

Summary The continuity, momentum, and energy equations comprise the governing equations for an incompressible flow. The solver for these equations uses a finite difference method with second-order accuracy in both time and space. The discretized equations are integrated via the 4th-order Runge-Kutta method in pseudo-time and computed on overset, structured grids. To enhance the computation, techniques including matrix-valued artificial viscosity, local time stepping, implicit residual smoothing, and V-cycle multigrid acceleration are utilized. Grid connectivity is described in detail, including its computation and accuracy, and the SI algorithm is presented to handle grid interfaces. Two MFBI algorithms, one based on Cartesian coordinates and the other on curvilinear coordinates, are introduced to facilitate the exchange of solutions at the interfaces. Following these, the procedure of the conventional Schwarz method is presented to compute the governing equations.

Numerical experiments are performed to test and illustrate the multiblock method for incompressible flows. First, experiments on the impulsive cavity flow are conducted to test and compare the two MFBI algorithms. The experiments show that the two algorithms produce solutions that are close to each other, both in good agreement with the results obtained on a single grid. They also demonstrate that the two MFBI algorithms yield similar convergence processes and residuals, as well as similar conservation errors and convergence rates, approximately 2. These findings suggest that the two MFBI algorithms yield similar results.

Further experiments illustrate the advantage of the MFBI over the SI for interface treatments. In the oscillatory cavity flow, solutions from the two subdomains exhibit apparent discrepancies in their overlapping region when the SI is adopted. Moreover, artificial oscillations are evident, particularly in the pressure values. Such discrepancies and oscillations become more pronounced as the difference between the spacings of the two meshes increases. When the MFBI is applied, they become minor or disappear altogether. Additionally, compared to the SI, the MFBI results in faster convergence of the computation. Moreover, the MFBI produces similar but more favorable conservation errors and interface numerical fluxes than those associated with the SI. Further experiments on a pipe-bend flow also confirm these advantages of the MFBI over the SI.

Finally, the methods presented in this chapter, particularly the flow solver and the multiblock method including the MFBI algorithm, are applied to the 3D flow past a cylinder with endplates. The simulation captures flow patterns, such as vortex shedding behind the cylinder, and relevant mechanisms, including those that lead to ascending secondary motion in the wake. In the simulation, flow structures, such as vortices, pass smoothly through the grid interface without artifacts. The computed frequencies of the lift and drag forces are consistent with those reported in the literature. A detailed discussion is presented on flow structures, behaviors, and mechanisms.

The content in this chapter deserves further study. A particular topic is the MFBI, whose advantages over the SI have been demonstrated. The MFBI has been constructed along a grid line or surface as a grid interface, but it has not been realized for an irregular grid interface, such as that surrounding the blanked zone associated with the H-type grid in Example 4.3.5. According to the same idea, the MFBI may be constructed cell by cell on such an interface. Additionally, the MFBI of this chapter is based on second-order, linear interpolation. To match its accuracy to that of the flow solver, one can construct MFBI using third-order, cubic interpolation. A candidate may be the BFECC technique, which enhances the accuracy of linear interpolation to third-order accuracy [22]. There are more interesting topics, such as coupling via machine learning, to be discussed in Chap. 6.

Questions The following questions are provided to aid in understanding, discussing, and investigating the content. Their answers and clues are embedded within the chapter.

1. The flow solver marches forward in physical time via an implicit discretization, see Eq. (4.7). Could it adopt another temporal discretization instead, and what are its advantages and disadvantages in comparison to those in Eq. (4.7)?
2. The discretized governing equations are solved with the AC method, Eq. (4.9), and the physical time derivative is treated implicitly as in Eq. (4.11). What is the purpose of this implicit treatment?
3. What is an interface node, its host cell, and the grid connectivity? How is the host cell searched for, and how is the location of the node within the cell determined? Discuss alternative methods for the host-cell search and location determination.
4. In the Schwarz iteration among subdomains, multiple iterations are performed, for example, $M = 20$ in the examples of this chapter. Such iteration is computationally expensive. Discuss how to reduce the number of iterations and accelerate convergence.
5. The formula for trilinear interpolation in this chapter, Eqs. (4.15) and (4.16), assumes that the curvilinear coordinate system (ξ^1, ξ^2, ξ^3) is orthogonal. How would the formula change if the system were not orthogonal, and would a similar one still be available?

6. The idea behind the MFBI is that numerical fluxes along an interface computed from two adjacent grids should be the same. What are the physical and mathematical bases for this idea?
7. The solver for the governing equations is globally second-order accurate in space and time, while the interface algorithms, both the adopted SI and MFBI, are globally first-order accurate only. Discuss how the lower-order accuracy at an interface affects the overall accuracy of the computation.
8. Continuing from the previous question, one might apply a locally third-order accurate interface algorithm to achieve uniform global second-order accuracy for the overall computation. Discuss how to construct such a third-order accurate interface algorithm, for example, as an extension of the SI or MFBI.
9. The MFBI algorithms in this chapter are designed along a grid line, e.g., $i = 1$, in Fig. 4.1. It is anticipated that MFBI algorithms could also be designed on a grid-cell-by-grid-cell basis and thus applied to an interface surrounding a blanked zone, such as that in Fig. 4.3b. Think about the construction of such an MFBI algorithm.
10. Theoretically, the conservation error on a grid is zero when a conservative scheme is applied. However, in actual computations, such as the oscillatory cavity simulation in Sect. 4.3.1, it is not zero. Discuss the reasons why this occurs.
11. The numerical flux at the grid interface approximates the mass flow rates across the interface. Discuss the values of such fluxes, e.g., $IntF_{3/2}$ and $IntF_{K-1/2}$, as the grid spacing tends to zero in the simulation of Sect. 4.3.2.
12. In the flow past the cylinder, the computed average lift coefficient is 0.075, which is close to, but differs from, values reported in other studies, for example, 0.047 in [27]. Discuss potential reasons for this discrepancy. Ideally, should the coefficient be zero due to the symmetry of the time-averaged flow about $y = 0$?

References

1. R.L. Meakin, R.L. Street, Simulation of environmental flow problems in geometrically complex domains. Part 2: A domain-splitting method. Comput. Meths. Appl. Mech. Eng. **68**, 311–331 (1988)
2. J.C. Strikwerdat, C.D. Scarbnick, A domain decomposition method for incompressible viscous flow. SIAM J. Sci. Comput. **14**(1), 49–67 (1993)
3. H.S. Tang, R.D. Haynes, G. Houzeaux, A review of domain decomposition methods for simulation of fluid flows: Concepts, algorithms, and applications. Arch. Comput. Methods Eng. **28**, 841–873 (2021)
4. P.K. Kundu, *Fluid Mechanics* (Academic Press, 1990)
5. H.S. Tang, K. Qu, X.G. Wu, An overset grid method for integration of fully 3D fluid dynamics and geophysical fluid dynamics models to simulate multiphysics coastal ocean flows. J. Comput. Phys. **273**, 548–571 (2014)
6. L.H. Shih, J.R. Koseff, G.N. Ivey, J.H. Ferziger, Parametrization of turbulent fluxes and scales using homogeneous sheared stably stratified turbulence simulations. J. Fluid Mech. **525**, 193–214 (2005)

7. J. Paik, F. Sotiropoulos, Coherent structure dynamics upstream of a long rectangular block at the side of a large aspect ratio channel. Phys. Fluids **17**, 115104 (2005)
8. H.S. Tang, J. Paik, F. Sotiropoulos, T. Khangaokar, Three-dimensional numerical modeling of initial mixing of thermal discharges at real-life configurations. ASCE J. Hydr. Eng. **134**, 1210–1224 (2008)
9. F.B. Lin, F. Sotiropoulos, Assessment of artificial dissipation models for three-dimensional incompressible flows. Trans. ASME J. Fluids Eng. **119**, 331–340 (1997)
10. A. Jameson, *Multigrid Algorithm for Compressible Flow Calculations*, MAE Report, Princeton University, 1743 (1985)
11. F. Sotiropoulos, Y. Ventikos, Transition from bubble-type vortex breakdown to columnar vortex in a confined swirling flow. Int. J. Heat Fluid Flow **19**, 446–458 (1998)
12. A. Jameson, W. Schmit, E. Turkel, Numerical solutions of the Euler equations by finite volume methods using Runge-Kutta time-stepping schemes. AIAA paper 81–1259, 1981
13. W.D. Henshaw, G. Chesshire, Multigrid on composite meshes. SIAM J. Sci. Statist. Comput. **8**, 914–923 (1987)
14. J.L. Steger, F.C. Dougherly, J.A. Benek, A Chimera grid scheme. ASME Mini-Symposium on Advance in Grid Generation, Houston, Texas, June 1983
15. W.H. Press, S.A. Flannery, B.P. Teukolsky, W.T. Vetterling, *Numerical Recipes in C: The Art of Scientific Computing, 2nd edition* (Cambridge University Press, 1992)
16. H.S. Tang, Study on a grid interface algorithm for solutions of incompressible Navier-Stokes equations. Comput. Fluids **35**, 1372–1383 (2006)
17. H.S. Tang, C. Jones, F. Sotiropoulos, An overset-grid method for 3D unsteady incompressible flows. J. Comput. Phys. **191**, 567–600 (2003)
18. B. Gustafsson, The convergence rate for difference approximations to general mixed initial boundary value problem. SIAM J. Numer. Anal. **18**, 179–190 (1981)
19. K. Xu, G. Sun, *Assessment of an interface conservative algorithm MFBI in a Chimera grid flow solver for multi-element airfoils. Proceedings of the world congress on engineering*, London, UK, 2009
20. A. Skillen, *The overset grid method, applied to the solution of the incompressible Navier-Stokes equations in two and three spatial dimensions*. Ph.D. thesis, School of Mechanical, Aerospace and Civil Engineering, University of Manchester, 2011
21. T. Craft, H. Iacovides, A. Skillen, A new overset grid algorithm applied to the simulation of flows involving complex geometries, in *Conference on Modelling Fluid Flow (CMFF'12). The 15th International Conference on Fluid Flow Technologies*, Budapest, Hungary, September 4–7, 2012
22. W. Dong, Y. Liu, H.S. Tang, Accuracy enhancing interface treatment algorithm: The back and forth error compensation and correction method. J. Scientific Comput. **93**, 1–25 (2022)
23. H.S. Tang, *Numerical simulation of unsteady three dimensional incompressible flows in complex geometries*. Ph.D. thesis, School of Civil and Enviro. Eng., Georgia Tech., 2001
24. H.S. Tang, F. Sotiropoulos, Fractional step artificial compressibility schemes for the unsteady incompressible Navier-Stokes equations. Comput. Fluids **36**, 974–986 (2007)
25. P.H.M. Bovendeerd, A.A. Steenhoven, F.A. Vosse, G. Vossers, Steady entry flow in a curved pipe. J. Fluid Mech. **177**, 233–246 (1987)
26. Zdravkovich, *Flow Around Circular Cylinder, Vol. 1* (Oxford Universtiy Press, New York, 1997)
27. S. Mittal, Computation of three-dimensional flows past cylinder of low aspect ratio. Phys. Fluids **13**, 177–191 (2001)
28. J.C. Wu, Theory for aerodynamics force and moment in viscous flow. AIAA J. **19**, 432–441 (1981)
29. C. Norberg, An experimental investigation of the flow around a circular cylinder: influnce of aspect ratio. J. Fluid Mech. **258**, 287–316 (1994)
30. M.R. Visbal, Structure of laminar juncture flows. AIAA J. **29**, 1273–1282 (1991)

Chapter 5
Coastal Ocean Flow

5.1 Background and Status

It has become necessary to advance our capability to simulate many emerging, complex flows in the ocean. Such ocean flows involve distinct types of physics spanning a vast range of temporal and spatial scales. Examples of such flows occurred in the 2010 Gulf of Mexico oil spill [1], the 2011 Tohoku Tsunami [2], and the fast-growing practice of ocean energy extraction [3]. In general, these flows are multiscale and multiphysics in nature. For instance, in the oil spill event, the crude oil gushed as a jet with initial mixing at $\mathcal{O}(10)$ m in length scales from a broken pipeline, and then it developed into patches of oil film drifting on the water surface of the ocean and spreading horizontally at $\mathcal{O}(10^2)$ mile in length scales [4]. Although both are fluid flows, the oil jet and the oil drift are distinct not only in terms of temporal and spatial scales but also in the patterns of physical phenomena described by different sets of PDEs. These phenomena and similar ones have been referred to as multiscale, multiphysics flows [5–7].

The aforementioned emerging flows challenge our modeling capability; conventionally, they have been split into subproblems and simulated separately by distinct sets of PDEs and computer models from different communities. For instance, for the 2010 Gulf of Mexico oil spill, the small-scale initial mixing of the jet is modeled widely in the engineering community using NS models [8]. Whereas, the large-scale drift of the floating oil is simulated intensively in the ocean science community using geophysical fluid dynamics (GFD) models [9]. However, there is a lack of numerical methods and computer models that can efficiently simulate the small-scale mixing and the large-scale drifting within a unified framework. Here and hereafter, as a flow solver, a model refers to a set of governing equations, computational methods, and the corresponding computer software package.

H. Tang, *Multiblock Method for Fluid Flow*, Nonlinear Systems and Complexity 40,
https://doi.org/10.1007/978-3-031-78568-9_5

In the past few decades, many NS models , i.e., the models built on the NS equations, particularly the continuity and momentum equations for fluid flow [10], have been designed for various small-scale physical phenomena in the engineering community (e.g., hydraulic engineering and aerospace engineering). Examples of NS models are FLUENT [11], OpenFoam [12], and Overture [13]. These small-scale flow models can satisfactorily predict fully 3D, complicated, local coastal ocean flows at scales $\mathcal{O}(1)$ s - $\mathcal{O}$ (1) hr in time and $\mathcal{O}$ (1) m - $\mathcal{O}$ (10) km in space with high fidelity, such as the jet in the above oil spill. The models are primarily restricted to small flow zones, and they are prohibitively expensive in terms of computational time and memory for simulating realistic, large-scale coastal ocean flows. They are also ineffective for such flows because of their governing equations and computational techniques.

At the same time, many GFD models , i.e., models based on the GFD equations, particularly those simplified from the NS equations and frequently based on the hydrostatic assumption (to be discussed in Sect. 5.5.2), have been developed for large-scale phenomena by the ocean science community. Examples of these GFD models are POM [14], FVCOM [15], ROMS [16], and SWAN [17]. These models are designed to predict current speed, sea level, water temperature, surface waves, etc., in the ocean but are restricted to periods of time as long as months, $\mathcal{O}(1)$ hr - $\mathcal{O}$ (1) month, and over regions with horizontal sizes as large as mesoscales, $\mathcal{O}(10)$ – $\mathcal{O}$ (10, 000) km. Occasionally, they provide limited dynamics at scales as small as $\mathcal{O}(100)$ m. Their restrictions come from their governing equations, numerical techniques, and parameterization. Because of the adoption of the hydrostatic assumption, these GFD models cannot directly deal with strongly 3D, local flows, such as the oil gushing at the bottom of the ocean in the above spill.

In view that computer modeling has reached the point where flow models have become mature for specific flow phenomena and scales, integration of different models, such as an NS model and a GFD model, via the multiblock method is considered as a natural, promising, and also currently most feasible approach for simulation of multiscale and multiphysics coastal ocean flows (e.g., [5, 18]). Efforts have been made to couple different models. These efforts include coupling of 1D/2D flows [19], 2D/2D flows [20], 2D/3D flows [21], and 3D/3D flows [22]. In these efforts, simplifications such as the Boussinesq assumption, potential flows, and one-way coupling are used, and the governing equations are not those best available in practice, such as the aforementioned NS and GFD models, resulting in that only limited multiphysics phenomena can be simulated. As a unique effort, the author and coworkers have explored and developed a modeling system that integrates an NS model and a GFD model using the multiblock method [5, 6, 23–26]. The NS model and the GFD model are state-of-the-art models in the engineering community and ocean science community, respectively. The resulting multiblock modeling system, to be introduced as follows, can capture multiscale, multiphysics flow phenomena in the ocean that are beyond the reach of the conventional models [27, 28].

5.2 Governing Equation and Model

The motion of an unsteady, 3D, incompressible water flow in a near-field ocean, e.g., the oil jet in the Gulf of Mexico oil spill, is described by the continuity equation, the NS equations, and the energy equation, which are respectively Eqs. (4.1), (4.2), and (4.3). The NS equations will be transformed for better coupling of the near-field equations to the governing equations for a far field. Particularly, inspired by an idea of Casulli [29], we introduce pressure split; the pressure, p, in the equations is split into two parts [23]:

$$p = p_h' + p_d' \tag{5.1}$$

With the above split, the NS equations, i.e., (4.2), are transformed as

$$\begin{aligned} \mathbf{u}_t + \nabla \cdot \mathbf{u}\mathbf{u} = & -\frac{1}{\rho_0}\nabla p_d' + \nabla \cdot ((\nu + \nu_t)\nabla \mathbf{u}) \\ & -g\nabla_H \eta + \alpha g(T - T_0)\mathbf{k} \end{aligned} \tag{5.2}$$

together with

$$-\frac{1}{\rho_0}\nabla_z p_h' = g\mathbf{k}$$

which is a balance of the gradient of p_h' and the gravity. In the aforementioned, subscript H indicates the horizontal direction. Integration of the last equation yields

$$\begin{aligned} p_h' &= \rho_0 g(\eta - z) \\ &= p_h + \rho_0 g \int_z^\eta \alpha(T - T_0)d\zeta \end{aligned} \tag{5.3}$$

in which η is the water surface elevation, and p_h is the hydrostatic pressure corresponding to the hydrostatic assumption, see Eq. (5.16). By (5.1) and (5.3), one has

$$\begin{aligned} p_d' &= p - p_h' \\ &= p - p_h - \rho_0 g \int_z^\eta \alpha(T - T_0)d\zeta \end{aligned} \tag{5.4}$$

The aforementioned pressure split, (5.1), is different from that adopted in the literature [29], see (5.17) and the discussion in Sect. 5.5.2. By the split in literature, the NS equations will be transformed into an integro-differential equation, (5.18). Such an integro-differential equation is more expensive to compute than (5.2). Indeed, numerical experiments confirm this, yet they also show that, in the presence of buoyancy, the two equations may lead to similar solution accuracies [25].

The advantage of the pressure-splitting approach over the one without it will be illustrated numerically in Sect. 5.5.1. However, as there is no buoyancy, i.e., $T = T_0$, the aforementioned split is the same as that in the literature. In this situation, p'_h and p'_d become the hydrostatic pressure, p_h, and the dynamic pressure, p_d $(= p - p_h)$, respectively.

A near field is simulated by solving continuity equation (4.1), NS equations (5.2), and energy equation (4.3). Particularly, these equations can be solved by an NS model called the Solver of Incompressible Flow on Overset Meshes (SIFOM). The SIFOM is based on the curvilinear coordinates, non-staggered, structured grids, which are described in Sect. 4.1, and its discretization techniques are those in that section, plus certain modifications [30, 31]. Particularly, it adopts a second-order accurate implicit finite difference method; the convective terms are discretized using the QUICK scheme, the other terms in space are approximated by central difference, and the time derivative term is discretized using an implicit, backward difference scheme. A third-order, fourth-difference artificial dissipation method is employed to eliminate odd-even decoupling of the pressure field. Chimera grids and the alternating Schwarz method for iteration among solutions on different grids are adopted. Available turbulence closures in SIFOM are a mixing length model, a $k-\varepsilon$ model, and a detached eddy simulation model [32, 33].

As a common practice, the geophysical fluid dynamics (GFD) equations are adopted as the governing equations for far-field ocean flows, e.g., the ocean currents in the Gulf of Mexico oil spill. The GFD equations are derived from the equations for the near field, particularly Eqs. (4.1), (4.2), and (4.3), plus simplification and parameterization. They consist of an external and an internal mode [34, 35]. The external mode is [34]:

$$\begin{aligned} &\eta_t + \nabla_H \cdot (\mathbf{V}D) = 0 \\ &(\mathbf{V}D)_t + \nabla_H \cdot (\mathbf{V}\mathbf{V}D) = -gD\nabla_H \eta + \frac{\boldsymbol{\tau}_s - \boldsymbol{\tau}_b}{\rho_o} + \mathbf{G} \end{aligned} \tag{5.5}$$

which are vertically averaged continuity and momentum equations. In the GFD equations, the internal mode consists of the 3D continuity and momentum equations associated with the hydrostatic assumption:

$$\begin{aligned} &\eta_t + \nabla_H \cdot (\mathbf{v}D) + \omega_\sigma = 0, \\ &(\mathbf{v}D)_t + \nabla_H \cdot (\mathbf{v}\mathbf{v}D) + (\mathbf{v}\omega)_\sigma = -gD\nabla_H \eta + \nabla_H \cdot (\kappa \mathbf{e}) + \frac{1}{D}(\lambda \mathbf{v}_\sigma)_\sigma \\ &\qquad - \frac{gD}{\rho_0}\left(\nabla_H (D \int_\sigma^0 \rho d\sigma') + \sigma \rho \nabla_H D\right) + \mathbf{H} \end{aligned} \tag{5.6}$$

together with the energy equation:

$$(TD)_t + \nabla_H \cdot (TD\mathbf{v}) + (T\omega)_\sigma = \nabla_H \cdot (\theta \nabla_H T) + \frac{1}{D}(\vartheta T_\sigma)_\sigma + \mathbf{I} \tag{5.7}$$

In the external mode, $\mathbf{V}$ is the depth-averaged velocity vector, with $\mathbf{V} = \mathbf{V}(x, y, t)$, D the water depth, with $D = D(x, y, t)$, $\boldsymbol{\tau}_s$ and $\boldsymbol{\tau}_b$ the shear stresses on the water surface and seabed, respectively, and **G** the rest terms such as the Coriolis force. In the internal mode, σ is a terrain-following vertical coordinate, called σ- coordinate ($0 < \sigma < 1$, with $\sigma = 0, 1$ at the seabed and the water surface, respectively). $\mathbf{v}$ is the horizontal velocity, ω the vertical velocity in the σ- coordinate, T the temperature, $\mathbf{e}$ the strain rate, and **H** and **I** the rest terms. κ, λ, θ, and ϑ are coefficients. Subscript σ stands for the derivative over σ. The Mellor and Yamada level-2.5 turbulent closure is used [36]. In the GFD equations, pressure p does not occur. Actually, due to the hydrostatic assumption, the pressure is the hydrostatic pressure, that is, $p = p_h$, and it can be recovered from η via Eq. (5.16).

The GFD equations, i.e., Eqs. (5.5), (5.6), and (5.7), will be solved by a GFD model, particularly the Finite Volume Method Coastal Ocean Model (FVCOM) [34]. In the FVCOM, the GFD equations are discretized using a finite volume method on a triangular mesh in the horizontal plane and a σ-grid in the vertical direction. In both its external and internal modes, the convection terms are discretized using upwind schemes, and the time derivative terms are discretized by Runge–Kutta methods. The external and internal modes may have different time steps. The reader is referred to [15, 34, 35] for further details of the FVCOM.

The NS and GFD models differ in governing equations, numerical algorithms, and computational grids; the former is designed for fully 3D flows at small scales, whereas the latter is developed primarily for 2D or 3D (with weak motion in the vertical direction), large-scale coastal ocean flows. To predict coastal flows with distinct physical phenomena spanning different scales, a model system has been developed by integrating the two models (e.g., [5, 23]). This system, together with its analysis and extension (e.g., [20]), is presented in the following sections.

5.3 Coupling of NS and GFD Equations

5.3.1 Integration of Governing Equations

An NS-GFD model system based on the overset-grid method is developed to simulate multiscale, multiphysics ocean flows. With the NS-GFD model system, the ocean is divided into a far field that covers the background, large-scale flow, Ω_{GFD}, and a near field that wraps a local, small-scale, complex flow of interest, Ω_{NS}, see Fig. 5.1a. As needed, there may be multiple near fields. In the system, the GFD equations are applied in subdomain Ω_{GFD}, and the NS equations are used in subdomain Ω_{NS}. Ω_{GFD} and Ω_{NS} overlap, and the solutions to the GFD and NS equations undergo a transition to each other over the overlapping region. The overlapping region will be designed and allocated (e.g., regarding its size and location) in such a way that the flow within it is hydrostatic, and thus, both the NS equations and the GFD equations are applicable. In Fig. 5.1, $\partial\Omega_{GFD}$ is an interface of Ω_{GFD} with Ω_{NS}, while $\partial\Omega_{NS}$ is an interface of Ω_{NS} with Ω_{GFD}. Moreover, as shown in Fig. 5.1b, the region enclosed by $\partial\Omega_{GFD}$ and the seabed line is a blanked

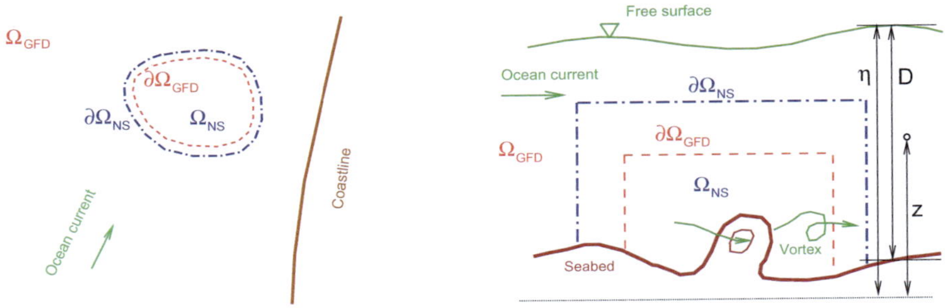

(a) Top view (b) Side view

Fig. 5.1 A schematic representation of the decomposition of the flow domain in the NS-GFD model system

zone of the GFD subdomain. This blanked zone is covered by Ω_{NS} but excluded by Ω_{GFD}, being denoted as $\Omega_{NS} \backslash \Omega_{GFD}$.

The variables in the NS equations, i.e., (4.1), (5.2), and (4.3), are ($\mathbf{u}$, p'_d, T). Note that p'_d, together with p'_h, leads to p. While those in the internal mode of the GFD equations, i.e., (5.6) and (5.7), are ($\mathbf{v}$, ω, η, T). The GFD equations have $\mathbf{v}$, plus ω that can be transferred from w, and thus they actually have $\mathbf{u}$. Additionally, these equations also contain the information of p'_d in view of (5.4) and p ($=p_h$, because of the hydrostatic assumption). As a result, under the hydrostatic assumption, essentially, the NS equations and GFD equations (precisely speaking, the internal mode of the GFD equations) have the same variables. Therefore, their integration will be made via an NS-GFD coupling, particularly the coupling between the NS equations, i.e., (4.1), (5.2), and (4.3), and the equations for the internal mode of the GFD equations, i.e., (5.6) and (5.7).

In the NS-GFD model system, the NS and GFD equations exchange their common variables at their interfaces. In the computation of either the NS equations or the GFD equations, various methods are used to impose boundary conditions. These conditions take into account both the well-posedness of each set of equations and the underlying physics. As a practical approach, the interface condition for either the NS equations or the GFD equations is prescribed solely based on the requirements of their respective boundary conditions. In other words, the interface condition is formulated to satisfy the boundary condition requirements of each set of governing equations independently, without jointly considering the two sets of equations in the coupled system. In such a consideration, the following interface condition, referred to as the NS-GFD interface condition I hereafter, is adopted:

$$\begin{aligned} &\mathbf{u}|_{NS} = \mathbf{u}|_{GFD}, \ \partial p'_d/\partial n = 0, \ T|_{NS} = T|_{GFD}, \quad &\text{on } \partial\Omega_{NS} \\ &\mathbf{u}|_{GFD} = \mathbf{u}|_{NS}, \ T|_{GFD} = T|_{NS}, \quad &\text{on } \partial\Omega_{GFD} \end{aligned} \tag{5.8}$$

Here, the condition imposes continuity of velocity across an interface.

The GFD solver does not perform the computation within its blanked zone, but it needs a solution there for its solution redistribution [15]. The solution in the blanked zone will be interpolated from the solution of the near field to maintain the structure of the software for the GFD solver. Particularly, as a special treatment of the solution exchange between the NS and GFD solvers, the NS solution is transferred to the GFD solver in its blanked zone as

$$\mathbf{u}_{GFD} = \mathbf{u}_{NS}, \qquad on \quad \Omega_{NS} \backslash \Omega_{GFD} \tag{5.9}$$

It is noted that, besides the aforementioned coupling at the interfaces and the blanked zone, the NS and GFD equations are also coupled through the gradient of water surface elevation as seen in Eq. (5.2); the water surface elevation computed from the GFD equations, for example at each time step, is transferred to the computation of the NS equations. In addition, extrapolation is used for other variables, such as those of turbulence, at the interfaces of the two solvers. Moreover, in the above NS-GFD model system, the NS subdomain has to be under the water surface, or submerged in water, since the NS solver, i.e., SIFOM, cannot deal with a free surface. To overcome this shortage, by following the same techniques in this chapter and employing the volume fraction method, another NS-GFD model system has been developed, wherein the NS solver is capable of handling free surfaces [24].

5.3.2 Grid Connectivity

The solver for the NS equations, i.e., SIFOM, adopts composite structured meshes. While the solver for the GFD equations, i.e., FVCOM, uses an unstructured mesh; it adopts a triangular mesh in the horizontal plane and a σ-grid, which is a layered grid, in the vertical direction. The grids of the two solvers overlap arbitrarily with each other in the horizontal plane, as shown in Fig. 5.2a, and they are overset grids, particularly Chimera grids, in the plane. In the figure, lines a-b-c and d-e-f correspond to interfaces $\partial\Omega_{NS}$ and $\partial\Omega_{GFD}$, respectively, in Fig. 5.1. Grid nodes on the interfaces, such as those marked as squares and triangles, are grids′ interface nodes, and interface condition I (5.8) will be applied. Figure 5.2b shows the meshes in a vertical plane. In this figure, line g-h-i is an interface of the NS equations, and interface condition I (5.8) will be applied also on the line, or at the nodes marked with squares on it. Moreover, in this figure, the region of the GFD equations filled with the nodes marked with triangles, yet covered by the region of the NS equations, is a blanked zone of the GFD subdomain, which corresponds to $\Omega_{NS} \backslash \Omega_{GFD}$ in Fig. 5.1b. Condition (5.9) will be used at those nodes marked with triangles. The blanked zone is introduced because of the structure of the GFD solver (i.e., FVCOM); for consistency between its external and internal mode, a redistribution of solution from the seabed to the water surface in the vertical direction is performed at each time step, and thus solution within the blanked zone, $\Omega_{NS} \backslash \Omega_{GFD}$, is needed

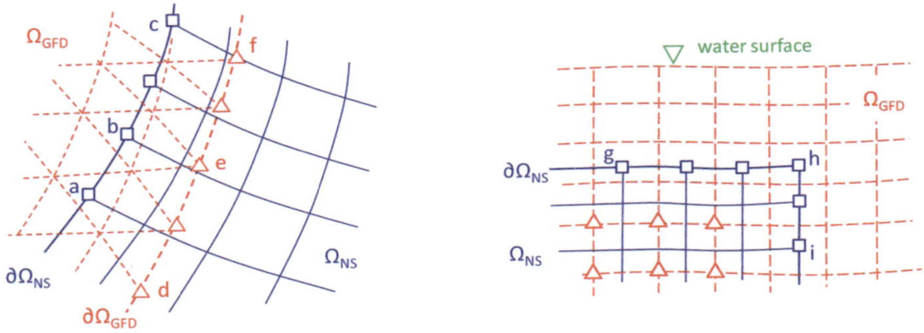

(a) Top view (b) Side view

Fig. 5.2 Overset meshes of NS and GFD solvers. Lines a-b-c and g-h-i are NS interfaces, and Line d-e-f is a GFD interface. The nodes marked with squares are the NS interface nodes. The grid nodes marked with triangles are the GFD nodes on the interface and inside the blanked zone of the GFD subdomain.

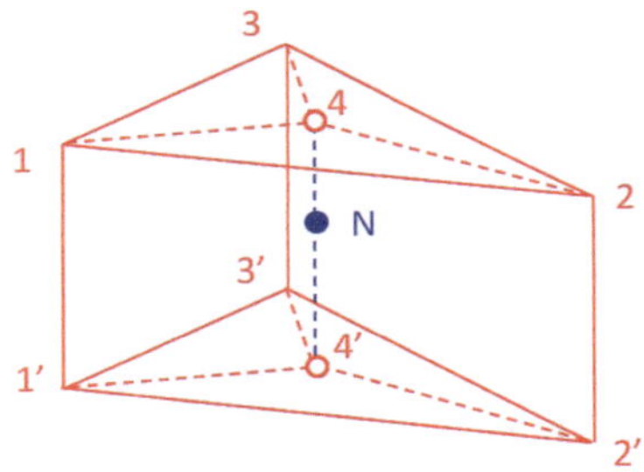

Fig. 5.3 An interface node of the NS mesh and its GFD host cell

for the GFD solver [15]. Since condition (5.9) for a blanked zone is part of condition (5.8) for an interface, a node of GFD within the blanked zone may also be viewed as an interface node (Fig. 5.2b). In the NS solver, all variables (e.g., $\mathbf{u}$ and T) are defined at grid nodes. In the GFD solver, variables are located at element centers, grid nodes, and the middle points of vertical edges of elements. More details are available in [15, 23, 34].

In the solution exchange between the NS and GFD solvers, it is necessary to determine the connectivity between their grids, or the relative locations of the grids. Let us first discuss the connectivity from the GFD grid to the NS grid; consider an interface node of the NS grid and its host cell, which is an element of the GFD grid, see Fig. 5.3. In the figure, the interface node is N, and the host cell is the triangular prism, i.e., the cell spanned by GFD grid nodes 1, 2, 3, $1'$, $2'$, and $3'$. Nodes 1, 2, and 3 are on a layer of the σ-grid, and nodes $1'$, $2'$, and $3'$ are on another. In this figure, 4 and $4'$ are the z-direction projections of N onto the plane spanned by triangles $\widehat{123}$ and $\widehat{1'2'3'}$, respectively. The search for the host cell of N is made in two steps as follows.

Step 1: search for the host cell of node N in the GFD mesh in the horizontal plane. A triangle is associated with the host cell if it satisfies the following conditions:

$$\mathbf{r}_{31} \times \mathbf{r}_{34} > 0, \quad \mathbf{r}_{12} \times \mathbf{r}_{14} > 0, \quad \mathbf{r}_{23} \times \mathbf{r}_{24} > 0$$

where $\mathbf{r}$ is a vector connecting the two points denoted by its subscripts. For instance, $\mathbf{r}_{31}$ is the vector pointing form node 3 to node 1.

Step 2: search in the vertical direction, i.e., in the σ-grid. The z-coordinates of 4 and $4'$ are determined by interpolation from the z-coordinates of the vertices of the horizontal triangle in which they reside. For instance,

$$z_4 = \frac{A_{\widehat{234}} z_1 + A_{\widehat{314}} z_2 + A_{\widehat{124}} z_3}{A_{\widehat{234}} + A_{\widehat{314}} + A_{\widehat{124}}}$$

where A represents the area of the triangle $\widehat{234}$, etc. Then, the σ-layer of the host cell is searched and identified as

$$z_{4'} < z_N \leq z_4$$

where z_N is the z-coordinate of interface node N. The GFD mesh will move up and down since it uses the σ-coordinate that changes in the vertical direction as the water surface sloshes, leading to a moving mesh problem. As a result, in general, the host cell of N and its position in the GFD mesh have to be determined at each time step. In view of the fact that the GFD mesh moves in the vertical direction only, luckily, the triangle of the host cell and x- and y-coordinates of interface node N remain the same. Therefore, the search for its host cell and the position is necessary only in the z-direction, and such a search will be made from a few cells below its prior host cell to a few cells above it, since the interface node moves gradually in the vertical direction.

The connectivity from the NS grid to the GFD grid is the same as that discussed in Sect. 4.2.1. Particularly, in Fig. 4.1, assume grid G_2 is the NS grid, replace grid G_1 with the GFD grid, and let N be its interface node. Then, the position of interface node N in grid G_2, or the connectivity from the NS grid to the GFD grid, can be determined in the way same as that discussed in Sect. 4.2.1.

5.3.3 Interface Algorithm

In solution exchange at the interfaces between the NS and GFD equations, particularly between SIFOM and FVCOM, it is desired to ensure that flow structures, such as vortices and internal waves, seamlessly pass the interfaces without distortion and artifacts. The interface algorithms are crucially important to achieve this goal, and conservative treatments are preferred, as discussed in the last two chapters. However, developing conservative interface treatments to couple the NS and GFD equations on overset grids is challenging, and so far, no such a treatment is available. A fact for the challenge is that, in general, it is difficult to realize strict conservation properties at interfaces between overset grids, and the discrepancy in the NS and GFD equations adds more challenge. In this chapter, as an interface algorithm, interpolation will be used to exchange the solutions of the NS and GFD equations.

Based on the grid connectivity described earlier, the GFD solver's solution can be mapped onto the interface nodes of the NS solver. For the interface node N in Fig. 5.3, the mapping is implemented by interpolation of the solution from its host cell:

$$f_4 = \frac{A_{\widehat{234}} f_1 + A_{\widehat{314}} f_2 + A_{\widehat{124}} f_3}{A_{\widehat{234}} + A_{\widehat{314}} + A_{\widehat{124}}}$$

$$f_{4'} = \frac{A_{\widehat{2'3'4'}} f_{1'} + A_{\widehat{3'1'4'}} f_{2'} + A_{\widehat{1'2'4'}} f_{3'}}{A_{\widehat{2'3'4'}} + A_{\widehat{3'1'4'}} + A_{\widehat{1'2'4'}}}$$

$$f_N = \frac{z_S - z_{4'}}{z_4 - z_{4'}} (f_4 - f_{4'}) + f_{4'}$$

where f represents a solution variable, such as u. The solution exchange from the NS solver to the GFD solver is determined precisely as in Sect. 4.2.1. Notably, as shown in Fig. 4.1, suppose N is an interface node on a GFD interface. Its interpolation from the solution at the host cell in the NS grid has been described in Sect. 4.2.1, and no further discussion is necessary.

When marching in time, the computation adopts the conventional method introduced in Sect. 1.2, and it follows algorithm (4.26). Note that application of a waveform relaxation algorithm, such as that depicted by (1.13), is not straightforward for real-world flow problems, and it may need a substantial change in their solvers, which are complex and stand-alone. The application of the conventional method is not only optional but also actually realistic if we utilize the existing GFD and NS solvers. Additionally, when following algorithm (4.26), the subdomains have different solvers; $\mathcal{S}_1$ is the GFD solver, and $\mathcal{S}_i$ ($i \geq 2, 3, ...$) are the NS solver.

The GFD and NS solvers will march in time simultaneously but may proceed with different time steps due to their own stability requirements. Particularly, the NS-GFD model system progresses with time step Δt, while the GFD solver and NS solver advance at their own time steps Δt_{GFD} and Δt_{NS}, respectively, as follows (Fig. 5.4):

$$\Delta t = n_{GFD} \Delta t_{GFD} = n_{NS} \Delta t_{NS} \tag{5.10}$$

where n_{NS} and n_{GFD} are integers. The time step of the GFD solver, Δt_{GFD}, is set to the time step of its internal mode, Δt_I. Its external mode may use another time step, Δt_E. For simplicity, in the simulations to follow, $\Delta t = max(\Delta t_{GFD}, \Delta t_{NS}), n_{GFD} = 1$, and/or $n_{NS} = 1$. That is, the time step of the

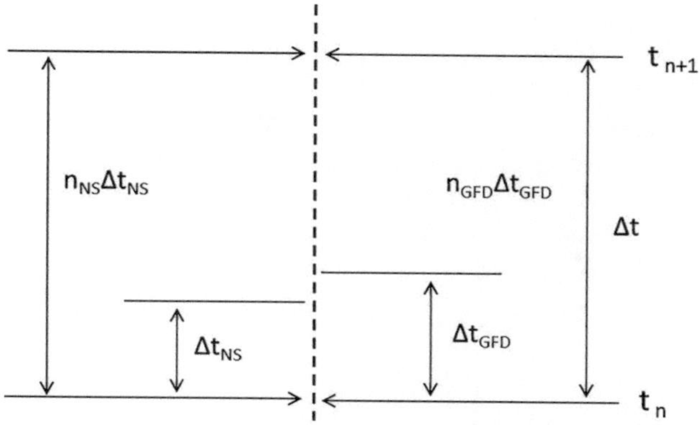

Fig. 5.4 The algorithm for marching in time in the NS-GFD model system. The NS and GFD solvers exchange solutions at t_n and t_{n+1}

NS-GFD model system is set to the larger of the time step of the GFD solver and that of the NS solver. A natural practice is to let $\Delta t_{GFD} = \Delta t_I = n_E \Delta t_E, n_E$ being an integer, usually, 10 [15]. After marching from time level t_n to $t_{n+1} = t_n + \Delta t$, the NS solver and the GFD solver will exchange solutions at interface nodes as shown in Fig. 5.4.

Since the NS solver and the GFD solver (actually, its internal mode) use implicit schemes, Schwarz iteration between the two solvers becomes necessary when they exchange solutions via implicit treatments at the interfaces. For computation over time interval $[0, T]$, the conventional Schwarz method introduced in Sect. 1.2 will be adopted. Particularly, the parallel Schwarz method/iteration, as illustrated by (1.12), is employed for solution exchange at model interfaces when marching in time. In reference to algorithm (4.26), the computational procedure of the conventional Schwarz method to compute the NS-GFD model system becomes

Do n=0,1, ..., N-1

Step 1: Iteration initialization

m=0

$(\mathbf{u}, p_d', T)^m_{NS} = (\mathbf{u}, p_d', T)^n_{NS}$, in Ω_{NS}, on $\partial\Omega_{NS}$

$(\mathbf{v}, \omega, \eta, T)^m_{GFD} = (\mathbf{v}, \omega, \eta, T)^n_{GFD}$, in Ω_{GFD}, on $\partial\Omega_{GFD}$

Step 2: Exchange of solutions at model interfaces by interpolation '$\Longleftarrow$'

$(\mathbf{u}, T)^{m+1}_{NS} \Longleftarrow (\mathbf{v}, \omega, T)^m_{GFD}$, on $\partial\Omega_{NS}$

$(\mathbf{v}, \omega, T)^{m+1}_{GFD} \Longleftarrow (\mathbf{u}, T)^m_{NS}$, on $\partial\Omega_{GFD}$, in $\Omega_{NS}\backslash\Omega_{GFD}$

Step 3: Solutions of NS and GFD solvers

NS solver produces $(\mathbf{u}, p_d', T)^{m+1}_{NS}$, in Ω_{NS}

GFD solver produces $(\mathbf{v}, \omega, \eta, T)^{m+1}_{GFD}$, in Ω_{GFD}

Step 4: Convergence check

$||(\mathbf{u}, p'_d, T)^{m+1}_{NS} - (\mathbf{u}, p'_d, T)^{m}_{NS}|| < \varepsilon, \quad$ in Ω_{NS}, on $\partial\Omega_{NS}$

$||(\mathbf{v}, \omega, \eta, T)^{m+1}_{GFD} - (\mathbf{v}, \omega, \eta, T)^{m}_{GFD}|| < \varepsilon,$ in Ω_{GFD}, on $\partial\Omega_{GFD}$

m $\longleftarrow$ m+1

Go to Step 5 if convergent or $m = M$. If not, return to Step 2.

Step 5: Solutions at time level $n + 1$

$(\mathbf{u}, p'_d, T)^{n+1}_{NS} = (\mathbf{u}, p'_d, T)^{m}_{NS}, \quad$ in Ω_{NS}, on $\partial\Omega_{NS}$

$(\mathbf{v}, \omega, \eta, T)^{n+1}_{GFD} = (\mathbf{v}, \omega, \eta, T)^{m}_{GFD}, \quad$ in Ω_{GFD}, on $\partial\Omega_{GFD}$

End Do

Here, m is the iteration index, M is a prescribed maximum iteration number, and ε is a prescribed residual tolerance. Note that further details on the exchange of the solutions at the model interfaces in Step 2–such as η in the subdomains and $\mathbf{u}$ in the blanked zone–have been explained previously.

The NS-GFD coupling presented in this chapter adopts linear interpolation at the NS-GFD interfaces, and it is second-order accurate locally (or first-order accurate globally) when full convergence is achieved. Besides, in practice, a prescribed M is usually set up to control the number of iterations, between the NS and GFD solvers, at which the full convergence is assumed. In the examples of the following section, $M = 1$ is adopted. In this situation, discretion is needed; a full convergence may not be achieved yet, and the accuracy of the interface treatment could deteriorate (e.g., [37]). Moreover, when $M = 1$, the GFD-NS model system reduces to one-way coupling (from GFD to NS) if the second line in Step 2 of the computational procedure is disabled.

In the NS-GFD coupling, the exchange of solutions between the NS and GFD solvers is performed through interpolation of all required variables. This is similar to the SI algorithm used in NS-NS coupling, as described in Eqs. (4.18) and (4.19). In this sense, and the NS-GFD coupling in this chapter adopts an SI algorithm, the standard interpolation, which is a nonconservative interface algorithm. In the next section, numerical examples—primarily drawn from [23]—will be presented to test and demonstrate the performance of the aforementioned NS-GFD model system.

5.4 Numerical Example

5.4.1 *Sill Flow*

The first example of applying the NS-GFD model system is about the flow over a sill, a benchmark problem for ocean flows. The flow exhibits rich physical phenomena, including non-hydrostatic effects. To simulate its physical processes, it is necessary to adequately resolve these effects (e.g., [38]).

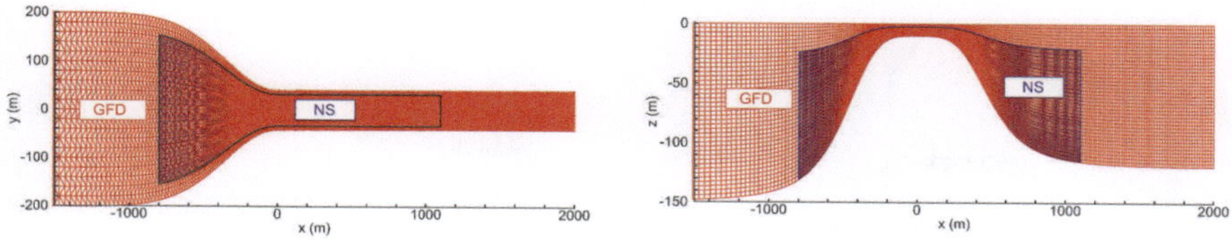

(a) Top view (b) Side view

Fig. 5.5 Meshes for the sill flow

Example 5.4.1 A sill is a typical form of topography for the seabed and shoreline, and its configuration is depicted by

$$\begin{cases} -1500 < x < 2000 \\ y = \pm 200(1 - 0.8e^{-4\times 10^{-6}x^2}), \; x < 0; \quad y = \pm 40, \qquad\qquad\qquad x > 0 \\ z = -150 + \dfrac{140}{1 + (x/500)^4}, \qquad x < 0; \quad z = -120 + \dfrac{110}{1 + (x/500)^4}, \; x > 0 \end{cases}$$

For the flow, the initial and boundary conditions are set up to

$$\mathbf{u} = 0, \eta = 0, \;\; p = \gamma(\eta - z), \;\; t = 0$$
$$u = a(1 - e^{-0.01t}), \;\; x = -1500, \;\; t > 0; \;\; \eta = 0, \;\; x = 2000, \;\; t > 0$$

In the above expressions, a is a constant, and $a = 0.006$ in this example. Additionally, the length is in m, the time in s, and the velocity in m/s. The flow is an initial-boundary value problem.

In the computation by the overset-grid method, the GFD grid extends over the whole flow domain except a blanked zone over the top of the sill, and it covers the far field. To adequately resolve its non-hydrostatic effects, the NS grid covers the top of the sill, the near field, see Fig. 5.5. The GFD grid has 10,400 triangle elements in the horizontal plane and 41 σ-layers in the vertical direction. The NS grid is 161×33×49 in the longitudinal, lateral, and vertical directions, respectively. The time steps for the GFD model are $\Delta t_{GFD} = 0.5$ s and $\Delta t_E = 0.125$ s, and that for the NS model is $\Delta t_{NS} = 0.5$ s. The interface condition for the coupling of NS and GFD equations is the NS-GFD interface condition I, i.e., (5.8), and the blanked zone is handled via (5.9). The computational procedure is described in the above section. As a reference, a run is also made using the GFD model alone, and another is conducted using the NS model alone. For the run with GFD alone, the grid is identical to the GFD grid in the coupling run. For the run with NS alone, the grid is 321×41×65, which is stretching and has a resolution similar to that in the coupling run. All other parameters remain the same as those in the coupling run.

A simulated instantaneous solution for the flow is shown in Fig. 5.6. At this instant, the solution is not steady but at an equilibrium state. As expected, the

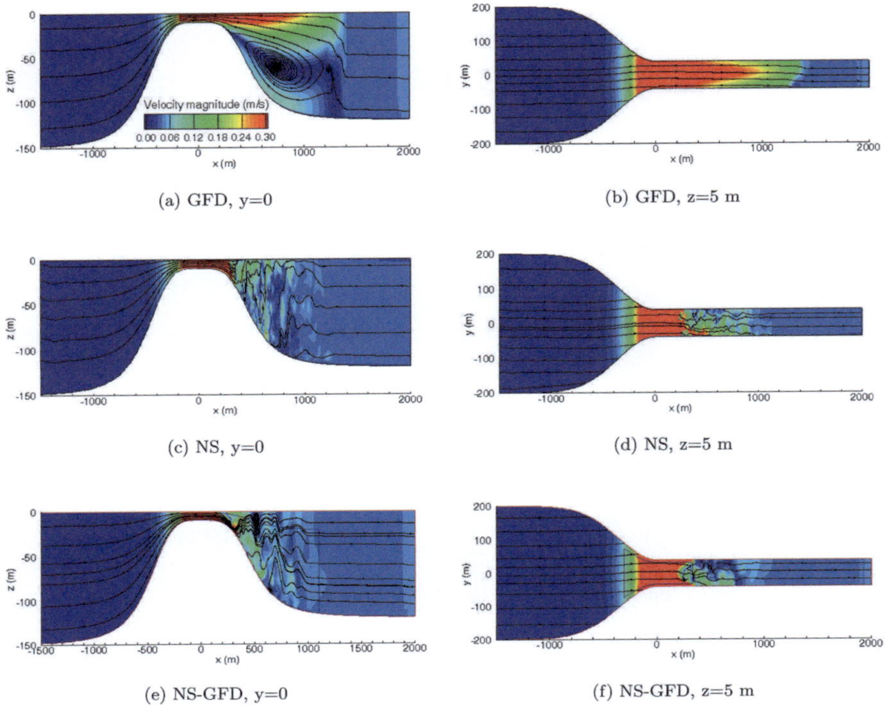

(a) GFD, y=0 (b) GFD, z=5 m

(c) NS, y=0 (d) NS, z=5 m

(e) NS-GFD, y=0 (f) NS-GFD, z=5 m

Fig. 5.6 Simulation of sill flow at $t = 10{,}000$ s

GFD model and the NS model do not produce the same solution for the flow, see Figs. 5.6a, b, c, and d. The figures indicate that both NS and GFD models capture similar large-scale patterns in the convergent section ($x < 0$). However, a large vortex is developing in the GFD solution within the divergent section, in the vertical plane, at this instant. In contrast, many small-scale structures, in the form of patches of velocity magnitude, are present in the NS solution for the divergence section. These small-scale structures are speculated to be associated with non-hydrostatic effects. In simulation by the NS-GFD model system, these structures are captured at the divergent section (Figs. 5.6e and f), and the flow pattern becomes similar to those obtained with NS alone (Figs. 5.6c and d).

Now, we look at the accuracy of the numerical solution associated with the NS-GFD model system. It is well known that a linear second-order scheme for a linear problem will yield a second-order accurate solution if it converges [39]. There is no such guarantee in the case of the NS-GFD model system since both the NS and GFD models are second-order methods but involve strong nonlinearity. But, the system's solutions are anticipated to be approximately second-order accurate if the impact of the interface algorithm accuracy (globally first-order) is minor. Still, the estimate of accuracy order formula (4.28) is adopted. As indicated in its derivation, the formula only leads to an appropriate estimate when the flow is relatively simple in structures.

Table 5.1 Estimated accuracy order γ of solution obtained with the NS-GFD model system

Solution	p	η	u	v	w
NS	0.96		1.75	2.12	1.63
GFD		1.08	1.53	2.73	1.68

Example 5.4.2 This example computes the sill flow as described above. Three sets of grid spacings and time steps are adopted:

I : NS grid: $41 \times 9 \times 13$, GFD grid: 650 triangle elements, 11, σ – layers; $\Delta t = 2\,s$

II : NS grid: $81 \times 17 \times 25$, GFD grid: 2,600 triangle elements, 21, σ – layers; $\Delta t = 1\,s$

III : NS grid: $161 \times 33 \times 49$, GFD grid: 10,400 triangle elements, 41, σ – layers; $\Delta t = 0.5\,s$

Here, the grid of set II is refined from that of set I, and the grid of set III is refined from that of set II. The grid-spacing ratio of set I, II, and III is about 4:2:1. As seen in the previous example, the flow is complicated in behaviors and structures. To facilitate the accuracy estimate via formula (4.28), the solutions of the three sets of meshes and time steps are selected at a moment shortly after the fluid enters the divergent section. At this moment, the flow remains simple in structures, roughly as smooth layers. The formula is applied to the NS and GFD solutions, and the estimate is summarised in Table 5.1. The table shows that the model system performs as expected; the accuracy order for velocity in both the NS and GFD solutions is close to 2, while that for surface elevation and pressure is about 1. These indicate that the numerical solutions for all variables converge as the grid spacing and time step decrease. As discussed in the previous section, the global accuracy of the interface algorithm is 1, one order lower than that of the NS and GFD models. The deduction of algorithm accuracy at the interface may impact, particularly decrease, the accuracy of the overall solution. The deduction may contribute to the discrepancy from the expected accuracy order 2 in the table. However, the result in the table shows that such a discrepancy is not pronounced. This test is an indication that the NS-GFD model system works appropriately.

The aforementioned two examples illustrate the necessity of coupling an NS model to a GFD model for resolving small-scale flow structures and show its potential to simulate multiphysics, multiscale flows. They also demonstrate the performance of the proposed coupling system.

5.4.2 *Current Past Bridge Pier*

This is an engineering flow problem. A laboratory study for the flow is made, and an NS simulation study for it has also been conducted [40]. It is noted that, although

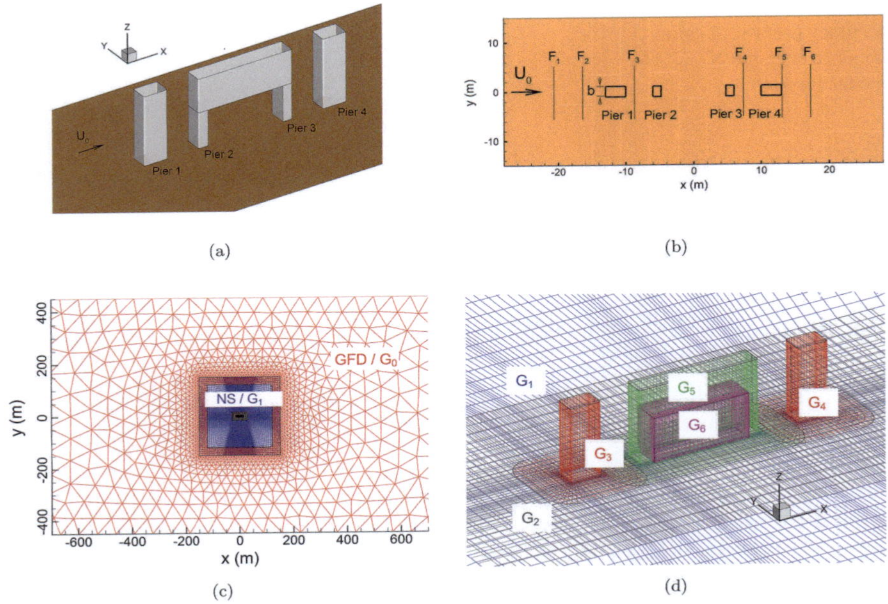

Fig. 5.7 Configuration of the bridge pier and computational meshes. (**a**) Bridge pier. (**b**) Top view of the pier. F_i are horizontal locations of the measurement (i = 1, ..., 6). (**c**) Meshes for the NS-GFD model system. (**d**) Meshes for the NS model

a constant flow rate is set as the upstream inflow condition, the flow cannot reach a steady state but keeps sloshing.

Example 5.4.3 The flow past a bridge pier in a channel is shown in Fig. 5.7a. Its Reynolds number is 28,516, defined by the incoming flow velocity, U_0, and the width of the bridge pier, b. The bridge height, H, is 15 m. The laboratory experiment measures velocity profiles along multiple horizontal lines at different heights above the ground, as shown in Fig. 5.7b.

In the simulation, a flow rate condition determined by U_0 is prescribed at the upstream boundary (U_0 is calculated according to the Reynolds number), water depth, h, is fixed as 8.9 m at the downstream boundary (the bridge height is 15 m), and a symmetry boundary condition is assumed at the two lateral sides. On the horizontal plane, the computational domain is a rectangle with 3,000 m in length and 900 m in width. The computational meshes are shown in Figs. 5.7c and d. In the figures, the GFD mesh overlaps with the outer mesh of the NS model, and it uses 3,500 grid nodes in the horizontal plane and 21 σ-layers in the vertical direction (Fig. 5.7c). The NS meshes consist of several overset grids wrapping the bridge piers (Fig. 5.7d), and they have a total of 765,000 grid nodes. The time steps are $\Delta t_{NS} = 3.6$ s, $\Delta t_{GFD} = 0.6$ s, and $\Delta t_E = 0.1$ s. For comparison, a run of the NS model alone is also made with the NS grids (Fig. 5.7d). In this run, the top surface

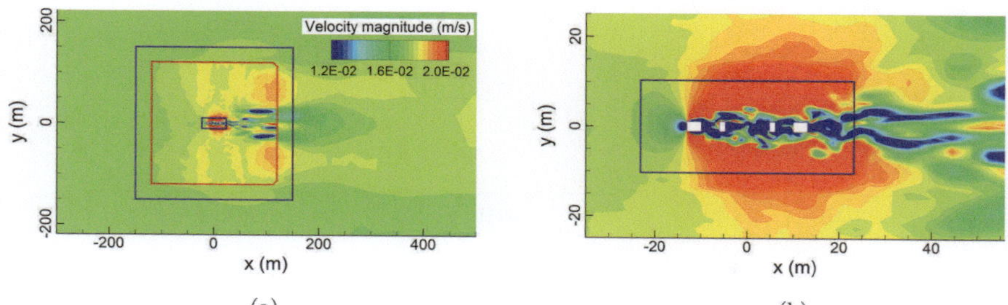

Fig. 5.8 Computed velocity magnitude at 3.56 m above ground. Blue line—NS interface (the blue square. The smaller blue rectangle is an interface between NS grids). Red line—GFD interface

is treated as a rigid wall associated with a symmetry condition, since the NS model cannot deal with a free surface.

Figure 5.8 depicts an instantaneous solution. It shows that the flow smoothly passes the model interfaces without apparent artifacts. The piers affect the flow field within the NS zone and exhibit disturbance in the GFD domain (Fig. 5.8a). The flow is strongly unsteady near the bridge piers and generates vortices (Fig. 5.8b). A comparison of the computed u-velocity and its measurement is shown in Fig. 5.9. The results shown in the figure are temporal averages of the solution since the computed flow is unsteady and keeps changing in time. Overall, it can be seen that there is a reasonable agreement between the simulation and the measurement. The computed velocity deviates the most from the laboratory data along F_3 and F_4 at a height of $0.2H$ above the ground. In addition, there is a difference between the solutions of the NS-only run and the coupling run, although they are close.

The deviation of the NS-GFD solution from the measurement data could be attributed to the strong vortex shedding and influence of the boundary layer at the bottom wall, plus a lack of tuning such as for mesh resolution. Additionally, the discrepancy between the solutions of NS-GFD and NS alone may be due to the latter's smaller computational domain and exclusion of a free surface. This discrepancy could also arise from the coupling of the NS and GFD models. All of these factors require further clarification. Despite these, the aforementioned results indicate that the NS-GFD model system works with reasonable accuracy.

5.4.3 *Thermal Effluent Discharge*

Effluent discharge into ambient flows is a common engineering practice, and examples include emission of treated wastewater, desalination plant brine release, and oil spill. The ocean flows with the discharge are generally multiscale and multiphysics, as indicated in Sect. 5.1. Until now, there has been a shortage of modeling packages that can simulate such multiscale and multiphysics flows. For

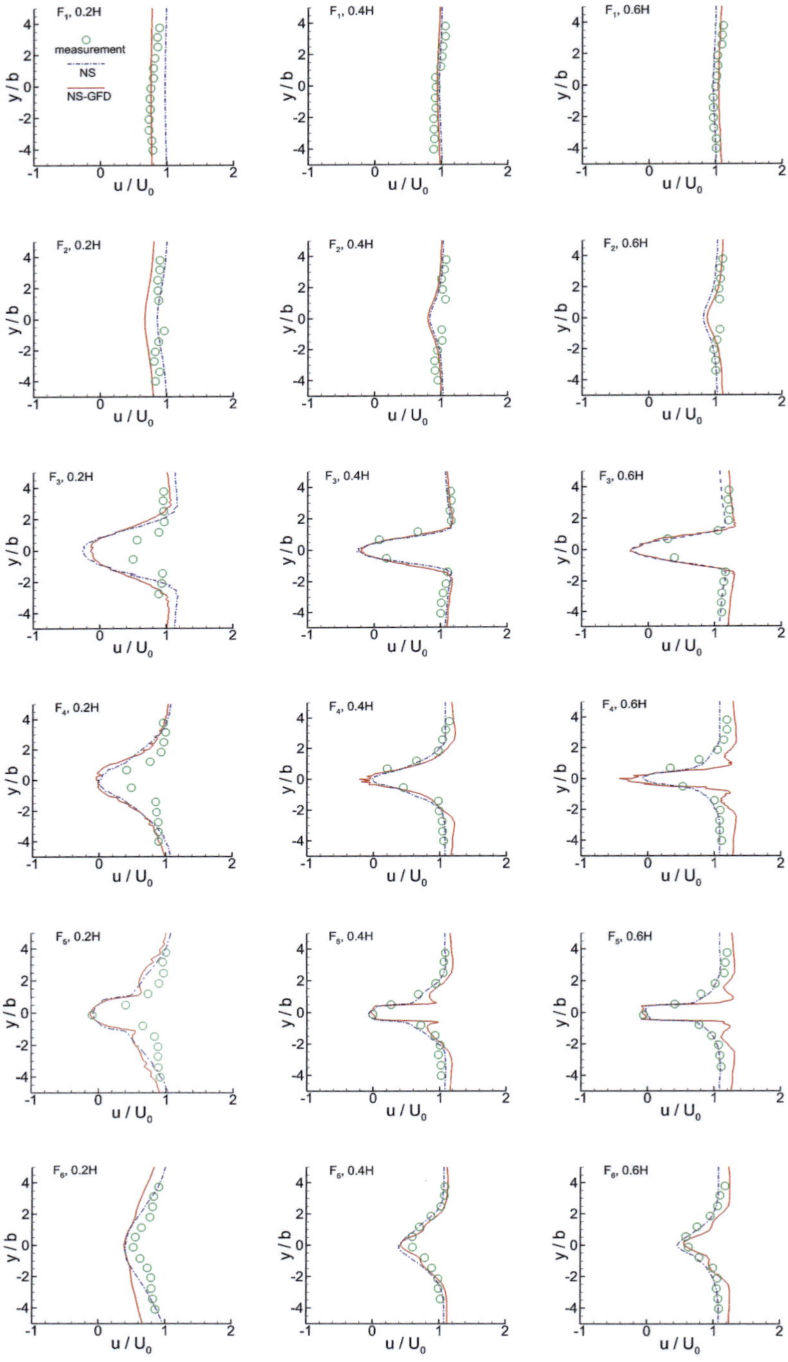

Fig. 5.9 Comparison between the computed velocity and measurement. For locations of F_i ($i = 1$, 2, ..., 6), refer to Fig. 5.7b

instance, there is a lack of computer packages to directly and simultaneously model the background ocean flows and the effluent discharged from the ocean bottom in the Gulf of Mexico oil spill. The NS-GFD model system sets up a platform to simulate such types of multiscale and multiphysics ocean flows, and this is illustrated by the following examples of thermal effluent discharge from diffusers into the ambient flow in a channel as well as in the ocean.

Example 5.4.4 This example is of a thermal effluent discharge flow in a channel. The velocity and temperature are respectively 0.3 m/s and 20.5 °C for the ambient flow, and they are respectively 3.92 m/s and 32 °C for the effluent flow. The diffuser consists of a horizontal pipeline, which is 1.32 m in diameter and lies on the ground at an angle of 110° to the ambient flow direction, and 10 discharge ports, which are 0.175 m in diameter and at different upward angles (Fig. 5.10a). More details, such as those for the diffuser, are available in [23, 33].

The computational domain is 3,000 m in length and 300 m in width. A flow rate condition corresponding to 0.3 m^3/s (note that FVCOM takes a flow rate as a boundary condition) is prescribed at the upstream boundary, the water depth is fixed

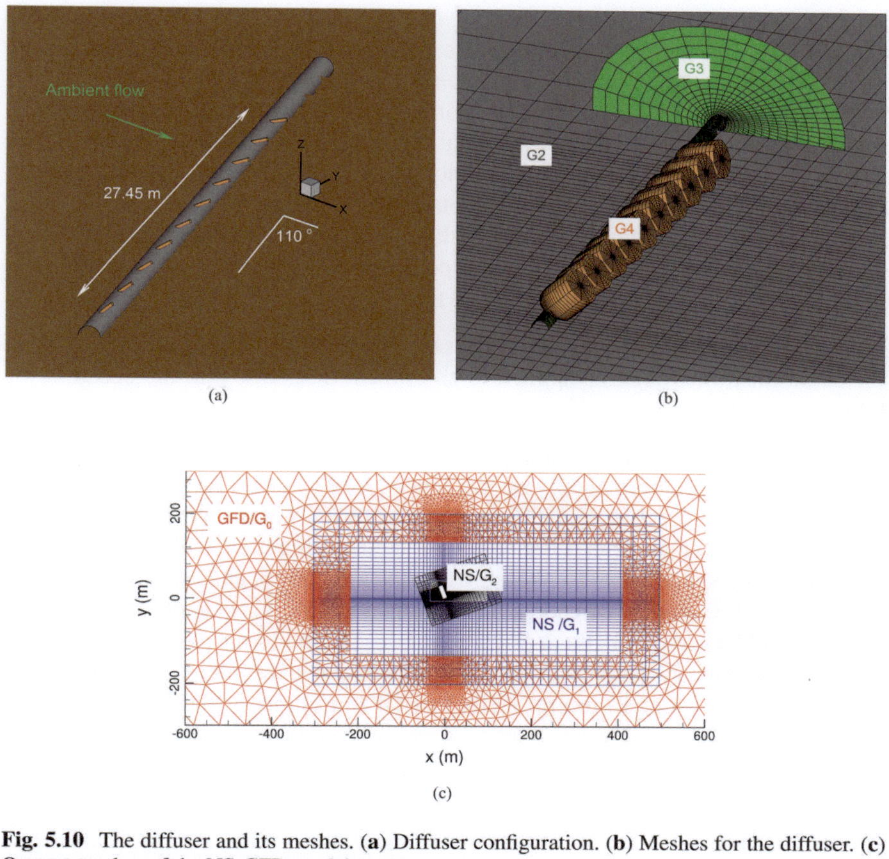

Fig. 5.10 The diffuser and its meshes. (**a**) Diffuser configuration. (**b**) Meshes for the diffuser. (**c**) Overset meshes of the NS-GFD model system

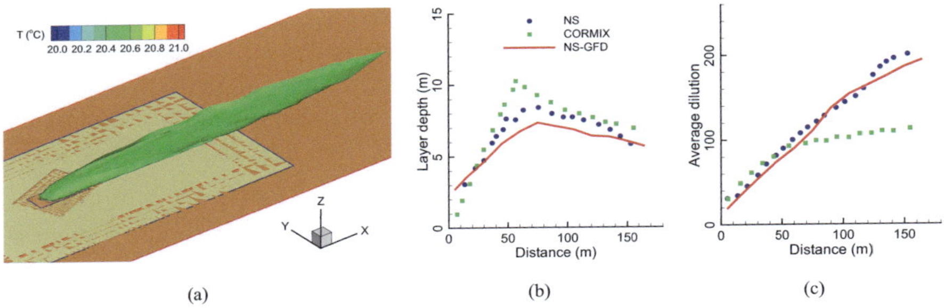

Fig. 5.11 Simulated thermal discharge into a channel. (**a**) Iso-temperature surface of the plume. The blue line is the interface of the NS model. (**b**) Vertical thickness of the plume along the plume center line. (**c**) Cross-section dilution along the plume center line

as 12.3 m at the downstream boundary, and a slip boundary condition is given at the two lateral sides. The NS model has 23 overset and patched grids to wrap the ports and the pipe (Fig. 5.10b), a total of 210,000 grid nodes. The GFD model has 1,200 grid elements on the horizontal plane and 11 σ-layers in the vertical direction (Fig. 5.10c). The grid of the GFD model, G_0, and the outer mesh of the NS model, G_1, overlap, and the two models couple on these two grids. The time steps are $\Delta t_{NS} = 3.6$ s, $\Delta t_{GFD} = 0.6$ s, and $\Delta t_E = 0.1$ s.

Figure 5.11a shows a 3D view of the computed instantaneous thermal plume. The figure indicates that the computed thermal plume passes through the interface from the NS into the GFD region without obvious distortion. A quantitative comparison of the NS-GFD solution with those obtained by other approaches is shown in Figs. 5.11b and c. The figures show that the computed solution compares favorably to those predicted by CORMIX, a popular empirical model for effluent flow problems [41]. Additionally, it is close to that predicted by the NS model alone.

The aforementioned example validates the NS-GFD model system and its setup. Based on this, the system is applied to an effluent discharge into the ocean in the following example.

Example 5.4.5 In this simulation, the diffuser in the aforementioned example, see Fig. 5.10a, is placed horizontally on the seabed in the ocean, particularly the New York Bight, see Fig. 5.12a. The diffuser pipeline extends in the northwest direction. The pipeline has an angle of 110° regarding the east direction, and its south end is set as the origins of x, y, and z, which are in the east, north, and upward directions, respectively. The effluent discharge has a temperature of 32 °C and a speed of 3.92 m/s. The GFD mesh covers the far field. Figure 5.12a shows the mesh, together with the ocean bathymetry. The NS meshes wrap the near field as in Fig. 5.10b. The GFD and NS meshes overlap as in Fig. 5.12b. The GFD mesh uses 10,000 grid elements in the horizontal plane with 15 σ-layers in the vertical direction, and the NS meshes adopt 630,000 grid nodes. The time steps are $\Delta t_{NS} = 2.4$ s, $\Delta t_{GFD} = 0.6$ s, and $\Delta t_E = 0.1$ s. At the open boundary of the deep ocean in Fig. 5.12a, the astronomical

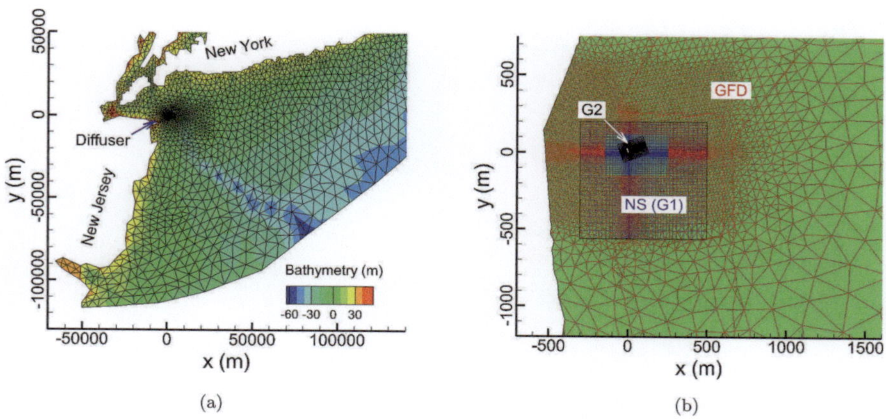

Fig. 5.12 Setup of NS-GFD model system for thermal plume in the ocean. (**a**) Bathymetry and GFD mesh. (**b**) GFD and NS meshes

tidal flow conditions are presented as the forcing for the flow. Moreover, winds are not included.

The simulation results are presented in Fig. 5.13. The simulation not only captures the far-field, background ocean flow but also resolves the thermal plume flow. Figure 5.13a shows an instant velocity field of the ocean. The patterns of the ambient flow's streamlines in the figure indicate that the flow experiences a flood tide at this instant. Figure 5.13b visualizes a 3D view of the thermal discharge from the diffuser. The computed individual thermal jets merge with each other after their release from the ports. To further illustrate the motion of the thermal effluent, a zoom-out view of the plume is presented in Figs. 5.13c and d. In these two figures, the temperature values on the iso-surfaces are lower than those in Fig. 5.13b, and thus, the size of the plume is larger. The two figures show that the plume swings forth and back with the tides. They also indicate that the plume goes smoothly from the NS zone into the GFD zone.

The two examples in this subsection show that not only can the NS-GFD model system deal with the far field, but it can also resolve the near field of a thermal discharge flow, modeling its plume from the near-field initial mixing all the way to the far-field dispersion.

5.5 Discussion on Coupling Method

5.5.1 *Pressure Split*

The coupling of NS and GFD equations is complex, and techniques for their seamless coupling are topics of research and exploration. In the previous section, the NS equations are transformed from (4.2) into (5.2) via the pressure split, i.e.,

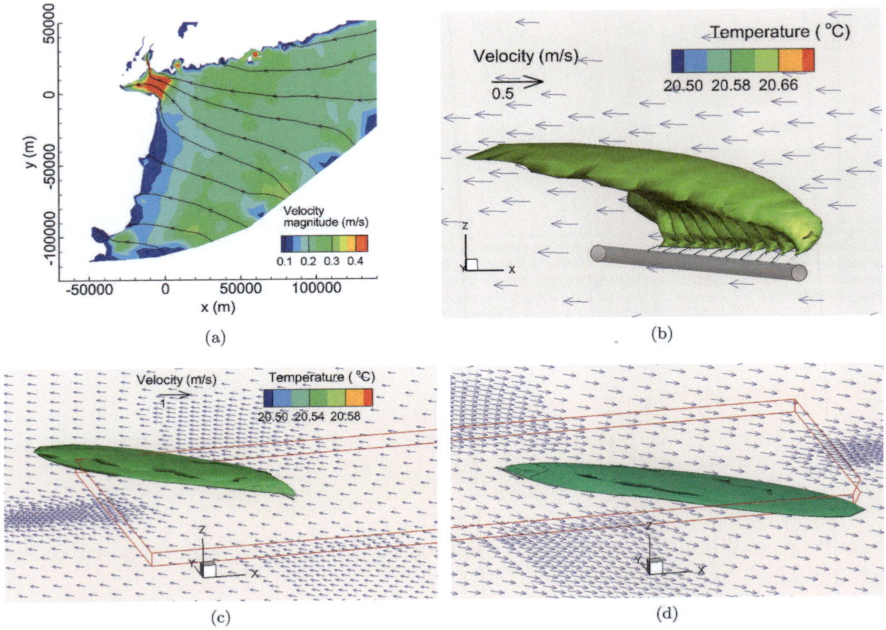

Fig. 5.13 Simulation of the thermal plume in tides. (**a**) Velocity field at flood tide, $z = 17$ m. (**b**) Zoom in effluent plume at flood tide. (**c**) Effluent plume at flood tide. (**d**) Effluent plume at ebb tide

(5.1). The pressure split is proven effective in better coupling the far- and near-field flows. To illustrate its effectiveness, a numerical experiment is presented, again on the sill flow in Sect. 5.4.1, at a highly transient stage during which the water body changes from a static to an equilibrium state.

Example 5.5.1 The sill flow is introduced in Sect. 5.4.1, and the simulation details are described there. To illustrate the effectiveness of the pressure split technique in the NS-GFD coupling, the NS equations are computed according to their form in (4.2), which uses no pressure split, also according to (5.2), which adopts it. In the simulation, the entrance speed is greater, with $a = 0.9175$ for the entrance speed.

Computed results for instantaneous solutions are presented in Fig. 5.14. In the computed results, non-physical solutions take place; it is seen in Figs. 5.14a and e that, at the front of the current, where water bodies in motion and at rest meet and the flow is highly transient, the solution of the NS model cannot react simultaneously to that obtained with the GFD model but instead exhibits a delay. There is a pronounced difference between the velocity solutions provided by the two models in their overlapping regions. In Fig. 5.14c, near the top of the sill, the GFD model provides a forward flow in the upper layer, while the NS model produces a backward flow in the lower layer. There is no expected physical mechanism to generate such a reverse flow. Moreover, as seen in the NS and GFD overlapping region, the NS and

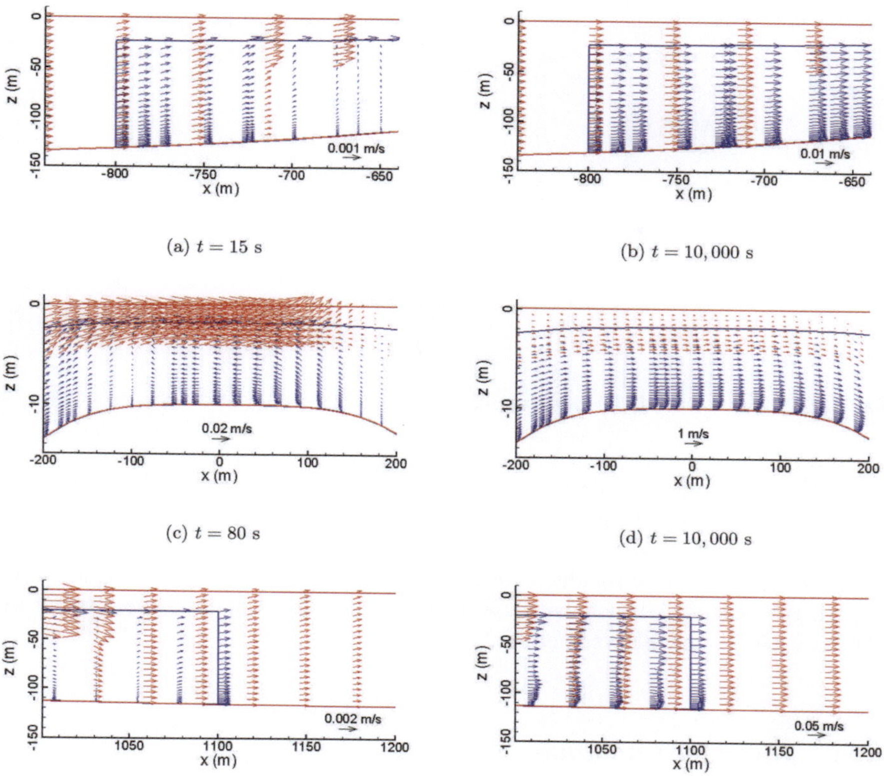

(a) $t = 15$ s

(b) $t = 10,000$ s

(c) $t = 80$ s

(d) $t = 10,000$ s

(e) $t = 65$ s

(f) $t = 10,000$ s

Fig. 5.14 Simulation of sill flow without pressure split, $y = 0$. Blue—NS, red—GFD

GFD solutions are distinct, or there are multiple solutions. The delay in response, the reverse flow, and multiple solutions are non-physical behaviors.

A long time after the front passes the interface region and as the flow becomes quasi-steady, the solutions become normal, and all the aforementioned problems disappear, see Figs. 5.14b, d, and f. Now, the solution at the interface region by the GFD model and that by the NS model are close in patterns and magnitudes, and there is a smooth transition between them in the region. This indicates that non-physical solutions tend to occur primarily where the flow is highly transient.

When the modeling system adopts the pressure split, the NS equations are computed as (5.2), the artifacts discussed in the above paragraph, i.e., those in Figs. 5.14a, c, and e, are essentially not presented. Figure 5.15 is a sample of simulated instantaneous solutions at the transient stage, and it shows that the GFD and NS models present similar patterns and magnitudes in the interface region.

This example indicates that, without the pressure split, the NS-GFD model system encounters difficulties in resolving the flow at the current front, which tends

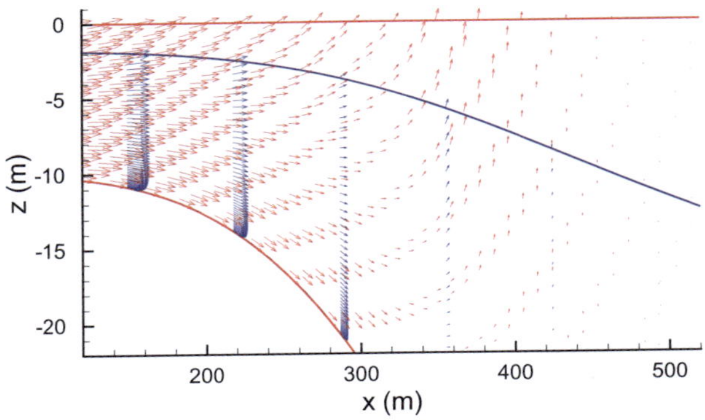

Fig. 5.15 Simulation of sill flow with pressure split, $t = 100$ s, $y = 0$. Blue—NS, red—GFD

to be highly transient and involves strong non-hydrostatic effects. The difficulties occur in the form of numerical oscillations and non-physical solutions. Such difficulties disappear as the flow reaches a quasi-steady state. When the pressure split is adopted, these difficulties are not present essentially, even though the flow is highly transient. All of these show that the coupling techniques are crucial in coupling the NS and GFD equations.

5.5.2 *Interface Condition*

NS Equations - NS Equations The interface condition, or the transmission condition, directly determines how NS and GFD solutions connect with each other at their interfaces. As illustrated in the previous content, such a condition is critical to the correctness and accuracy of the coupled solutions. The NS-GFD interface condition I adopted in the previous sections, i.e., (5.8), is based on individually considering the computation of the NS equations and that of the GFD equations. This subsection examines and explores an alternative interface condition that jointly considers the computations of the NS and GFD equations.

Let us start with coupling the NS equations in subdomains. Suppose an interface Γ divides a flow domain into two subdomains, as shown in Fig. 3.1. In the two subdomains, (4.1) and (4.2) hold. Consider the weak form of continuity equation (4.1) across the interface:

$$\int_{\omega} \nabla\phi \cdot \mathbf{u} d\omega = 0 \tag{5.11}$$

Here, $\phi(\mathbf{x}) \in C_0^\infty(\omega)$, and ω is an arbitrary region across the interface. The solution $\mathbf{u}$ to the weak form is a weak solution of the continuity equation. In the tensor form, the weak form leads to

$$
\begin{aligned}
0 &= \int_{\omega} \frac{\partial \phi}{\partial x^l} u^l d\omega \\
&= \int_{\omega_1 \cup \omega_2} \left(\frac{\partial \phi u^l}{\partial x^l} - \phi \frac{\partial u^l}{\partial x^l} \right) d\omega \\
&= \oint_{\partial \omega_1 \cup \partial \omega_2} \phi u^l ds_l \\
&= \int_{\Gamma} \phi (u^{l,-} - u^{l,+}) ds_l
\end{aligned}
$$

Here, $l = 1, 2, 3$, and repeat indices imply summation. Superscripts $-$ and $+$ represent the left and right sides of Γ, respectively. In the aforementioned derivation, after the second equal sign, the derivative in the second term is actually the divergence of $\mathbf{u}$, and it is zero according to continuity equation (4.1). Additionally, the Gauss divergence theorem is applied from the second to the third equal sign, and $\phi = 0$ on $\partial\omega$ is utilized from the third to the fourth equal sign. Given the arbitrariness of ϕ, in the vector form, the aforementioned yields an interface condition:

$$
\mathbf{u}_n|_{\Gamma^-} = \mathbf{u}_n|_{\Gamma^+} \tag{5.12}
$$

Here n means the normal direction of Γ pointing from ω_1 to ω_2. The interface condition permits the situation where the tangential velocity of the weak solution is discontinuous across the interface.

Now, consider a weak form of the NS equations (4.2) across the interface:

$$
\int_{\omega} \left(\phi_t \mathbf{u} + \nabla \phi \mathbf{u}\mathbf{u} + \frac{1}{\rho_0} \nabla \phi p - (\nu + \nu_t) \nabla \phi \cdot \nabla \mathbf{u} + \phi \mathbf{g} \right) d\omega = 0 \tag{5.13}
$$

Here, $\mathbf{g} = (0, 0, g)$, the acceleration of gravity, and $\phi(\mathbf{x}, t) \in C_0^{\infty}(\omega)$. A solution to the weak form is a weak solution for the NS equations. In the tensor form, the formulation yields

$$
\begin{aligned}
0 &= \int_{\omega} \left(\frac{\partial \phi}{\partial t} u^i + \frac{\partial \phi}{\partial x^l} u^l u^i + \frac{1}{\rho_0} \frac{\partial \phi}{\partial x^i} p - (\nu + \nu_t) \frac{\partial \phi}{\partial x^l} \frac{\partial u^i}{\partial x^l} + \phi g_i \right) d\omega \\
&= \int_{\omega} \left(\frac{\partial \phi u^i}{\partial t} + \frac{\partial}{\partial x^l} \left(\phi u^l u^i + \frac{1}{\rho_0} \phi p - (\nu + \nu_t) \phi \frac{\partial u^i}{\partial x^l} \right) \right) d\omega \\
&\quad - \int_{\omega} \phi \left(\frac{\partial u^i}{\partial t} + \frac{\partial u^l u^i}{\partial x^l} + \frac{1}{\rho_0} \frac{\partial p}{\partial x^i} - (\nu + \nu_t) \frac{\partial^2 u^i}{\partial x^{l2}} - g_i \right) d\omega \\
&= \int_{\omega_1 \cap \omega_2} \left(\frac{\partial \phi u^i}{\partial t} + \frac{\partial}{\partial x^l} \left(\phi u^l u^i + \frac{1}{\rho_0} \phi p - (\nu + \nu_t) \phi \frac{\partial u^i}{\partial x^l} \right) \right) d\omega
\end{aligned}
$$

In the aforementioned derivation, ν, $\nu_t = const$. Moreover, the second integral after the second equal sign contains the left-hand side of the NS equations, i.e., the second equation in (4.2), and thus, it is zero. Application of the Green formula to the integral after the third equal sign gives rise to

$$
\begin{aligned}
0 &= \oint_{\partial\omega_1 \cap \partial\omega_2} \left(-\left(\phi u^l u^i + \frac{1}{\rho_0}\phi p - (\nu + \nu_t)\phi \frac{\partial u^i}{\partial x^l} \right) + \phi u^i \frac{ds_l}{dt} \right) dt \\
&= -\oint_{\partial\omega_1 \cap \partial\omega_2} \phi \left(u^l u^i + \frac{p}{\rho_0} - (\nu + \nu_t)\frac{\partial u^i}{\partial x^l} \right) dt \\
&= -\int_\Gamma \phi \left[u^l u^i + \frac{p}{\rho_0} - (\nu + \nu_t)\frac{\partial u^i}{\partial x^l} \right] dt
\end{aligned}
$$

In the above, s_l is an area on Γ, and in view that the interface Γ is stationary, $ds_l/dt = 0$. Moreover, the bracket $[\cdot]$ indicates difference across the interface. Since ϕ is smooth and arbitrary, the difference is zero across the interface. In the vector form, this leads to the following interface condition:

$$
(\mathbf{u}_n \mathbf{u} + p/\rho_0 - (\nu + \nu_t)\partial \mathbf{u}/\partial n)_{\Gamma^-} = (\mathbf{u}_n \mathbf{u} + p/\rho_0 - (\nu + \nu_t)\partial \mathbf{u}/\partial n)_{\Gamma^+} \tag{5.14}
$$

which requires the flux of momentum to be continuous across the interface in its normal direction. This condition has no restriction on the continuity of the flux in the interface's tangential direction, or it permits discontinuity of the flux in this direction regarding the weak solution.

NS Equations - Hydrostatic NS Equations As discussed in Sect. 5.3, the NS-GFD coupling is actually made between the NS equations and the internal mode of the GFD equations, particularly between (4.1), (5.2), and (4.3) for the NS equations and (5.6) and (5.7) for the GFD equations. The first equation in Eq. (5.6) is the continuity equation, and it is derived from Eq. (4.1). The second equation in Eq. (5.6) is actually a hydrostatic version of the NS equations, particularly the momentum equations in Eq. (4.2). Now, let us derive the hydrostatic NS equations using the hydrostatic assumption, which states that the gravity balances the pressure gradient, or the acceleration and diffusion are negligible, in the vertical direction [10]. In this situation, pressure, p, is referred to as the hydrostatic pressure, p_h, that is, $p = p_h$, and NS equations (4.2) in the vertical direction are simplified as

$$
-\frac{1}{\rho_0}\nabla_z p_h = g\big(1 - \alpha(T - T_0)\big)\mathbf{k} \tag{5.15}
$$

Integration of the aforementioned equation, one has

$$p_h = \rho_0 g\big(\eta - z - \alpha \int_z^{\eta} (T - T_0)\, d\zeta\big) \tag{5.16}$$

For a non-hydrostatic flow, its pressure differs from the hydrostatic pressure, and it can be expressed as

$$p = p_h + p_d \tag{5.17}$$

in which p_d is the dynamic pressure. Equation (5.17) is a pressure split in the existing literature [29]. It is seen that, under the hydrostatic assumption, $p = p_h$, $p_d = 0$, that is, the dynamic pressure becomes zero. Adoption of pressure split (5.17), NS equations (4.2) for a non-hydrostatic flow become

$$\begin{aligned} \mathbf{u}_t + \nabla \cdot \mathbf{uu} = & -\frac{1}{\rho_0} \nabla p_d + \nabla \cdot ((\nu + \nu_t)\nabla \mathbf{u}) \\ & - g \nabla_H \left(\eta - \int_z^{\eta} (\alpha (T - T_0))\, d\zeta \right) \end{aligned} \tag{5.18}$$

This is an integro-differential equation, and it is expensive to compute since the integral introduces an additional computational load that is not cheap; at every spatial point z in a flow, the computation involves an integration from this point to the water surface elevation above it. Under the hydrostatic assumption and in the situation of constant temperature, Eq. (5.18) reduces to

$$\mathbf{v}_t + \nabla \cdot \mathbf{uv} = \nabla \cdot ((\nu + \nu_t)\nabla \mathbf{v}) - g \nabla_H \eta \tag{5.19}$$

in the horizontal plane. The above equation is the hydrostatic NS equations. Equation (5.16) complements Eq. (5.19) in the vertical direction, and it becomes

$$p_h = \rho_0 g\big(\eta - z\big) \tag{5.20}$$

in the situation of constant temperature.

To facilitate the NS-GFD coupling, an interface condition is derived for the NS equations, i.e., (5.2), and the hydrostatic NS equations, i.e., (5.19), at a constant temperature. Consider an interface Γ, and its left and right are, respectively, the NS subdomain, in which (5.2) holds, and the hydrostatic NS subdomain, in which (5.19) holds. In the NS subdomain, the pressure split technique, i.e., (5.17) and (5.20), leads to

$$p = \rho_0 g(\eta - z) + p_d \tag{5.21}$$

Plugging (5.20) and (5.21) into (5.14) leads to an interface condition between the NS and the hydrostatic NS equations:

$$(\mathbf{u}_n \mathbf{u} + p_d/\rho_0 - (\nu + \nu_t)\partial \mathbf{u}/\partial n)_{\Gamma-} = (\mathbf{u}_n \mathbf{u} - (\nu + \nu_t)\partial \mathbf{u}/\partial n)_{\Gamma+} \tag{5.22}$$

In deriving the condition, the continuity of water surface elevation and position is applied, that is, $\eta_{\Gamma^-}=\eta_{\Gamma^+}$, $z_{\Gamma^-}=z_{\Gamma^+}$. In an NS-GFD computation, conditions (5.12) and (5.22) may serve as interface conditions. According to them, a set of interface conditions is

$$\begin{aligned}
&(\mathbf{u}_n\mathbf{u} + p_d/\rho - (\nu+\nu_t)\partial\mathbf{u}/\partial n)\,|_{NS} = (\mathbf{u}_n\mathbf{u} - (\nu+\nu_t)\partial\mathbf{u}/\partial n)\,|_{GFD},\\
&\partial p_d/\partial n = 0, \qquad\qquad\qquad\qquad\qquad \text{on } \partial\Omega_{NS}\\
&\mathbf{u}_n|_{GFD} = \mathbf{u}_n|_{NS}, \qquad\qquad\qquad\qquad \text{on } \partial\Omega_{GFD}
\end{aligned} \tag{5.23}$$

Condition (5.23) is referred to as interface condition II. This interface condition is similar to interface condition I, i.e., (5.8). The difference is that, on an NS interface, the former requires continuity of the momentum flux in the normal direction, while the latter imposes continuity of the velocity. Additionally, on a GFD interface, the former imposes continuity of the normal velocity, whereas the latter requires continuity of the total velocity. Condition II accounts for the coupling between the two sets of NS equations, which is not directly addressed by condition I. Condition II, or a similar one, has been proposed to couple the NS equations and the hydrostatic NS equations in previous investigations [42, 43].

Numerical Example A theoretical analysis on the performance of both interface conditions I and II is not straightforward. In the following, a numerical experiment is made to explore an understanding of their performance.

Example 5.5.2 This experiment is about a transient water flow in a curved channel, commonly referred to as a riverbend, as shown in Fig. 5.16. In the channel, the water is static initially, with a depth of 15 m, and then it flows because of a velocity imposed at the channel's entrance. The following are the initial and boundary conditions:

$$\begin{aligned}
&\mathbf{u} = 0,\ \eta = 0,\ \ p = \gamma(\eta - z),\ \ t = 0\\
&u = 0.25(1 - e^{-0.01t}),\ \ \text{at entrance},\ t > 0;\ \eta = 0,\ \ \text{at exit},\ t > 0
\end{aligned}$$

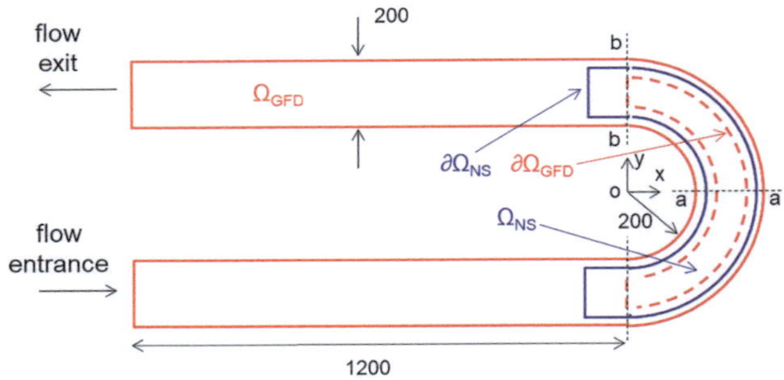

Fig. 5.16 Riverbend configuration and subdomains. Top view. $z = 0$ is at the initial water surface, and the channel bottom is at $z = -15$ m

Here, the length, time, and velocity are in m, s, and m/s, respectively. As seen in the figure, the GFD subdomain covers the entire channel, and its grid has 10,000 triangular cells in the horizontal plane and 20 layers in the vertical direction. The NS subdomain occupies the bend section, and its grid has a grid of $111 \times 13 \times 13$ in the streamwise, lateral, and vertical directions, respectively. The time step is set to 0.01 s.

Figure 5.17 depicts the simulated instantaneous water surface elevation by the GFD simulation and the NS-GFD simulations. For the NS-GFD simulations, interface condition I, i.e., (5.8), and interface condition II, i.e., (5.23), are adopted. The figure shows solutions at $t = 60$ s, when the flow is within the transient stage, and at $t = 1{,}200$ s, when it is roughly at a quasi-steady stage. It is seen that, at the transient stage, the GFD and NS-GFD simulations for elevations are largely identical in the scale of the figure. However, they exhibit differences at the quasi-steady stage. Additionally, the NS-GFD simulations with the two interface conditions are similar in main patterns but present a certain difference.

To further examine their differences, Fig. 5.18 shows the simulated velocities at cross-section a-a within the bend (Fig. 5.16). The GFD simulation has a single vortex covering the whole cross-section (Fig. 5.18a). However, the NS-GFD simulations contain multiple vortices: two counter-rotation vortices in the middle of the cross-section, plus one at the left lower corner, one at the right lower corner, and one at the upper right corner (Figs. 5.18b and c). Such multiple vortices have been reported in an earlier NS simulation [44]. This difference in the GFD and NS-GFD simulations at the cross-section flow could depict the necessity for the NS-GFD coupling. It is also seen that, in NS-GFD simulations with the two interface conditions, the relative positions of the vortices differ to a certain extent, although both lead to these multiple vortices at the cross-section.

The vertically averaged velocities at two cross-sections are plotted in Fig. 5.19. The figure shows that the solutions for both streamwise and lateral velocities obtained with the GFD simulation are distinct from those computed by the NS-GFD simulations. Note that the streamwise velocity is v-velocity and u-velocity at cross-sections a-a and b-b, respectively. In addition, interface conditions I and II result in similar streamwise velocities but distinct lateral velocities. To illustrate this with more details, Table 5.2 presents a quantification of the difference in the solutions obtained with the two interface conditions. The table shows that the lateral velocity has a more pronounced difference, indicating that the two interface conditions lead to distinct simulations for the secondary flow in the channel's curved section. More details for the simulations are available in [45] (for instance, a modified condition II is applied in the actual computation).

The aforementioned numerical experiment illustrates that the GFD simulation and the NS-GFD simulation present distinct solutions. Furthermore, interface conditions I and II produce similar flow patterns in a large picture but lead to differences, especially in the secondary flow within the bend. However, further investigations and tests against benchmarks are necessary to understand the interface conditions clearly.

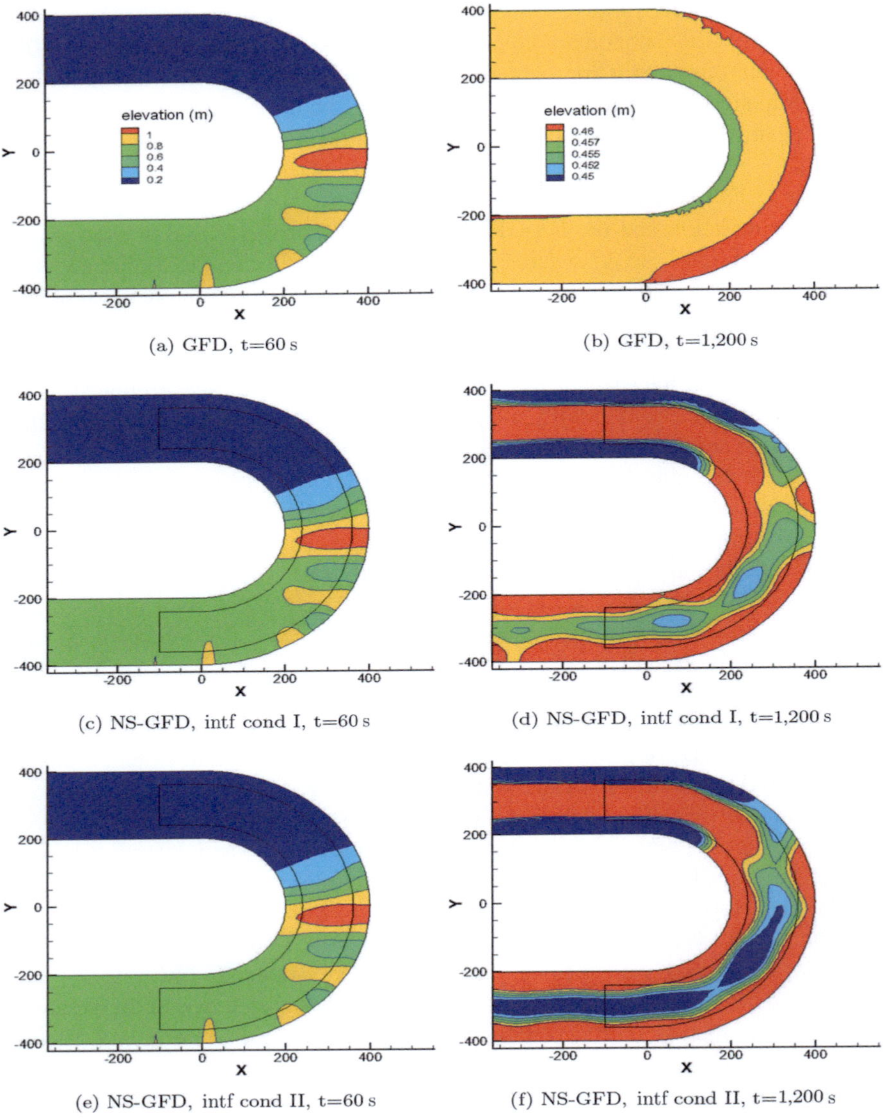

(a) GFD, t=60 s

(b) GFD, t=1,200 s

(c) NS-GFD, intf cond I, t=60 s

(d) NS-GFD, intf cond I, t=1,200 s

(e) NS-GFD, intf cond II, t=60 s

(f) NS-GFD, intf cond II, t=1,200 s

Fig. 5.17 Water surface elevations obtained with the GFD simulation and the NS-GFD simulations with interface condition I, (5.8), and interface condition II, (5.23)

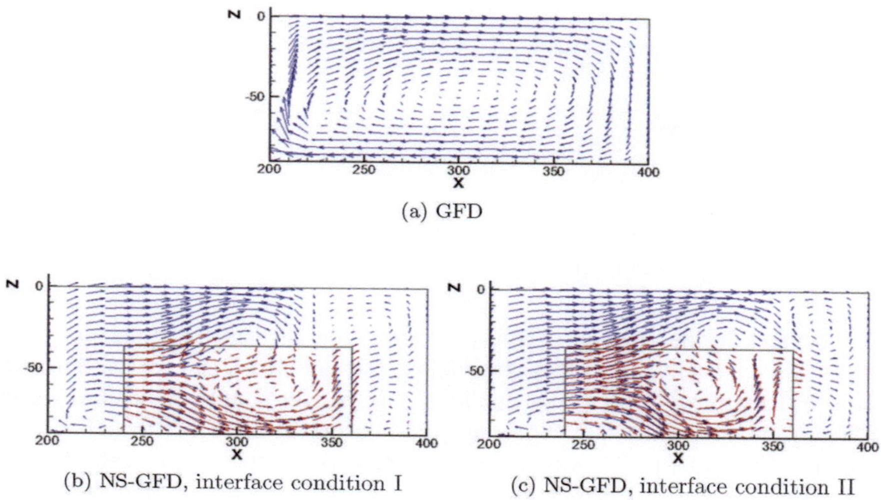

(a) GFD

(b) NS-GFD, interface condition I

(c) NS-GFD, interface condition II

Fig. 5.18 Simulated flow field, cross-section a-a, $t = 1{,}200$ s

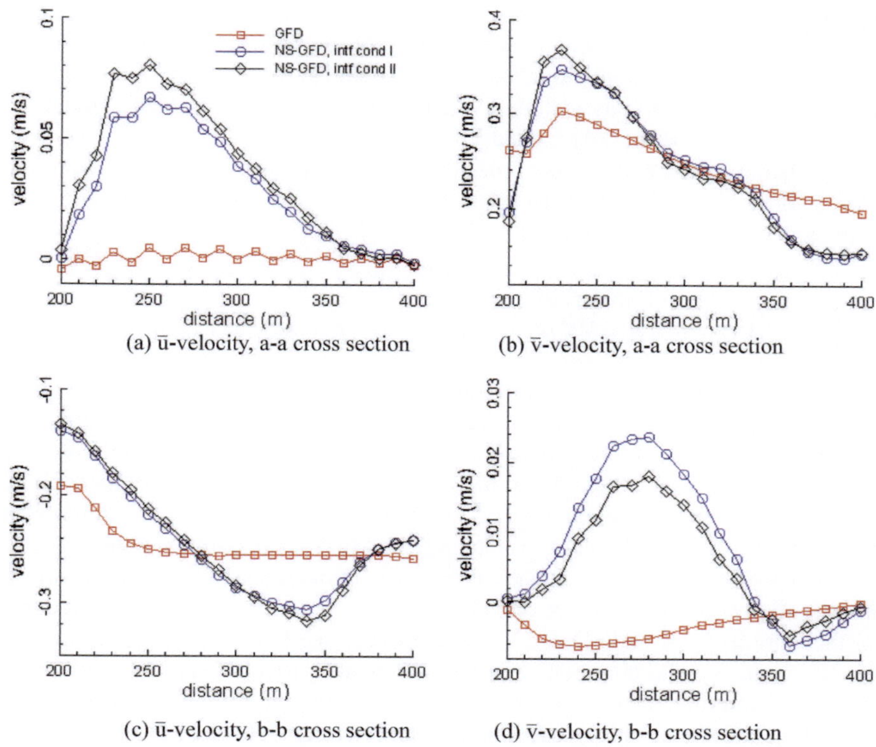

(a) ū-velocity, a-a cross section

(b) v̄-velocity, a-a cross section

(c) ū-velocity, b-b cross section

(d) v̄-velocity, b-b cross section

Fig. 5.19 Vertically averaged velocity at cross-sections a-a and b-b, $t = 1{,}200$ s

Table 5.2 The NS-GFD solutions obtained with interface conditions I and II. $t = 1{,}200$ s. $\bar{u}$ and $\bar{v}$ are vertically averaged u-velocity and v-velocity, respectively, and subscripts I and II depict interface conditions (5.8) and (5.23), respectively

Cross-section	$\|\|\bar{u}_{II} - \bar{u}_I\|\|_2/\|\bar{u}_I\|_{max}$	$\|\|\bar{v}_{II} - \bar{v}_I\|\|_2/\|\bar{v}_I\|_{max}$
a-a	1.809	0.385
b-b	0.268	2.196

5.6 Coupling of SW and GFD Equations

The methods in the previous sections to couple the NS and GFD equations are extendable to coupling other sets of equations for fluid flows, and coastal flooding is one of these flows. This section applies the methods to couple the shallow water (SW) equations and the GFD equations, introducing an SW-GFD model system [20]. The SW equations provide an ideal framework to model flows with surface waves such as surge bores (a surge bore is a surface wave with a step shape). In the past two decades, many mature algorithms have been developed to solve the SW equations and have been implemented in software packages, such as GeoClaw [46], to simulate these surface waves. Additionally, the SW equations are cheaper to compute than the GFD equations. Therefore, an SW-GFD model system may capture flow phenomena, such as surge bores, more accurately than the GFD model while requiring less computational cost. In the model system, the SW model depicts a near-field ocean that may contain a surge bore, and the GFD model simulates a far-field ocean that is the background ocean.

The shallow water equations are the governing equations for shallow water flows, whose horizontal scales are much bigger than those in the vertical direction. The shallow water equations read as [20, 47]

$$\begin{aligned} &\eta_t + \nabla_H \cdot (\mathbf{V}D) = 0 \\ &(\mathbf{V}D)_t + \nabla_H \cdot (\mathbf{V}\mathbf{V}D) = -gD\nabla_H\eta + gD(\boldsymbol{\tau}_s - \boldsymbol{s}_b) \end{aligned} \tag{5.24}$$

In the SW equations, $\boldsymbol{\tau}_s$ is the water-surface stress. $\boldsymbol{s}_b$ is the bottom friction slope, and it is determined as $\boldsymbol{s}_b = n^2\mathbf{V}|\mathbf{V}|/D^{4/3}$, with n being the Manning coefficient. Note that the SW equations are also considered as a type of GFD equations. The SW equations will be solved by an in-house SW solver, which adopts an explicit, Godunov-type, finite volume method on a triangular mesh and the third-order Runge–Kutta method for the time integration. More details are available in [20].

The SW equations and the equations for the external mode of the GFD equations are essentially the same, except that the latter has an additional term that accounts for the Coriolis force and 3D and viscosity effects, see Eqs. (5.24) and (5.5). They are 2D equations with common variables, i.e., water surface elevation, η, and flow speed, $\mathbf{V}$. Therefore, a natural approach to constructing the SW-GFD model system is to couple the SW equations with those of the external mode of GFD and exchange their solutions for the common variables. In the SW-GFD coupling, a flow domain is split into an SW subdomain and a GFD subdomain (Fig. 5.20a). On an SW interface, $\partial\Omega_{SW}$, and a GFD interface, $\partial\Omega_{GFD}$, the two solvers exchange the solutions:

$$\begin{aligned} (\eta, \mathbf{V})_{SW} &= (\eta, \mathbf{V})_{GFD}, \qquad on \quad \partial\Omega_{SW} \\ (\eta, \mathbf{V})_{GFD} &= (\eta, \mathbf{V})_{SW}, \qquad on \quad \partial\Omega_{GFD} \end{aligned} \tag{5.25}$$

Since both the solver for the GFD equations, i.e., FVCOM, and the solver for the SW equations, i.e., the in-house solver, adopt finite volume methods on triangular meshes, it is natural to let their grids overlap. Particularly, the two solvers' meshes share elements and nodes in their overlapping region, see Fig. 5.20a. A grid node on the interface of the SW solver is its interface node, and a grid element of the solver with a node on the interface is an interface element. Similar terms are used for the GFD solver.

In the SW solver, η and $\mathbf{V}$ are defined at element centroids of its grid (Fig. 5.20b). Consequently, the SW solver needs boundary conditions for η and $\mathbf{V}$ at the centroid of each interface element of its grid. These conditions will come from the solutions

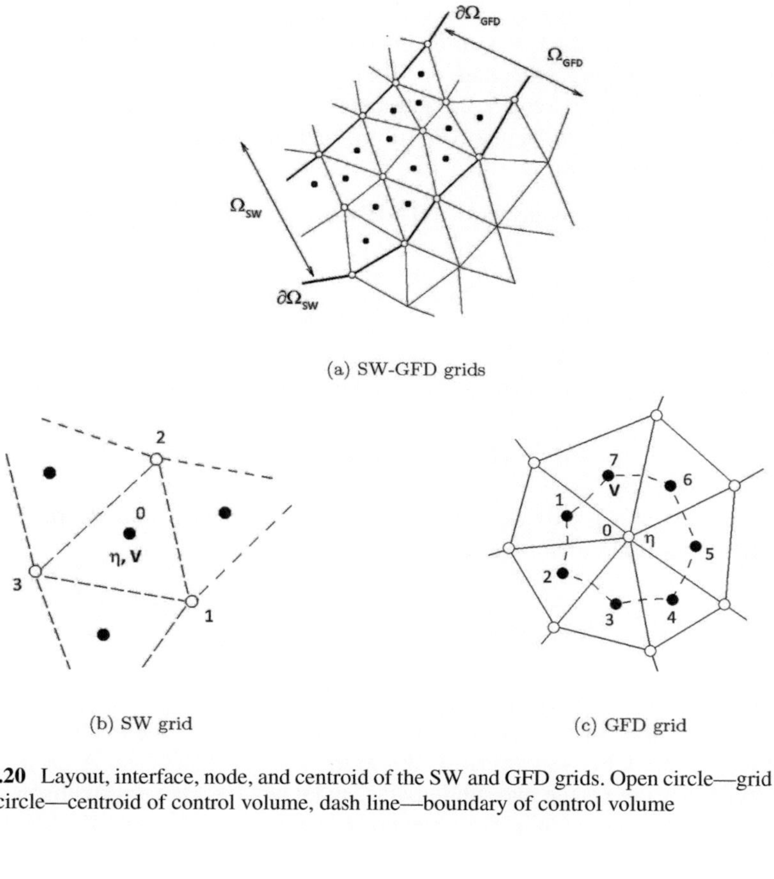

(a) SW-GFD grids

(b) SW grid

(c) GFD grid

Fig. 5.20 Layout, interface, node, and centroid of the SW and GFD grids. Open circle—grid node, filled circle—centroid of control volume, dash line—boundary of control volume of the GFD solver for these variables in these elements. For the GFD solver, η is located as grid nodes, and $\mathbf{V}$ is located at grid element centroids (Fig. 5.20c). In return, the GFD solver needs the SW solutions at GFD interface elements. Therefore, interpolating solutions between the nodes and centroids of elements is necessary to implement interface condition (5.25).

In the computation of the SW solver, the interpolation from the nodes of a grid element to its centroid may be implemented via previous techniques. For instance, in the situation of Fig. 5.20b, the interpolation is

$$\mathbf{V}_0 = \frac{A_{\widehat{023}}\mathbf{V}_1 + A_{\widehat{031}}\mathbf{V}_2 + A_{\widehat{012}}\mathbf{V}_3}{A_{\widehat{023}} + A_{\widehat{031}} + A_{\widehat{012}}}$$

In the GFD solver, the solutions at centroids can also be transferred to a grid node. For instance, in Fig. 5.20c, the transferring can be implemented as follows:

$$\eta_0 = \frac{\eta_1/d_{10} + \eta_2/d_{20} + \cdots + \eta_7/d_{70}}{1/d_{10} + 1/d_{20} + \cdots + 1/d_{70}}$$

in which d_{i0} is the distance between centroid i and node 0 ($i = 1, 2, \ldots 7$).

When marching in time, the computation of the SW-GFD model system may follow the methods and procedure for the NS-GFD model system illustrated in Fig. 5.4 and described in Sect. 5.3.3. Particularly, in the figure and the description, the NS solver is replaced by the SW solver, and the computational methods and procedure remain the same.

Example 5.6.1 This is an unsteady, 1D tidal flow in an open channel, whose bottom elevation is

$$h_b = 50.5 - \frac{40x}{L} - 10sin\left(\frac{4\pi x}{L} - \frac{\pi}{2}\right)$$

Here, L is the channel length, and $L = 14{,}000$ m. Initially, the water is stationary:

$$(D_{t=0}, u_{t=0}) = (h_b, 0)$$

where D is the depth, u is the velocity, measured respectively in m and m/s. As $t > 0$, the water flows under the action of the boundary condition:

$$(D_{x=0}, u_{x=L}) = \left(h_{bx=0} + 4 - 4sin(\frac{4\pi t}{86400} + \frac{\pi}{2}), 0\right)$$

An analytical solution to this problem is available [48].

In the application of the SW-GFD model system, the SW subdomain is $0 < x <$ 9,000 m, and the GFD subdomain is 3,600 m $< x <$ 14,000 m. A mesh in the horizontal plane covers the two subdomains, and it has 100 elements with $\Delta x = 140$ m. It is noted that the mesh is 2D in the horizontal plane, since both the SW model

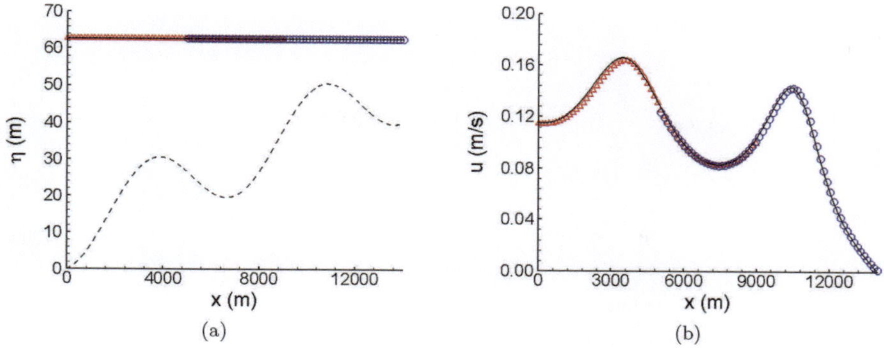

Fig. 5.21 Solution of the 1D tidal flow. $t = 7{,}500$ s. Dash line—bed elevation, triangle—SW solution, circle—GFD solution, solid line—analytical solution

and the GFD model's external mode are designed for 2D flows in the plane. The two models adopt a portion of the horizontal mesh in their subdomains. As a note, the 2D mesh consists of triangular elements, and the layout of these elements is made symmetric with the two lateral sides. Otherwise, the GFD simulation for a 1D flow tends to become effectively 2D, with its direction turning toward one lateral side rather than pointing strictly forward. This is a numerical issue in the GFD model. Additionally, the GFD model has five layers in the vertical direction. Time steps are set as Δt_{SW}, Δt_E, and $\Delta t_{GFD} = 1.5$ s. The numerical solution and its corresponding analytical solution are presented in Fig. 5.21. In terms of quantification, $||\eta_{numer} - \eta_{anal}||_2 = 2.2$, $||u_{numer} - u_{anal}||_2 = 0.28$. Here, subscripts "numer" and "anal" indicate numerical and analytical solutions, respectively.

Example 5.6.2 This example concerns a dam-break flow [49]. Initially, two stationary water bodies with different elevations are separated by a dam, see Fig. 5.22a. Then, the water flows due to the sudden removal of the dam section between the two dashed lines in the figure. The initial conditions in terms of a 3D flow are

$$(D, u, v, w) = \begin{cases} (10, 0, 0, 0), & x < 0 \\ (5, 0, 0, 0), & x > 0 \end{cases}$$

Here the length is in m, and the velocity is in m/s. A surge bore occurs right after the dam is partially removed, moving toward the right. At the same time, a rarefaction wave (a wave of expansion) takes place in the flow, propagating to the left, and vortices appear at the tips of the breaking wall. In this process, the flow near the removed dam section, including the bore and the vortices, is a near field, and the rest is a far field.

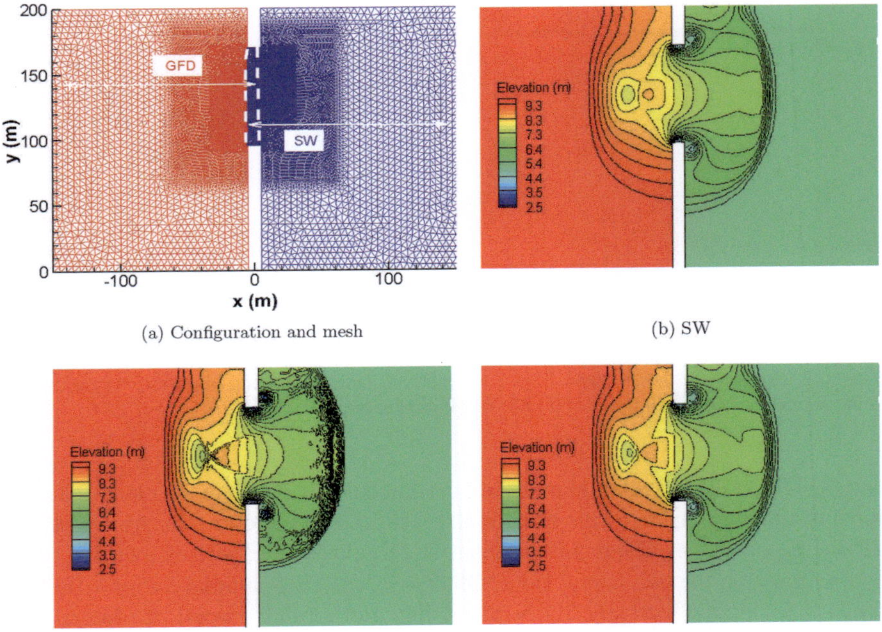

(a) Configuration and mesh (b) SW

(c) GFD (d) SW-GFD

Fig. 5.22 Meshes and simulated partial dam-break flow. $t = 7.2$ s

The computational domain covers $-195 < x < 195$ m and $0 < y < 200$ m. A mesh in the horizontal plane with elements of 50,071 covers the whole flow domain, and the SW model and the GFD model use 26,760 elements and 24,491 elements, respectively, of the horizontal mesh, see Fig. 5.22a. The SW and GFD meshes overlap, particularly in the region between the two dashed lines and the dam in the figure. Additionally, the GFD model has five layers in the vertical direction. The time steps are Δt_{SW}, Δt_E, and $\Delta t_{GFD} = 0.005$ s. For comparison, simulations are also made using the SW model alone and the GFD model alone, on the horizontal mesh.

Figure 5.22b shows that the SW model captures the surge bore with a clear front and presents no spurious oscillations. In Fig. 5.22c, the GFD model resolves fine structures of the vortices at the wall tips and within the rarefaction wave; however, it produces a substantial amount of artificial oscillations at the surge bore, as shown by the oscillatory or noisy contours at its wavefront. In Fig. 5.22d, the SW-GFD solution quality for the surge bore is basically the same as that computed by the SW model alone, and its resolution for the vorticies and rarefaction wave is similar to that obtained by the GFD model alone. These illustrate that the SW-GFD model system retains the two models' advantages while avoiding their disadvantages.

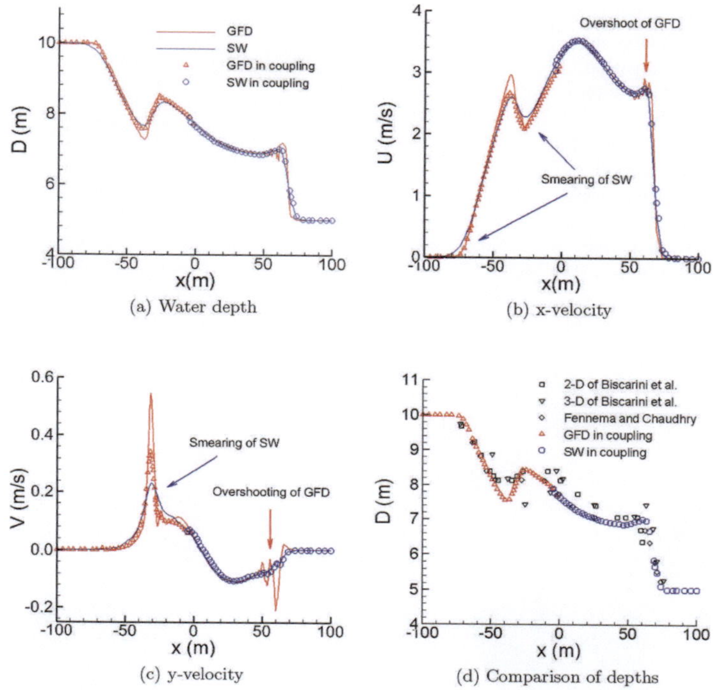

Fig. 5.23 Solution for the partial dam-break flow. $y = 130$ m. U and V stand for velocities in x- and y-directions, respectively. In (**d**), diamonds show 2D modeling of Fennema and Chaudhry [49], and squares and gradients respectively represent the 2D and 3D modeling by Biscarini et al. [50]

Additionally, the solution transition at the interfaces between the two models is smooth, and no non-physical phenomena occur. The computed solution and these observations are further illustrated in Figs. 5.23a, b, and c. These figures show that the run of the GFD model alone exhibits artificial oscillations at the surge bore, around $x = 70$ m. A comparison between the computed solution and those obtained by other investigators is shown in Fig. 5.23d, confirming the solution's correctness by the SW-GFD model system. Regarding the computational time, the SWM-only run uses 1 min of CPU time, the GFD-only run takes 44 min of CPU time, and the coupling run takes 31 min of CPU time. Compared with the GFD-only run, the coupling approach saves 31%. in CPU time.

Example 5.6.3 This example is about an ocean tidal flow in Delaware Bay during April 10–14, 2010. The bay is primarily a shallow water body with a mean depth of 7 m and a maximum depth of approximately 30 m near its mouth [51]. In applying the SW-GFD model system, the GFD model covers most of the bay, and the SW model includes an eastern coastal zone next to Cape May County, see Fig. 5.24. A mesh in the horizontal plane with 39,976 elements covers the whole computational

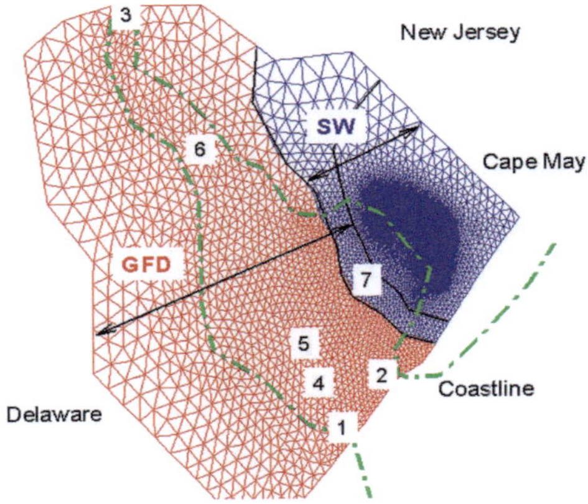

Fig. 5.24 Computational domain, subdomains, their meshes, and observation stations. The observation stations are at six locations: Location 1—Lewes, location 2—Cape May, location 3—Reedy Point, location 4—Brown Shoal Light, location 5—Brandywine Shoal, and location 6—Ship John Shoal. Location 7 is within the overlapping region of the SW and GFD subdomains

(a) Flood tide (b) Ebb tide

Fig. 5.25 SW-GFD simulation for the instantaneous velocity field and surface elevation

domain. In this mesh, the GFD model adopts 3,325 elements, and the SW model uses 37,353 elements. The SW zone has a high resolution, as fine as below 50 m. Bathymetric data with about 100 m resolution and the DEM data with 30 m resolution are respectively obtained from the NOAA NGDC and the USGS seamless data warehouse. The boundary conditions of water surface elevation observed by NOAA NDBC are applied at Lewes and Cape May for the southern boundary, i.e., locations 1 and 2 in Fig. 5.24, and Reedy Point for the northern boundary, i.e., location 3. $\Delta t_{SWM} = 4.5$ s, $\Delta t_E = 1.5$ s, and $\Delta t_{GFD} = 4.5$ s. For comparison, a run of the GFD model alone on the horizontal-plane mesh is also made to further examine the accuracy and efficiency of the coupling approach.

Figure 5.25 presents the computed flow at a flood-tide instant and an ebb-tide instant in the bay. The figure shows a smooth transition between the SW and

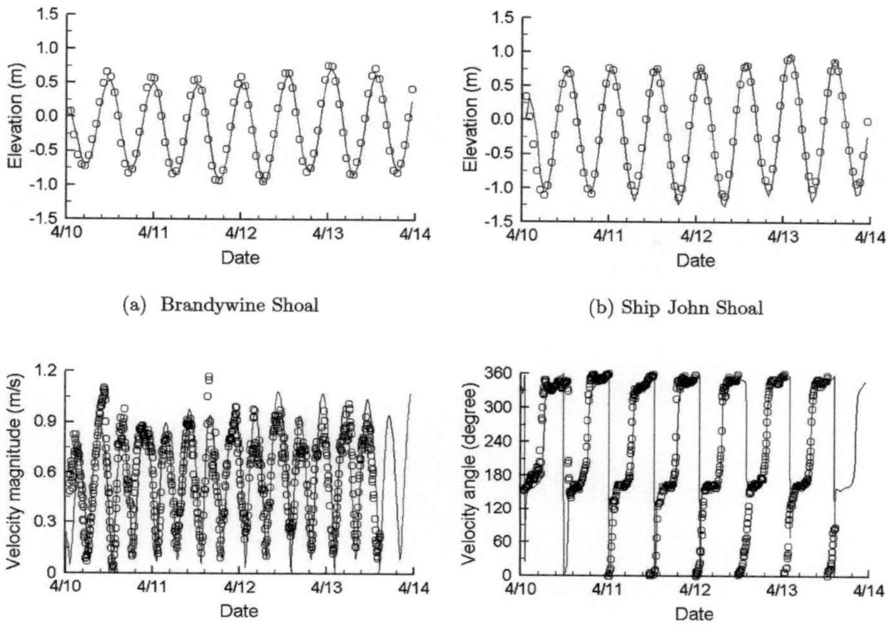

(a) Brandywine Shoal (b) Ship John Shoal

(c) Brown Shoal Light (d) Brown Shoal Light

Fig. 5.26 SW-GFD and GFD simulations at specific locations and its comparison with observation. Solid line—simulations, circle—observation

GFD solutions. A comparison between the simulations by the SW-GFD model system and the GFD model alone against field measurements at three locations by NOAA NDBC is presented in Figs. 5.26. It is noted that the simulations by the SW-GFD model system and the GFD model overlap at these locations. That is, their simulations are essentially indistinct in the plot scales. The figure shows that the simulations match the measurements, serving as validation of them. As a prediction, the simulations of the model system and the GFD model at another location are depicted in Fig. 5.27. Again, at the location, the simulations by the two approaches are indistinct at the scales of the figure.

In these simulations, the SW-GFD run saves computing time, using 1 hr 29 min of CPU time, and the GFD-only run uses 5 hr 40 min of CPU time. Thus, the SW-GFD run saves 74% in CPU time. It should be noted that an even higher CPU time saving is expected if more grid elements with fine spacing are added to the high-resolution zone.

The aforementioned three examples show that the SW-GFD model system works well. It can handle multiphysics flows and save computational time. More simulation results are available for coastal flooding in sea-level-rise conditions [20, 52].

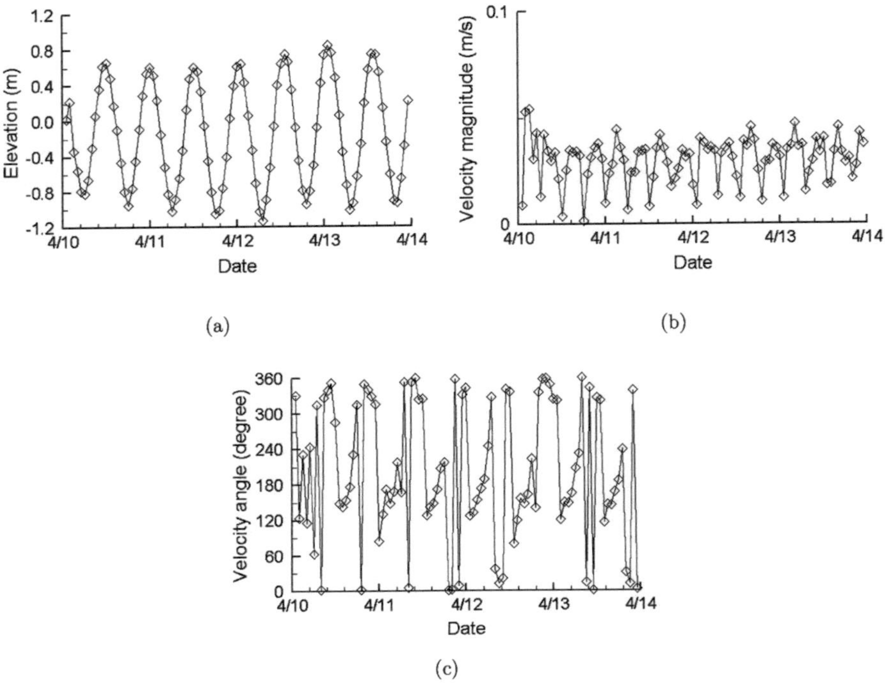

Fig. 5.27 Prediction by the SW-GFD model system and the GFD model, location 7. Solid line—SW-GFD, diamond—GFD

Summary and Question

Summary The chapter begins with an overview of emerging ocean flow problems, highlighting their multiscale and multiphysics nature. It then presents the need for new modeling capabilities to address these emerging problems, identifying the multiblock method as the most feasible and promising approach. A multiblock model system is introduced that couples the NS equations for near-field ocean flows with the GFD equations for far-field flows. The governing equations of the NS-GFD model system are presented, followed by a discussion on the NS and GFD solvers. The discussion also involves coupling techniques, including grid connectivity, interface algorithms, their numerical accuracy, and computational procedures. The interface algorithm based on interface condition I is essentially the same as the SI algorithm discussed in Chap. 4.

A series of numerical examples is then presented to demonstrate the performance of the multiblock method. Through a sill flow simulation, the NS-GFD model system illustrates both the necessity of the NS-GFD coupling and its solution accuracy. The system is further validated through simulations of a flow past bridge piers and a thermal discharge, showing agreement with measurement data. Its capability to handle realistic ocean problems is demonstrated by an application to effluent discharge into the New York Bight, covering spatial scales from $\mathcal{O}(10)$ m to $\mathcal{O}(100)$ km.

The chapter continues with a deeper examination of coupling techniques. First, it introduces a pressure-splitting method in the NS equations designed for the NS-GFD coupling. Although a similar split has appeared in the literature, it results in integro-differential equations. In contrast, the proposed split avoids such equations while reducing to the known version in the absence of buoyancy. Numerical experiments demonstrate that this pressure split is effective in mitigating numerical artifacts and non-physical solutions, which frequently arise without such a split. Second, from the perspective of weak solutions and coupling the NS with the hydrostatic NS equations, the chapter introduces interface condition II. Unlike interface condition I, which requires velocity continuity across an interface, interface condition II enforces momentum continuity across the interface. Numerical simulations of riverbend flows illustrate differences between the two interface conditions, particularly in the prediction of secondary flows.

The chapter then addresses the coupling of the SW equations with the GFD equations to capture surface wave dynamics. The SW solver is based on a Godunov-type finite volume method and is applied to resolve local phenomena such as surge bores. The SW-GFD model system follows the structural and methodological framework of the NS-GFD model system, connecting the SW model with the external mode of the GFD model. Since both models employ triangular elements in the horizontal plane, their meshes may overlap, allowing for the construction of conservative interface treatments. Numerical examples of tidal flow and dam-break flow are presented to demonstrate the performance of the SW-GFD model system. Additionally, the system is applied to simulate coastal flooding in the Delaware Bay.

The multiblock method discussed in this chapter integrates state-of-the-art models from different communities to capture emerging multiscale and multiphysics ocean flows. The developed multiblock systems offer a powerful tool for studying complex oceanic processes that are difficult to resolve using existing standalone models. For example, the NS-GFD system can be used to study near-field wakes behind infrastructure (e.g., offshore wind energy foundation) and their influence on far-field ocean mixing and stratification. However, further testing of the systems, especially with data from the ocean, is still needed. While these systems employ advanced techniques, several challenges persist. The interface conditions, such as conditions I and II, deserve deeper theoretical investigation to assess their mathematical and physical validity. Furthermore, the development of conservative interface treatments is expected to substantially improve the accuracy and reliability of these simulations.

Questions The following questions are provided to aid in understanding, discussing, and investigating the content. Their answers and clues are embedded within the chapter.

1. In general, multiphysics refers to multiple physical phenomena and their interactions across disciplines, such as fluid-structure interaction. Why does this chapter classify certain ocean flows as multiphysics?
2. This chapter presents a pressure-splitting technique in the NS equations, which differs from the one proposed in earlier literature. Discuss the relation between the two split techniques and the advantages of the former.
3. Why does the NS model couple with the internal mode of the GFD model in the NS-GFD system, while the SW model couples with its external mode in the SW-GFD system?
4. What is a blanked zone in the GFD subdomain, and why is it needed in the NS-GFD model system?
5. The GFD mesh responds to water surface sloshing, causing its connectivity with the NS mesh to change over time. However, the two meshes' connectivity updates are required only in the vertical direction at each time step. Explain why.
6. Describe the time-marching schemes used in the NS-GFD and SW-GFD systems. In particular, explain how the component models exchange solutions at the interfaces in time and how the time steps are managed.
7. What are the accuracy orders of the NS and GFD solvers? What is the accuracy order of the interface algorithms presented in this chapter?
8. Both the NS-GFD and SW-GFD model systems use two-way coupling. However, one-way coupling, from GFD to NS or SW, can also be implemented. Consider how the computational procedure in Sect. 5.3.3 could be modified to enable the one-way coupling.
9. What are the erroneous and non-physical solutions observed in the NS-GFD simulation of the sill flow in Sect. 5.5.1? What technique is used to mitigate them?
10. Discuss the considerations for introducing interface conditions I and II from the perspectives of boundary conditions and computations. What is the key difference between these two conditions?
11. In the above summary of the chapter, it is noted that the overlapping meshes of the SW and GFD models enable the possibility of designing a conservative interface treatment. Propose how such a treatment could be constructed using the numerical flux enforcement strategy discussed in Chaps. 3 and 4.
12. The NS-GFD and SW-GFD model systems offer a promising framework for simulating multiscale, multiphysics ocean flows, but they require continued development. What are potential topics for the further study?

References

1. E.B. Kujawinski, M.C.K. Soule, D.L. Valentine, A.K. Boysen, K. Longnecker, M.C. Redmond, Fate of dispersants associated with the Deepwater Horizon oil spill. Environ. Sci. Technol. **45**, 1298–1306 (2011)

2. N. Mimura, K. Yasuhara, S. Kawagoe, H. Yokoki, S. Kazama, Damage from the great east Japan earthquake and tsunami - a quick report. Mitig. Adapt. Strateg. Glob. Change **16**, 803–818 (2011)
3. H.S. Tang, S. Kraatz, K. Qu, G.Q. Chen, N. Aboobaker, C.B. Jiang, High-resolution survey of tidal energy towards power generation and influence of sea-level-rise: A case study at coast of New Jersey, USA. Renew. Sustain. Energy Rev. **32**, 960–982 (2014)
4. A.J. Mariano, V.H. Kourafalou, A. Srinivasan, H. Kanga, G.R. Halliwell, E.H. Ryan, M. Roffer, On the modeling of the 2010 Gulf of Mexico oil spill. Dynam. Atmospheres Oceans **52**, 322–340 (2011)
5. H.S. Tang, X.G. Wu, Multi-scale coastal flow simulation using coupled CFD and GFD models, in *Modelling for Environment's Sake, Fifth Biennial Meeting*, July 5–8 2010, Ottawa, Canada, ed. by D.A. Swayne, W. Yang, A.A. Voinov, A. Rizzoli, T. Filatova (2010)
6. H.S. Tang, T. Keen, in *Hybrid Model Approaches to Predict Multiscale and Multiphysics Coastal Hydrodynamic and Sediment Transport Processes, Sediment Transport*, chap. 4, ed. by S.S. Ginsberg (InTech, Austria, 2011), pp. 71–90. ISBN: 978-953-307-189-3
7. A.S. Candy, An implicit wetting and drying approach for non-hydrostatic baroclinic flows in high aspect ratio domains. Adv. Water Resour. **102**, 188–205 (2017)
8. C. Peng, S.L. Xiao, D. Yang, Large-eddy simulation model for the effect of gas bubble dissolution on the dynamics of hydrocarbon plume from deep-water blowout. J. Geophys. Res. Ocean **125**, e2019JC016037 (2020)
9. M. Le Henaff, V.H. Kourafalou, C.B. Paris, J. Helgers, Z.M. Aman, P.J. Hogan, A. Srinivasan, Surface evolution of the deepwater horizon oil spill patch: Combined effects of circulation and wind-induced drift. Environ. Sci. Technol. **46**, 7267–7273 (2012)
10. P.K. Kundu, *Fluid Mechanics* (Academic Press, 1990)
11. ANSYS, Inc., ANSYS FLUENT User's Guide, release 14.0, 2011
12. OpenFOAM, The Open Source CFD Toolbox, version 2312. 2023
13. B. Henshaw, Overture short course: Solving PDE's on overlapping grids. 2011
14. A.F. Blumberg, G.L. Mellor, A description of a three-dimensional coastal ocean circulation model. *Three-Dimensional Coastal Ocean Models*, vol. 4, ed. by N.S. Heaps (American Geophysical Union, New York, 1987)
15. C.S. Chen, R.C. Beardsley, G. Cowles, FVCOM User Manual, 2nd edition. *SMAST/UMASSD-06-0602*, 2006
16. A.F. Shchepetkin, J.C. McWilliams, The regional oceanic modeling system (ROMS): A split-explicit, free-surface, topography-following-coordinate oceanic model. Ocean Modell. **9**(4), 347–404 (2005)
17. The SWAN team, SWAN technical documentation. SWAN Cycle III version 40.51, 2006
18. O.R. Fringer, J.C. McWilliams, R.L. Street, A new hybrid model for coastal simulation. Oceanography **19**, 64–77 (2006)
19. K.I. Sitanggang, P.J. Lynett, Multi-scale simulation with a hybrid Boussinesq-RANS hydrodynamic model. Int. J. Numer. Methods Fluids **62**, 1013–1046 (2010)
20. H.S. Tang, S. Kraatz, X.G. Wu, W.L. Cheng, K. Qu, J. Polly, Coupling of shallow water and circulation models for prediction of multiphysics coastal flows: method, implementation, and experiment. Ocean Eng. **62**, 56–67 (2013)
21. K. Fujima, K. Masamura, C. Goto, Development of the 2D/3D hybrid model for tsunami numerical simulation. Coastal Eng. J. **44**, 373–397 (2002)
22. B.T. Paulsen, H. Bredmose, H.B. Bingham, An efficient domain decomposition strategy for wave loads on surface piercing circular cylinders. Coastal Eng. **86**, 57–76 (2014)
23. H.S. Tang, K. Qu, X.G. Wu, An overset grid method for integration of fully 3D fluid dynamics and geophysical fluid dynamics models to simulate multiphysics coastal ocean flows. J. Comput. Phys. **273**, 548–571 (2014)
24. K. Qu, H.S. Tang, A. Agrawal, Integration of fully 3D fluid dynamics and geophysical fluid dynamics models for multiphysics coastal ocean flows: Simulation of local complex free-surface phenomena. Ocean Modell. **135**, 14–30 (2019)

25. K. Qu, H.S. Tang, A. Agrawal, C.B. Jiang, B. Deng, Evaluation of SIFOM-FVCOM system for high-fidelity simulation of small-scale coastal ocean flows. J. Hydrodynamics **28**(6), 994–1002 (2016)
26. X.G. Wu, H.S. Tang, Coupling of CFD model and FVCOM to predict small-scale coastal flows. J. Hydrodynamics **22**, 284–289 (2010)
27. H.S. Tang, R.D. Haynes, G. Houzeaux, A review of domain decomposition methods for simulation of fluid flows: Concepts, algorithms, and applications. Arch. Comput. Methods Eng. **28**(3), 841–873 (2021)
28. H.S. Tang, C.R. Nichols, L.D. Wright, D. Resio, Modeling multiscale and multiphysics coastal ocean processes: A discussion on necessity, status, and advances. J. Marine Sci. Eng. **9**, 847 (2021)
29. V. Casulli, A semi-implicit finite difference method for non-hydrostatic, free-surface flows. Int. J. Numer. Methods Fluids **30**, 425–440 (1999)
30. H.S. Tang, C. Jones, F. Sotiropoulos, An overset-grid method for 3D unsteady incompressible flows. J. Comput. Phys. **191**, 567–600 (2003)
31. L. Ge, F. Sotiropoulos, 3D unsteady RANS modeling of complex hydraulic engineering flows. I: numerical model. J. Hydraul. Eng. **131**, 800–808 (2005)
32. J. Paik, F. Sotiropoulos, Coherent structure dynamics upstream of a long rectangular block at the side of a large aspect ratio channel. Phys. Fluids **17**, 115104 (2005)
33. H.S. Tang, J. Paik, F. Sotiropoulos, T. Khangaokar, Three-dimensional numerical modeling of initial mixing of thermal discharges at real-life configurations. ASCE J. Hydr. Eng. **134**, 1210–1224 (2008)
34. C. Chen, H. Liu, R.C. Beardsley, An unstructured, finite-volume, three-dimensional, primitive equation ocean model: application to coastal ocean and estuaries. J. Atm. Oceanic Tech. **20**, 159–186 (2003)
35. Z. Lai, C. Chen, G. Cowles, R.C. Beardsley, A non-hydrostatic version of FVCOM, Part I: Validation experiments. J. Geophys. Res. Oceans **115**, C11010 (2010)
36. G.L. Mellor, T. Yamada, A hierarchy of turbulence closure models for planetary boundary layers. J. Atmos. Sci. **31**, 1791–1806 (1974)
37. B. Gustafsson, The convergence rate for difference approximations to general mixed initial boundary value problem. SIAM J. Numer. Anal. **18**, 179–190 (1981)
38. J. Xing, A.M. Davies, On the importance of non-hydrostatic processes in determining tidally induced mixing in sill regions. Continental Shelf Res. **27**, 2162–2185 (2007)
39. G.A. Sod, *Numerical Methods in Fluid Dynamics – Initial and Initial Boundary-Value Problems* (Cambridge University Press, Cambridge, 1985)
40. L. Ge, S. Lee, F. Sotiropoulos, T.W. Sturm, 3D unsteady RANS modeling of complex hydraulic engineering flows. II: model validation and flow physics. J. Hydraul. Eng. **131**, 809–820 (2005)
41. G.H. Jirka, R.L. Doneker, S.W. Hinton, User's Manual for CORMIX: A hydrodynamic mixing zone model and decision support system for pollutant discharges into surface waters. 1996
42. E. Blayo, About interface conditions for coupling hydrostatic and nonhydrostatic Navier-Stokes flows. Discrete Contin. Dynam. Syst. S **9**(5), 1565–1574 (2016)
43. H.S. Tang, Y.J. Liu, Coupling of Navier-Stokes equations and their hydrostatic versions for ocean flows: A discussion on algorithm and implementation, in *Domain Decomposition Methods in Science and Engineering*, ed. by R. Haynes, S. MacLachlan, X.C. Cai, L. Halpern, H.H. Kim, A. Klawonn, O. Widlund. XXV, Lecture Notes in Computational Science and Engineering, vol. 138 (Springer, 2020), pp. 236–333
44. B. Li, X. Zhang, H.S. Tang, R. Tsubaki, Influence of deflection angles on flow behaviours in open channel bends. J. Mountain Sci. **15**, 2292–2306 (2018)
45. W.B. Dong, H.S. Tang, and Y. J. Liu, Coupling of Navier-Stokes equations and their hydrostatic versions and simulation of riverbend Flow, in *Domain Decomposition Methods in Science and Engineering*, ed. by S.C. Brenner, E. Chung, A. Klawonn, F. Kwok, J. Xu, J. Zou. XXVI, Lecture Notes in Computational Science and Engineering, vol. 145 (Springer, 2022), pp. 531–538

46. M.J. Berger, D.L. George, R.J. LeVeque, K.M. Mandli, The GeoClaw software for depth-averaged flows with adaptive refinement. Adv. Water Resour. **34**, 1195–1206 (2011)
47. T.H. Yoon, S.-K. Kang, Finite volume model for two-dimensional shallow water flows on unstructured grids. J. Hydraul. Eng. **130**, 678–688 (2004)
48. A. Bermudez, M.E. Vazquez, Upwind methods for hyperbolic conservation laws with source terms. Comput. Fluids **23**, 1049–1071 (1994)
49. R.J. Fennema, M.H. Chaudhry, Explicit methods for two-dimensional transient free surface flows. ASCE J. Hydraul. Eng. **116**, 1013–1034 (1990)
50. C. Biscarini, S.D. Francesco, P. Manciola, CFD modelling approach for dam break flow studies. Hydrol. Earth Syst. Sci. Discuss. **6**, 6759–6793 (2009)
51. P.A. Muscarella, N.P. Barton, Lipphardt Jr. B.L., D.E. Veron, Wong K.C., A.D. Kirwan Jr., Surface currents and winds at the Delaware Bay mouth. Continental Shelf Res. **311**, 1282–1293 (2011)
52. H.S. Tang, SI.-Jy. Chien, M. Temimi, C.A. Blain, K. Qu, L.H. Zhao, S. Kraatz, Vulnerability of population and transportation systems at east bank of delaware bay to coastal flooding in climate change conditions. Natural Hazards **69**, 141–163 (2013)

Chapter 6
Special Topics

6.1 Odd–Even Oscillation

6.1.1 Background

In the computation of differential equations, artifacts are commonly observed, and frequently, they take place in the form of spurious oscillations in numerical solutions for fluid-related problems [1–3]. A typical oscillation is the odd–even oscillation, which is also referred to as a checkerboard pattern [3], a zig-zag form [4], a grid-scale oscillation [5], odd–even decoupling [6], etc. A numerical solution with such odd–even oscillation fluctuates alternately in value from one grid node to the next and exhibits a zig-zag pattern in plots. Not only is the fluctuation non-physical, but it also pollutes solution accuracy and even leads to numerical problems, such as instability in computing differential equations [2, 3, 7].

The prevention and suppression of the odd–even oscillation have been one of the central issues of compressible flow simulations, especially when the central difference is used for spatial discretization [1, 2]. An example of such oscillation in the simulation of compressible flows occurs at the right-moving shocks in Figs. 3.5c and d. Researchers have made substantial efforts to suppress and prevent the oscillation associated with compressible flows, first introducing artificial viscosity [1], and, after two decades, finally reaching a few pieces of classic work on high-resolution schemes for shocks, including MUSCL, TVD, and ENO schemes [8–10]. In calculating incompressible flows, checkerboard patterns, frequently in pressure, tend to happen as the central difference is used for the divergence-free condition, the advection term, and the pressure gradient term [6, 11]. An example of oscillation in the incompressible flow simulation occurs within the subdomain overlapping region in Fig. 4.10a. Actually, grid-scale oscillation is a universal issue, and it also appears in the computation of problems beyond fluid flows, such as heat transfer and mass transport [3], seismic wave propagation [12], and many more problems.

H. Tang, *Multiblock Method for Fluid Flow*, Nonlinear Systems and Complexity 40,
https://doi.org/10.1007/978-3-031-78568-9_6

Spurious oscillations and related artifacts have been observed frequently not only on a single grid but also on multiple grids associated with the multiblock method. As shown in Chap. 3, in the computation of an inviscid, compressible flow, an oscillation may take place at interfaces between two grids. Such an oscillation occurs no matter whether a nonconservative or a conservative interface algorithm is used, even if the numerical solution tends to be a weak solution [13, 14]. In the computation of an incompressible flow, fluctuation in a zig-zag form in both pressure and velocity tends to appear near grid interfaces when a Chimera grid method is used [15].

Although grid-scale oscillation has been a classical topic in many ways, the early work focuses primarily on conditions and criteria for its onset and the development of algorithms for its suppression. For instance, see the discussions on the domain decomposition computation in Sect. 4.2.2 and in the literature [15, 16]. So far, explanations for its onset and presence mechanisms remain heuristic, abstract, and fragmentary. It is unclear if the reasons for its onset in different situations are related or the same in a strict sense. In addition, it is unclear if they are connected to causes for other types of oscillation, such as the Runge phenomenon in interpolation [17]. Overall, systematic analysis of the generation and behaviors of the grid-scale oscillation is scarce.

For further understanding of the grid-scale oscillation, this section presents a study on its generation and behaviors in computation in a single domain as well as two subdomains [18]. This study gets to the very bottom of the problem and examines the oscillation associated with a simple ordinary differential equation. It presents a rigorous but plain and direct description of why such oscillation occurs and how it behaves. The study also gives criteria for its onset, in conjunction with numerical examples for verification and illustration.

6.1.2 Problem of Study

We consider the computation of a simplified model problem, particularly an initial value problem of an ordinary differential equation:

$$\begin{cases} u_x = f(x), & x > 0 \\ u = u_0, & x = 0 \end{cases} \tag{6.1}$$

where f is a prescribed function of x, $f \in C^0$, and u_0 is a constant. This problem's analytical solution is

$$u = u_0 + \int_0^x f(s)ds \tag{6.2}$$

Let us transform initial value problem (6.1) into an alternative problem:

$$\begin{cases} \hat{u}_x = 0, & x > 0 \\ \hat{u} = u_0, & x = 0 \end{cases} \tag{6.3}$$

where $\hat{u} = u - \int_0^x f(s)ds$. The transformed problem is discretized on a uniform grid ($0 = x_1, x_2, \ldots, x_I = L$), and its numerical solution is denoted as $\hat{v}_i$. The discretization adopts finite difference, particularly the central finite difference, at the interior nodes and the backward finite difference at the right end, and it reads as

$$\begin{cases} \hat{v}_{i+1} - \hat{v}_{i-1} = 0, & i = 2, \cdots, I-1 \\ \hat{v}_1 = u_0, & \hat{v}_I - \hat{v}_{I-1} = 0 \end{cases} \tag{6.4}$$

In the calculation, when I is odd, the computation proceeds as

$$\begin{cases} \text{Step I}: & \hat{v}_{i+1} - \hat{v}_{i-1} = 0, \quad i = 2, 4, \cdots, I-1 \\ \text{Step II}: & \hat{v}_{I-1} = \hat{v}_I \\ \text{Step III}: & \hat{v}_{i+1} - \hat{v}_{i-1} = 0, \quad i = I-2, I-4, \cdots, 5, 3 \end{cases}$$

While I is even, it proceeds as

$$\begin{cases} \text{Step I}: & \hat{v}_{i+1} - \hat{v}_{i-1} = 0, \quad i = 2, 4, \cdots, I-2 \\ \text{Step II}: & \hat{v}_I = \hat{v}_{I-1} \\ \text{Step III}: & \hat{v}_{i+1} - \hat{v}_{i-1} = 0, \quad i = I-1, I-3, \cdots, 5, 3 \end{cases}$$

A straightforward derivation from the aforementioned, regardless of whether I is odd or even, yields the solution: $\hat{v}_i = u_0$. Then, the solution for $\hat{v}_i$, together with the trapezoid rule of integration, leads to the numerical solution of (6.1), v_i, as follows:

$$v_i = u_0 + \frac{\Delta x}{2} \sum_{k=2}^{i} (f_{k-1} + f_k), \quad i = 2, 3, \cdots, I \tag{6.5}$$

in which Δx is the grid spacing. Here, the derivation utilizes $v_i = \hat{v}_i + \int_0^x f(s)ds$. Obviously, the numerical solution converges to the analytical solution of the problem, i.e., (6.2), as Δx tends to zero.

Two definitions are presented as follows to facilitate the discussion on the oscillation.

Definition 6.1.1 Let v_i ($i = 1, 2, \cdots, I$) be a numerical solution for problem (6.1). The concaveness of the numerical solution is defined as

$$e_i \equiv v_{i-1} + v_{i+1} - 2v_i, \quad 2 \le i \le I-1 \tag{6.6}$$

The solution is called concave up at node i if $e_i > 0$, and it is called concave down at the node if $e_i < 0$.

Definition 6.1.2 A numerical solution v_i ($i = 1, 2, \cdots$, I) is said to exhibit odd–even oscillation at node i if its concaveness' sign at node i is different from those at nodes $i \pm 1$:

$$\begin{aligned} sign(e_i) * sign(e_{i-1}) < 0 \\ sign(e_i) * sign(e_{i+1}) < 0 \end{aligned} \tag{6.7}$$

With the aforementioned two definitions, we say the odd–even oscillation is present in a numerical solution if it is concave up at node i and concave down at its neighbor nodes $i \pm 1$, or if it is concave down at node i and concave up at nodes $i \pm 1$. The magnitude of e_i measures the degree of such oscillation. As indicated previously, such oscillation is a kind of grid-scale fluctuation, and it is also called odd–even decoupling, a zig-zag form, etc., in the literature. In general, the odd–even oscillation is an artificial or numerical oscillation.

It is readily checked that, given $f(x) \in C^0$, numerical solution (6.5) exhibits no odd–even oscillations according to the aforementioned definitions. Unluckily, some methods of computation could lead to such oscillations. In the following subsections, the discussion first deals with the oscillations in the situation of a single grid, and then it proceeds to the situation of two grids.

6.1.3 Oscillation on Single Grid

Dual-Mode Solution Consider the computation of initial value problem (6.1) on grid ($0 = x_1, x_2, \ldots, x_I = L$). Using the central difference at all grid nodes except the last node, where the backward difference is adopted, one has

$$\begin{cases} v_{i+1} - v_{i-1} = 2\Delta x f_i, \quad i = 2, \cdots, I-1 \\ v_1 = u_0, \quad v_I - v_{I-1} = \Delta x f_I \end{cases} \tag{6.8}$$

The computation can be realized via three steps for odd and even I in the last subsection. For convenience of discussion, the discretization is formulated as a linear algebraic system of size $(I-1) \times (I-1)$:

$$AV = b \tag{6.9}$$

in which

$$A = \begin{pmatrix} 0 & 1 & 0 & 0 & \cdots & & & \\ -1 & 0 & 1 & 0 & \cdots & & & \\ 0 & -1 & 0 & 1 & \cdots & & & \\ \vdots & \vdots & \vdots & \ddots & & & & \\ & & & & \ddots & \vdots & \vdots & \vdots \\ & & & \cdots & -1 & 0 & 1 & 0 \\ & & & \cdots & 0 & -1 & 0 & 1 \\ & & & \cdots & 0 & 0 & -1 & 1 \end{pmatrix}, \quad V = \begin{pmatrix} v_2 \\ v_3 \\ v_4 \\ \vdots \\ \vdots \\ v_{I-2} \\ v_{I-1} \\ v_I \end{pmatrix}, \quad b = \begin{pmatrix} u_0 + \bar{f}_2 \\ \bar{f}_3 \\ \bar{f}_4 \\ \vdots \\ \vdots \\ \bar{f}_{I-2} \\ \bar{f}_{I-1} \\ \bar{f}_I \end{pmatrix}$$

Then, by the Gaussian elimination on the augmented matrix of A, one has

$$\tilde{A} = \begin{pmatrix} 0 & 1 & 0 & 0 & \cdots & & & & u_0 + \bar{f}_2 \\ -1 & 0 & 1 & 0 & \cdots & & & & \bar{f}_3 \\ 0 & -1 & 0 & 1 & \cdots & & & & \bar{f}_4 \\ \vdots & \vdots & \vdots & \ddots & & & & & \\ & & & & \ddots & \vdots & \vdots & \vdots & \vdots \\ & & & \cdots & -1 & 0 & 1 & 0 & \bar{f}_{I-2} \\ & & & \cdots & 0 & -1 & 0 & 1 & \bar{f}_{I-1} \\ & & & \cdots & 0 & 0 & -1 & 1 & \bar{f}_I \end{pmatrix} \tag{6.10}$$

$$\rightarrow \begin{pmatrix} -1 & 0 & 0 & 0 & \cdots & & & 2 & u_0 + \sum_{k=2}^{I} \bar{f}_k \\ -1 & -1 & 0 & 0 & \cdots & & & 2 & \sum_{k=3}^{I} \bar{f}_k \\ 0 & -1 & -1 & 0 & \cdots & & & 2 & \sum_{k=4}^{I} \bar{f}_k \\ \vdots & \vdots & \vdots & \ddots & & & & & \\ & & & & \ddots & \vdots & \vdots & \vdots & \vdots \\ & & & \cdots & -1 & -1 & 0 & 2 & \sum_{k=I-2}^{I} \bar{f}_k \\ & & & \cdots & 0 & -1 & -1 & 2 & \sum_{k=I-1}^{I} \bar{f}_k \\ & & & \cdots & 0 & 0 & -1 & 1 & \sum_{k=I}^{I} \bar{f}_k \end{pmatrix}$$

Here, $\tilde{A}$ is the augmented matrix, the right arrow means the Gaussian elimination operation, and

$$[\bar{f}_2, \bar{f}_3, \cdots, \bar{f}_{I-1}, \bar{f}_I] = [2\Delta x f_2, 2\Delta x f_3, \cdots, 2\Delta x f_{I-1}, \Delta x f_I] \tag{6.11}$$

When the total number of grid nodes, I, is even, a continuation of Gauss elimination from the aforementioned results in

$$\tilde{A} \to \left(\begin{array}{cccccccccc}
-1 & 0 & 0 & 0 & \cdots & & & 2 & & u_0 + \sum_{k=2}^{I} \bar{f}_k \\
0 & -1 & 0 & 0 & \cdots & & & 0 & & -u_0 - \bar{f}_2 \\
0 & 0 & -1 & 0 & \cdots & & & 2 & & u_0 + \bar{f}_2 + \sum_{k=4}^{I} \bar{f}_k \\
\vdots & \vdots & \vdots & \ddots & & & & & & \\
& & & & \ddots & \vdots & \vdots & \vdots & & \vdots \\
& & & \cdots & 0 & -1 & 0 & 2 & & u_0 + \sum_{k=2,4,\cdots}^{I-4} \bar{f}_k + \sum_{k=I-2}^{I} \bar{f}_k \\
& & & \cdots & 0 & 0 & -1 & 0 & & -u_0 - \sum_{k=2,4,\cdots}^{I-2} \bar{f}_k \\
& & & \cdots & 0 & 0 & 0 & -1 & & -u_0 - \sum_{k=2,4,\cdots}^{I} \bar{f}_k
\end{array}\right)$$

$$\to \left(\begin{array}{cccccccccc}
-1 & 0 & 0 & 0 & \cdots & & & 0 & & -u_0 - \sum_{k=2,4,\cdots}^{I} \bar{f}_k + \sum_{k=3,5,\cdots}^{I-1} \bar{f}_k \\
0 & -1 & 0 & 0 & \cdots & & & 0 & & -u_0 - \bar{f}_2 \\
0 & 0 & -1 & 0 & \cdots & & & 0 & & -u_0 - \sum_{k=2,4,\cdots}^{I} \bar{f}_k + \sum_{k=5,7,\cdots}^{I-1} \bar{f}_k \\
\vdots & \vdots & \vdots & \ddots & & & & & & \\
& & & & \ddots & \vdots & \vdots & \vdots & & \vdots \\
& & & \cdots & 0 & -1 & 0 & 0 & & -u_0 - \sum_{k=2,4,\cdots}^{I} \bar{f}_k + \sum_{k=I-1}^{I-1} \bar{f}_k \\
& & & \cdots & 0 & & -1 & 0 & & -u_0 - \sum_{k=2,4,\cdots}^{I-2} \bar{f}_k \\
& & & \cdots & 0 & 0 & 0 & -1 & & -u_0 - \sum_{k=2,4,\cdots}^{I} \bar{f}_k
\end{array}\right)$$

The last matrix leads to the numerical solution for (6.8) ($i = 2, 3, \ldots, I$):

$$v_i = \begin{cases} u_0 + \sum_{k=2,4,\cdots}^{i-1} \bar{f}_k, & i - odd \\ u_0 + \sum_{k=2,4,\cdots}^{I_e} \bar{f}_k - \sum_{k=i+1,i+3,\cdots}^{I_o} \bar{f}_k, & i - even \end{cases} \tag{6.12}$$

Here, I_o and I_e are respectively the largest even and odd integers no larger than I. With a similar derivation, it is known that when I is odd, the solution is (6.12), too.

Now let us examine numerical solution (6.12). This solution has two branches or two modes, one at the odd nodes and the other at the even nodes. In other words, the solution is a dual-mode solution. The two modes are distinct in expression, and, in general, they are different in values. The presence of the dual-mode solution at the odd and even nodes implies that the solution may exhibit odd–even oscillation. This is stated as follows.

Proposition 6.1.1 *(1) A necessary and sufficient condition for numerical solution (6.8) to exhibit odd–even oscillation is*

$$R_i(R_i + \bar{f}_{II} - \bar{f}_{II+1}) > 0,\ II = i-1, i \tag{6.13}$$

in which

$$R_i = \begin{cases} 2\sum\limits_{k=i+1,i+3,\cdots}^{I_e} \bar{f}_k - 2\sum\limits_{k=i,i+2,\cdots}^{I_o} \bar{f}_k + \bar{f}_i, & i-odd \\ -2\sum\limits_{k=i,i+2,\cdots}^{I_e} \bar{f}_k + 2\sum\limits_{k=i+1,i+3,\cdots}^{I_o} \bar{f}_k + \bar{f}_i, & i-even \end{cases} \tag{6.14}$$

(2) The concaveness of the numerical solution tends to zero as the grid spacing goes to zero, i.e., $\lim\limits_{\Delta x \to 0} e_i = 0$.

Proof At the grid node with an odd index i, application of solution (6.12) yields

$$\begin{aligned}
v_{i-2} &= u_0 + \sum_{k=2,4,\cdots}^{i-3} \bar{f}_k \\
v_{i-1} &= u_0 + \sum_{k=2,4,\cdots}^{I_e} \bar{f}_k - \sum_{k=i,i+2,\cdots}^{I_o} \bar{f}_k \\
v_i &= u_0 + \sum_{k=2,4,\cdots}^{i-1} \bar{f}_k \\
v_{i+1} &= u_0 + \sum_{k=2,4,\cdots}^{I_e} \bar{f}_k - \sum_{k=i+2,i+4,\cdots}^{I_o} \bar{f}_k \\
v_{i+2} &= u_0 + \sum_{k=2,4,\cdots}^{i+1} \bar{f}_k.
\end{aligned}$$

These give rise to

$$\begin{aligned}
e_{i-1} &= -2\sum_{k=i-1,i+1,\cdots}^{I_e} \bar{f}_k + 2\sum_{k=i,i+2,\cdots}^{I_o} \bar{f}_k + \bar{f}_{i-1} \\
e_i &= 2\sum_{k=i+1,i+3,\cdots}^{I_e} \bar{f}_k - 2\sum_{k=i,i+2,\cdots}^{I_o} \bar{f}_k + \bar{f}_i \\
e_{i+1} &= -2\sum_{k=i+1,i+3,\cdots}^{I_e} \bar{f}_k + 2\sum_{k=i+2,i+4,\cdots}^{I_o} \bar{f}_k + \bar{f}_{i+1}
\end{aligned}$$

From the above and (6.14), we see that R_i equals e_i and

$$e_{i-1} = -(e_i + \bar{f}_{i-1} - \bar{f}_i)$$
$$e_{i+1} = -(e_i + \bar{f}_i - \bar{f}_{i+1}))$$

Therefore, condition (6.13) leads to

$$e_i e_{i-1} < 0, \quad e_i e_{i+1} < 0$$

which is actually the same as (6.7). Similarly, we derive the following at the grid node with an even index i:

$$e_{i-1} = 2 \sum_{k=i,i+2,\cdots}^{I_e} \bar{f}_k - 2 \sum_{k=i+1,i+3,\cdots}^{I_o} \bar{f}_k - \bar{f}_{i-1}$$
$$e_i = -2 \sum_{k=i,i+2,\cdots}^{I_e} \bar{f}_k + 2 \sum_{k=i+1,i+3,\cdots}^{I_o} \bar{f}_k + \bar{f}_i$$
$$e_{i+1} = 2 \sum_{k=i+2,i+4,\cdots}^{I_e} \bar{f}_k - 2 \sum_{k=i+3,i+5,\cdots}^{I_o} \bar{f}_k - \bar{f}_{i+1}$$

With the aforementioned, it is readily shown that condition (6.13) is also the same as (6.7). In summary, condition (6.13) is equivalent to condition (6.7). According to Definition 6.1.2, it is concluded that condition (6.13) is a necessary and sufficient condition for the numerical solution to exhibit odd–even oscillation. This proves (1) in the proposition.

The proof for (2) in the proposition is straightforward. As shown earlier, $e_i = R_i$. Additionally, from (6.14) and (6.11), it is seen that $e_i = \mathcal{O}(\Delta x)$, which leads to (2) in the proposition. This completes the proof of Proposition 6.1.1. □

In discretization (6.8), the solution value at odd nodes and that at even nodes are not completely decoupled but related to each other by the backward difference at the right end of the domain. Therefore, the reason for the presence of odd–even oscillation in Proposition 6.1.1 is not same as the "decoupling" revealed in the literature [6]. Instead, such oscillation is attributed to dual modes in the solution because of this clear fact: solution (6.12) has two modes, or two patterns, one at odd nodes and the other at even nodes, and the two modes are different in value in general. It is the dual-mode behavior of the solution that leads to the zig-zag form. Such oscillation is called the dual-mode odd–even oscillation.

Example 6.1.1 This example illustrates Proposition 6.1.1, particularly, its oscillation criterion for the numerical solution to (6.8), i.e., numerical solution (6.12). With two functions of $f(x)$, a direct calculation according to (6.6), (6.12), and (6.13) leads to Table 6.1. As an illustration, for $f(x) = f^1(x)$ and at node $i = 3$, the calculation of (6.6) and (6.12) yields $e_3 = -35.6 < 0$, so e_3 is assigned with "−." In addition, for this node, the computed values for e_3 and $\bar{f}_2$, $\bar{f}_3$, and $\bar{f}_4$ result in $e_3(e_3 + \bar{f}_2 - \bar{f}_3) = 1311.4 > 0$, $e_3(e_3 + \bar{f}_3 - \bar{f}_4) = 1336.2 > 0$. Thus condition (6.13) is satisfied, and node $i = 3$ is assigned with "✓" in the table. To further illustrate the proposition, the solution to (6.8) is also computed numerically, and it is plotted in Fig. 6.1.

Table 6.1 Oscillation of solution (6.12) due to dual modes. $L = 1$. $f^1(x) = 5e^{5x} - 1$, $I = 11$. $f^2(x) = 2(x - 0.5)$, $I = 11$. "+," "−," and "zero" mean that the value of e_i is positive, negative, and zero, respectively. "✓," "×," and "n/a" indicate that condition (6.13) is satisfied, violated, and inapplicable (e.g., at ends of subdomains, $x = 0, L$), respectively

	i	1	2	3	4	5	6	7	8	9	10	11
$f^1(x)$	e_i	n/a	+	−	+	−	+	−	+	−	+	n/a
	(6.13)	n/a	n/a	✓	✓	✓	✓	✓	✓	✓	n/a	n/a
$f^2(x)$	e_i	n/a	+	0	+	0	+	0	+	0	+	n/a
	(6.13)	n/a	n/a	×	×	×	×	×	×	×	n/a	n/a

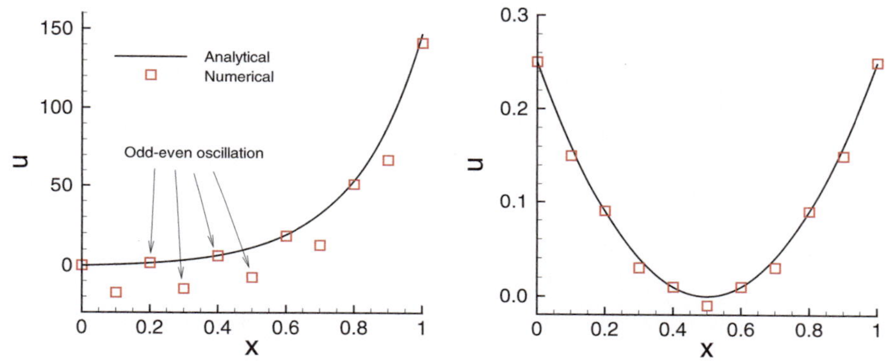

(a) $f^1(x)$ (b) $f^2(x)$

Fig. 6.1 Visualization of oscillation in solution (6.12) due to dual modes

The table shows that, for $f^1(x)$, condition (6.13) is true for all grid nodes (except the two end nodes of the domain). According to Proposition 6.1.1, solution (6.12) exhibits odd–even oscillation. This is validated by direct computation; e_i takes "+" and "−" alternately at these nodes. By Definition 6.1.2, it is known that the numerical solution presents odd-even oscillation, and this is illustrated in Fig. 6.1a. In contrast, for $f^2(x)$, condition (6.13) is violated at all grid nodes (except the two end nodes of the domain). Then, the proposition concludes that no oscillation will be present. Also, this is visualized in Fig. 6.1b.

Solution with Inconsistent Boundary Conditions Now, we examine another cause for the node-to-node oscillation. Introducing a modification to discretization (6.4) by imposing a boundary value at $x = L$, one has

$$\begin{cases} \hat{v}_{i+1} - \hat{v}_{i-1} = 0, \quad i = 2, \cdots, I - 1 \\ \hat{v}_1 = u_0, \quad \hat{v}_I = u_l \end{cases} \tag{6.15}$$

in which $u_l = u_L - (\Delta x/2) \sum_{k=2}^{I} (f_{k-1} + f_k)$, u_L being a prescribed value, and I is an even number. The modification transfers initial value problem (6.3) to a boundary value problem, which is not well-posed because of imposing the extra boundary condition at $x = x_I$. However, such a treatment is frequently adopted, such as in the computation of the continuity equation in Sect. 6.1.4.

One can solve the system (6.15) from its augmented matrix:

$$\tilde{A} = \begin{pmatrix} 0 & 1 & 0 & 0 & \cdots & & & & u_0 \\ -1 & 0 & 1 & 0 & \cdots & & & & 0 \\ 0 & -1 & 0 & 1 & \cdots & & & & 0 \\ \vdots & \vdots & \vdots & \ddots & & & & & \\ & & & & \ddots & \vdots & \vdots & \vdots & \vdots \\ & & & \cdots & -1 & 0 & 1 & 0 & 0 \\ & & & \cdots & 0 & -1 & 0 & 1 & 0 \\ & & & \cdots & 0 & 0 & -1 & 0 & -u_l \end{pmatrix}$$

As I is even, one has

$$\tilde{A} \rightarrow \begin{pmatrix} -1 & 0 & 0 & 0 & \cdots & & & 1 & u_0 - u_l \\ 0 & -1 & 0 & 0 & \cdots & & & 0 & -u_0 \\ 0 & 0 & -1 & 0 & \cdots & & & 1 & u_0 - u_l \\ \vdots & \vdots & \vdots & \ddots & & & & & \\ & & & & \ddots & \vdots & \vdots & \vdots & \vdots \\ & & & \cdots & 0 & -1 & 0 & 0 & -u_0 \\ & & & \cdots & 0 & -1 & -1 & 1 & -u_l \\ & & & \cdots & 0 & 0 & -1 & 0 & -u_l \end{pmatrix}$$

which leads to

$$\hat{v}_i = \begin{cases} u_0, & i - odd \\ u_l, & i - even \end{cases} \tag{6.16}$$

where $i = 1, 2, \cdots, I$. Utilizing the trapezoid rule to approximate the integral and rearranging solution (6.16), one obtains

$$v_i = \begin{cases} u_0 + \dfrac{\Delta x}{2} \sum_{k=2}^{i} (f_{k-1} + f_k), & i = 1, 3, \cdots, I-1 \\ u_L - \dfrac{\Delta x}{2} \sum_{k=i+1}^{I} (f_{k-1} + f_k), & i = 2, 4, \cdots, I \end{cases} \tag{6.17}$$

Here, I is even. The following is the conclusion for the solution.

Proposition 6.1.2 *A necessary and sufficient condition for solution (6.17) to exhibit odd–even oscillation is*

$$S_i\left(S_i+\frac{\Delta x}{2}\langle f_{II-2}+f_{II-1}-f_{II}-f_{II+1}\rangle\right)>0,\ II=i,i+1 \tag{6.18}$$

in which

$$S_i=\begin{cases}2(u_l-u_0)+\dfrac{\Delta x}{2}(f_{i+1}-f_{i-1}), & i-odd\\ 2(u_0-u_l)+\dfrac{\Delta x}{2}(f_{i+1}-f_{i-1}), & i-even\end{cases}$$

Proof From the second branch of solution (6.17), and notice that $u_l = u_L - (\Delta x/2)\sum_{k=2}^{I}(f_{k-1}+f_k)$, we have

$$v_i = u_l + \frac{\Delta x}{2}\sum_{k=2}^{i}(f_{k-1}+f_k)$$

for even i. Now, let i be odd. With the aid of the above and the first branch of solution (6.17), one has

$$\begin{aligned}e_{i-1}&=-2(u_l-u_0)+\frac{\Delta x}{2}(f_i-f_{i-2})\\ e_i&=2(u_l-u_0)+\frac{\Delta x}{2}(f_{i+1}-f_{i-1})\\ e_{i+1}&=-2(u_l-u_0)+\frac{\Delta x}{2}(f_{i+2}-f_i)\end{aligned}$$

Since S_i equals e_i and

$$\begin{aligned}e_{i-1}&=-\left(e_i+\frac{\Delta x}{2}(f_{i-2}+f_{i-1}-f_i-f_{i+1})\right)\\ e_{i+1}&=-\left(e_i+\frac{\Delta x}{2}(f_{i-1}+f_i-f_{i+1}-f_{i+2})\right)\end{aligned}$$

condition (6.18) is equivalent to inequality (6.7). This indicates that solution (6.17) fluctuates at the node with an odd index i. Similarly, we can prove that (6.17) has the same behavior at the node with an even index i. This completes the proof of Proposition 6.1.2. □

Discretization (6.15) exhibits a complete decoupling at odd and even nodes. It has two boundary conditions, one at $x = 0$ and the other at $x = x_L$. Therefore, precisely speaking, the odd–even oscillation stated in Proposition 6.1.2 results from inconsistent-boundary-conditions, particularly the inconsistency of the two boundary conditions at the two ends of the computational domain. Such oscillation is called the inconsistent-boundary-conditions odd–even oscillation. Actually, in (6.15), if the boundary condition $\hat{v}_I = u_l$ is imposed in a consistent manner with

$$u_L = u_0 + \frac{\Delta x}{2} \sum_{k=2}^{I} (f_{k-1} + f_k)$$

then, solution (6.17) becomes solution (6.5), which contains no fluctuation. Moreover, as seen in (6.16), such oscillation, in its transform $\bar{v}$, occurs alternately at odd and even grid nodes between mode u_0 and mode u_l. Distinct from the previous type of oscillation due to the dual-mode pattern, the oscillation in $\bar{v}$ does not decay with grid spacing in the sense of concaveness, as long as there is a difference in boundary values u_0 and u_l.

Example 6.1.2 Table 6.2 and Fig. 6.2 present numerical examples of the oscillation criterion of Proposition 6.1.2. In these examples, u_L is set to the value of analytical solution (6.2). As seen in Table 6.2, condition (6.18) reveals whether oscillations take place or not, and Fig. 6.2 visualizes this. It is worth noting that this example demonstrates that oscillations may still occur even when u_L is set to the analytical solution.

Table 6.2 Oscillation in solution (6.17) due to inconsistent boundary conditions. $L = 1$. $f^2(x) = 2(x - 0.5)$, $I = 10$. $f^3(x) = -1/x^2$, $I = 9$

	i	1	2	3	4	5	6	7	8	9	10
$f^2(x)$	e_i	n/a	+	+	+	+	+	+	+	+	n/a
	(6.18)	n/a	n/a	×	×	×	×	×	×	n/a	n/a
$f^3(x)$	e_i	n/a	+	−	+	−	+	−	+	n/a	
	(6.18)	n/a	n/a	✓	✓	✓	✓	✓	n/a	n/a	

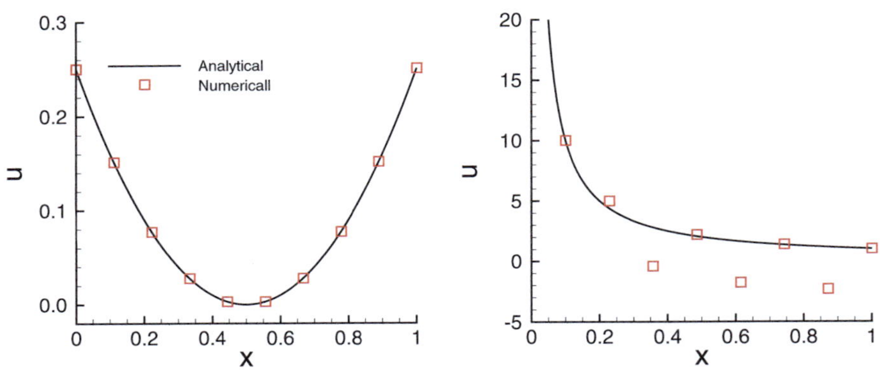

(a) $f^2(x)$ (b) $f^3(x)$

Fig. 6.2 Visualization of oscillation in solution (6.17) due to inconsistent boundary conditions

6.1.4 Oscillation on Multiblock Grids

Oscillation Due to Dual-Mode Solution Now, let us proceed to the oscillation on two grids. Consider the computation of problem (6.1) on two subdomains:

$$\begin{cases} u_{1x} = f(x), \ x < x_1 \\ u_1 = u_0, \quad\quad x = 0 \\ u_1 = u_2, \quad\quad x = x_1 \end{cases}$$
$$\begin{cases} u_{2x} = f(x), \ x > x_2 \\ u_2 = u_1, \quad\quad x = x_2 \end{cases} \tag{6.19}$$

The grids of the two subdomains overlap, as shown in Fig. 6.3. In the figure, K is the grid node in subdomain 1 that overlaps with node $I_1 + 1$ in subdomain 2, and $I_1 + K'$ is the grid node in subdomain 2 that overlaps with node I_1 in subdomain 1. I_1 and $I_1 + 1$ are the interface nodes in subdomains 1 and 2, respectively, corresponding to the points x_1 and x_2, respectively, in (6.19). Let v_i $(1 \le i \le I_1)$ be the numerical solution for u_1 in the grid of subdomain 1 and v_i $(I_1 + 1 \le i \le I_1 + I_2)$ be the numerical solution for u_2 in the grid of subdomain 2. The multiblock computation is formulated as

$$\begin{cases} Subdomain\ 1: \\ v_{i+1} - v_{i-1} = \bar{f}_i, \ i = 2, 3, \cdots, I_1 - 1 \\ v_1 = u_0, \ \ v_{I_1} = v_{I_1+K'} \\ Subdomain\ 2: \\ v_{I_1+i+1} - v_{I_1+i-1} = \bar{f}_{I_1+i}, \ i = 2, 3, \cdots, I_2 - 1 \\ v_{I_1+1} = v_K, \ \ v_{I_1+I_2} - v_{I_1+I_2-1} = \bar{f}_{I_1+I_2} \end{cases} \tag{6.20}$$

and

$$\begin{aligned} &[\bar{f}_2, \bar{f}_3, \cdots, \bar{f}_{I_1-1}, \bar{f}_{I_1+2}, \cdots, \bar{f}_{I_1+I_2}] \\ &= [2\Delta x_1 f_2, 2\Delta x_1 f_3, \cdots, 2\Delta x_1 f_{I_1-1}, 2\Delta x_2 f_{I_1+2}, \cdots, \Delta x_2 f_{I_1+I_2}] \end{aligned}$$

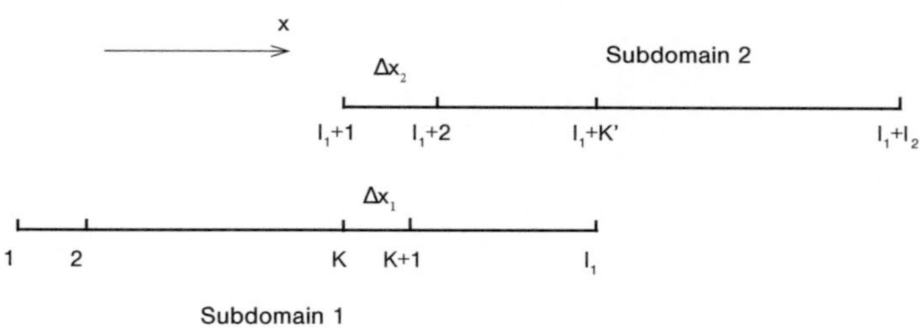

Fig. 6.3 Grid layout for multiblock computation

Here, Δx_1 and Δx_2 are the grid spacings in subdomains 1 and 2, respectively. The solutions in the two subdomains are coupled by a Dirichlet condition at I_1 and another at $I_1 + 1$.

Consider that the two subdomains have even numbers of grid nodes in the overlapping region; that is, $I_1 - K + 1$ and K' are even. Then, multiblock problem (6.20) can be transformed into a linear algebra system of size $(I_1 + I_2 - 1) \times (I_1 + I_2 - 1)$, in the form of (6.9) with an augmented matrix as

$$
\tilde{A} = \left(\begin{array}{cccccccccccccccccr}
0 & 1 & 0 & \cdots & & & & & & & & & & & & & u_0 + \bar{f}_2 \\
-1 & 0 & 1 & \cdots & & & & & & & & & & & & & \bar{f}_3 \\
\vdots & \vdots & \ddots & & & & & & & & & & & & & & \vdots \\
& & & & \ddots & \vdots & \vdots & \vdots & & & & & & & & & \\
& & & \cdots & -1 & 0 & 1 & 0 & & & & & & & & & \bar{f}_{I_1-2} \\
& & & \cdots & 0 & -1 & 0 & 1 & & & & & & & & & \bar{f}_{I_1-1} \\
& & & \cdots & 0 & 0 & 0 & 1 & & & & & -1 & & & & 0 \\
& & & -1 & & & & & 1 & 0 & 0 & 0 & \cdots & & & & 0 \\
& & & & & & & & -1 & 0 & 1 & 0 & \cdots & & & & \bar{f}_{I_1+2} \\
& & & & & & & & 0 & -1 & 0 & 1 & \cdots & & & & \bar{f}_{I_1+3} \\
& & & & & & & & \vdots & \vdots & \vdots & \ddots & & & & & \\
& & & & & & & & & & & & & \ddots & \vdots & \vdots & \vdots \\
& & & & & & & & & & & & \cdots & -1 & 0 & 1 & \bar{f}_{I_1+I_2-1} \\
& & & & & & & & & & & & \cdots & 0 & -1 & 1 & \bar{f}_{I_1+I_2}
\end{array}\right)
$$

Since $I_1 - K + 1$ and K' are even, we have

$$
\tilde{A} \rightarrow \left(\begin{array}{ccccccccccccccr}
0 & 1 & \cdots & & & & & & & & & & & & u_0 + \bar{f}_2 \\
-1 & 0 & \cdots & & & & & & & & & & & & \bar{f}_3 \\
\vdots & \vdots & & & & & & & & & & & & & \vdots \\
& & & \ddots & \vdots & \vdots & \vdots & & & & & & & & \\
& & \cdots & -1 & 0 & 1 & 0 & & & & & & & & \bar{f}_{I_1-2} \\
& & \cdots & 0 & -1 & 0 & 1 & & & & & & & & \bar{f}_{I_1-1} \\
& & \cdots & 0 & 0 & 0 & 1 & 0 & -1 & & & & & & -\bar{f}_{I_1+1} \\
& & & & & -1 & 0 & 1 & 0 & 0 & 0 & \cdots & & & \bar{f}_{I_1} \\
& & & & & & & -1 & 0 & 1 & 0 & \cdots & & & \bar{f}_{I_1+2} \\
& & & & & & & 0 & -1 & 0 & 1 & \cdots & & & \bar{f}_{I_1+3} \\
& & & & & & & \vdots & \vdots & \vdots & \ddots & & & & \\
& & & & & & & & & & & & \vdots & \vdots & \vdots \\
& & & & & & & & & & & \cdots & 0 & 1 & \bar{f}_{I_1+I_2-1} \\
& & & & & & & & & & & \cdots & -1 & 1 & \bar{f}_{I_1+I_2}
\end{array}\right)
$$

$$
\rightarrow \left(\begin{array}{ccccccccccccc}
0 & 1\cdots & & & & & & & & & & & u_0+\bar{f}_2 \\
-1 & 0\cdots & & & & & & & & & & & \bar{f}_3 \\
\vdots & \vdots & & & & & & & & & & & \vdots \\
 & & \ddots & \vdots & \vdots & \vdots & & & & & & & \\
 & & \cdots -1 & 0 & 1 & 0 & & & & & & & \bar{f}_{I_1-2} \\
 & & \cdots\ 0 & -1 & 0 & 1 & & & & & & & \bar{f}_{I_1-1} \\
 & & \cdots\ 0 & 0 & -1 & 0 & 1 & & & & & & \bar{f}_{I_1} \\
 & & & & & -1 & 0 & 1 & 0 & 0\cdots & & & \bar{f}_{I_1+1} \\
 & & & & & & -1 & 0 & 1 & 0\cdots & & & \bar{f}_{I_1+2} \\
 & & & & & & 0 & -1 & 0 & 1\cdots & & & \bar{f}_{I_1+3} \\
 & & & & & & \vdots & \vdots & \vdots\ddots & & & & \\
 & & & & & & & & & & \vdots & \vdots & \vdots \\
 & & & & & & & & & & \cdots\ 0 & 1 & \bar{f}_{I_1+I_2-1} \\
 & & & & & & & & & & \cdots -1 & 1 & \bar{f}_{I_1+I_2}
\end{array}\right) \tag{6.21}
$$

where

$$
\bar{f}_{I_1} = -\sum_{k=K+1,K+3,\cdots}^{I_1-2} \bar{f}_k, \quad \bar{f}_{I_1+1} = -\sum_{k=I_1+3,I_1+5,\cdots}^{I_1+K'-1} \bar{f}_k \tag{6.22}
$$

Matrix (6.21) is exactly the same as matrix (6.10) for the computation on a single grid, except that their last columns are different.

Then, with the same derivation from matrix (6.10) to solution (6.12) for a single-grid problem, a solution for the multiblock problem can be obtained from matrix (6.21) in the form of (6.12). Particularly, the solution in subdomain 1 is derived as ($i = 2, 3, \cdots, I_1$)

$$
v_i = \begin{cases} u_0 + \sum\limits_{\text{k=2,4,}\cdots}^{\text{i-1}} \bar{f}_k, & i - odd \\ u_0 + \sum\limits_{k=2,4,\cdots,\neq I_1, I_1+1}^{I_e^{12}} \bar{f}_k - \sum\limits_{k=i+1,i+3,\cdots,\neq I_1,I_1+1}^{I_o^{12}} \bar{f}_k & \\ \quad +(\delta(I_1) - \delta(I_1+1))(\bar{f}_{I_1} - \bar{f}_{I_1+1}), & i - even \end{cases} \tag{6.23}
$$

and the solution in subdomain 2 ($i = 1, 2, \cdots I_2$) is derived as

$$
v_{I_1+i} = \begin{cases} u_0 + \sum\limits_{k=2,4,\cdots,\neq I_1,I_1+1}^{i+I_1-1} \bar{f}_k & \\ \quad +\delta(I_1)\bar{f}_{I_1} + \delta(I_1+1)\bar{f}_{I_1+1}, & i + I_1 - odd \\ u_0 + \sum\limits_{k=2,4,\cdots,\neq I_1,I_1+1}^{I_e^{12}} \bar{f}_k - \sum\limits_{k=i+I+1,i+I_1+3,\cdots}^{I_o^{12}} \bar{f}_k & \\ \quad +\delta(I_1)\bar{f}_{I_1} + \delta(I_1+1)\bar{f}_{I_1+1}, & i + I_1 - even \end{cases} \tag{6.24}
$$

Here, I_o^{12} and I_e^{12} are respectively the largest odd and even integers no larger than $I_1 + I_2$. In addition, $\delta(i) = 0, i - odd; \ = 1, i - even$. Equations (6.23) and (6.24) show that, similar to that for single domain problem (6.8), the solution for two-subdomain problem (6.20) has a dual-mode pattern. Thus, it is anticipated that its solution may present oscillation, which is concluded in the following proposition.

Proposition 6.1.3 *(1) A necessary and sufficient condition for solution (6.23) and (6.24) to problem (6.20) to exhibit the odd–even oscillation is*

$$T_i(T_i + \bar{f}_{II} - \bar{f}_{II+1}) > 0, \ II = i - 1, i \tag{6.25}$$

in which

$$T_i = \begin{cases} 2 \sum\limits_{n=i+1,i+3,\cdots}^{I_e^{12}} \bar{f}_k - 2 \sum\limits_{k=i,i+2,\cdots}^{I_o^{12}} \bar{f}_k + \bar{f}_i, & i - odd \\ -2 \sum\limits_{k=i,i+2,\cdots}^{I_e^{12}} \bar{f}_k + 2 \sum\limits_{k=i+1,i+3,\cdots}^{I_o^{12}} \bar{f}_k + \bar{f}_i, & i - even \end{cases} \tag{6.26}$$

(2) In both subdomains 1 and 2, the concaveness tends to zero as the grid spacing goes to zero, i.e., $\lim\limits_{\Delta x_1, \Delta x_2 \to 0} e_i = 0$.

Proof The proof is the same as in the last subsection for the single grid situation. □

Since solution (6.23) and (6.24) is derived with the method for the solution on a single grid, i.e., (6.12), the odd–even oscillation in the former solution may be viewed as a manifestation of that inherent in the latter solution. However, adopting multiple grids plays a role in the former solution. As seen in (6.23) and (6.24), the multiblock effects come from the two terms in (6.22), which depend on the location of the interfaces of the two subdomains, the size of their overlapping region, etc.

Example 6.1.3 To illustrate the proposition, the odd–even oscillation in solution (6.23) and (6.24) is shown in two examples in Table 6.3 and Fig. 6.4. Both the table and the figure show that condition (6.25) predicts the oscillation. Note that in these examples, the numbers of grid nodes of the two subdomains are even in their overlapping region.

Oscillation Due to Inconsistent Boundary Conditions Let us compute problem (6.19) via a discretization modified from (6.20), as follows:

$$\begin{cases} Subdomain\ 1: \\ \hat{v}_{i+1} - \hat{v}_{i-1} = 0, \ i = 2, 3, \cdots, I_1 - 1 \\ \hat{v}_1 = u_0, \ \ \hat{v}_{I_1} = \hat{v}_{I_1+K'} \\ Subdomain\ 2: \\ \hat{v}_{I_1+i+1} - \hat{v}_{I_1+i-1} = 0, \ i = 2, 3, \cdots, I_2 - 1 \\ \hat{v}_{I_1+1} = \hat{v}_K, \ \ \hat{v}_{I_1+I_2} = u_l \end{cases} \tag{6.27}$$

Table 6.3 Oscillation in solution (6.23) and (6.24) due to the dual-mode pattern. $L = 1$. $f^1(x) = 5e^{5x} - 1$. $f^4(x) = -0.9/(1 - 0.9x)$. $I_1 = 10$, $I_2 = 10$, $K = 5$, and $K' = 8$

Subdomain 1	i	1	2	3	4	5	6	7	8	9	10
$f^1(x)$	e_i	n/a	−	+	−	+	−	+	−	+	n/a
	(6.25)	n/a	n/a	✓	✓	✓	✓	✓	✓	n/a	n/a
$f^4(x)$	e_i	n/a	+	−	+	−	+	−	+	−	n/a
	(6.25)	n/a	n/a	✓	✓	✓	✓	✓	✓	n/a	n/a
Subdomain 2	i	1	2	3	4	5	6	7	8	9	10
$f^1(x)$	e_i	n/a	−	+	−	+	−	+	−	+	n/a
	(6.25)	n/a	n/a	✓	✓	✓	✓	✓	✓	n/a	n/a
$f^4(x)$	e_i	n/a	+	−	+	−	+	−	+	−	n/a
	(6.25)	n/a	n/a	✓	✓	✓	✓	✓	✓	n/a	n/a

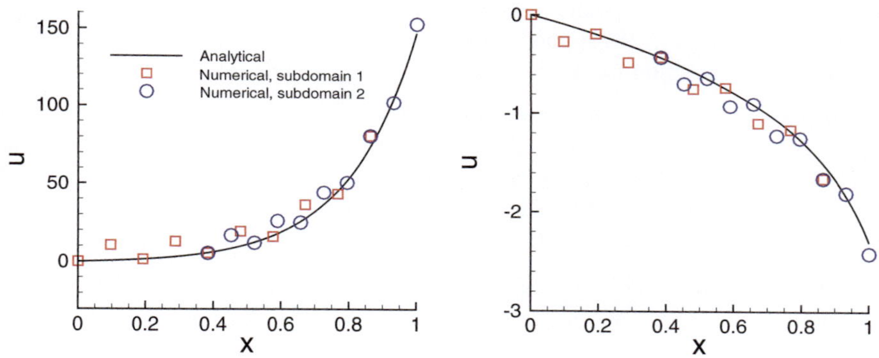

(a) $f^1(x)$ (b) $f^4(x)$

Fig. 6.4 Visualization of oscillation in solution (6.23) and (6.24) due to dual modes

The modification is to add a Dirichlet condition at node $I_1 + I_2$: $\hat{v}_{I_1+I_2} = u_l$, with $u_l = u_L - (\Delta x/2) \sum_{k=2}^{I_1+I_2} (f_{k-1} + f_k)$, u_L being a prescribed value. Furthermore, consider the situation in which the two grids have an even number of grid nodes and also an even number of grid nodes in their overlapping region. It is noted that the added condition at node $I_1 + I_2$ is redundant. However, such treatment happens in practical calculations.

With the steps to derive matrix (6.21) and the solutions (6.23) and (6.24), the solution to the aforementioned computation is derived as

$$v_i = \begin{cases} u_0 + \dfrac{\Delta x_1}{2} \sum_{k=2}^{i} (f_{k-1} + f_k), & i - odd \\ u_L - \dfrac{\Delta x_1}{2} \sum_{k=i+1}^{I_1} (f_{k-1} + f_k) \\ \quad - \dfrac{\Delta x_2}{2} \sum_{k=I_1+K'+1}^{I_1+I_2} (f_{k-1} + f_k), & i - even \end{cases} \tag{6.28}$$

in subdomain 1 ($i = 2, 3, \ldots, I_1$), and

$$v_{I_1+i} = \begin{cases} u_0 + \dfrac{\Delta x_1}{2} \sum_{k=2}^{K} (f_{k-1} + f_k) \\ \quad + \dfrac{\Delta x_2}{2} \sum_{k=I_1+2}^{I_1+i} (f_{k-1} + f_k), & I_1 + i - odd \\ u_L - \dfrac{\Delta x_2}{2} \sum_{k=I_1+i+1}^{I_1+I_2} (f_{k-1} + f_k), & I_1 + i - even \end{cases} \tag{6.29}$$

in subdomain 2 ($i = 1, 2, \ldots, I_2$).

Proposition 6.1.4 *Condition (6.18) is a necessary and sufficient condition for solution (6.28) and (6.29) to present odd–even oscillation.*

Proof Because the computation in each subdomain of (6.27) is of the same type as (6.15), condition (6.18) applies to the solution of each subdomain as a necessary and sufficient condition for the presence of oscillation. □

Example 6.1.4 To illustrate Proposition 6.1.4, numerical examples are given in Table 6.4 and Fig. 6.5. Again, The boundary value u_L is set to the value of analytical solution (6.2) (Fig. 6.5a). Additionally, it is set to a perturbed value of the solution, $1.1u_L$ (Fig. 6.5b). The examples show that the results concerning grid-scale fluctuation in the table match exactly those predicted by the proposition. □

Application to a Model Flow Problem Consider a steady, 1D, inviscid flow problem:

$$\begin{cases} u_x = f(x), \quad uu_x = -p_x, \; 0 < x \leq L \\ u = u_0, \; p = p_0, \; x = 0 \end{cases} \tag{6.30}$$

Here u is the flow velocity, p is the pressure, and $f(x)$ reflects an inflow from the lateral side. The first and the second ordinary differential equations in the problem

Table 6.4 Oscillation in solution (6.28) and (6.29) due to inconsistent boundary conditions. $L = 1$. $f^2(x) = 2(x - 0.5)$, $I_1 = 10$, $I_2 = 10$, $K = 5$, and $K' = 8$

Subdomain 1	node i	1	2	3	4	5	6	7	8	9	10
$f^2(x)$	e_i	n/a	+	+	+	+	+	+	+	+	n/a
	(6.18)	n/a	n/a	×	×	×	×	×	×	n/a	n/a
$f^2(x)$	e_i	n/a	−	+	−	+	−	+	−	+	n/a
with $1.1u_L$	(6.18)	n/a	n/a	✓	✓	✓	✓	✓	✓	n/a	n/a
Subdomain 2	node i	1	2	3	4	5	6	7	8	9	10
$f^2(x)$	e_i	n/a	+	+	+	+	+	+	+	+	n/a
	(6.18)	n/a	n/a	×	×	×	×	×	×	n/a	n/a
$f^2(x)$	e_i	n/a	−	+	−	+	−	+	−	+	n/a
with $1.1u_L$	(6.18)	n/a	n/a	✓	✓	✓	✓	✓	✓	n/a	n/a

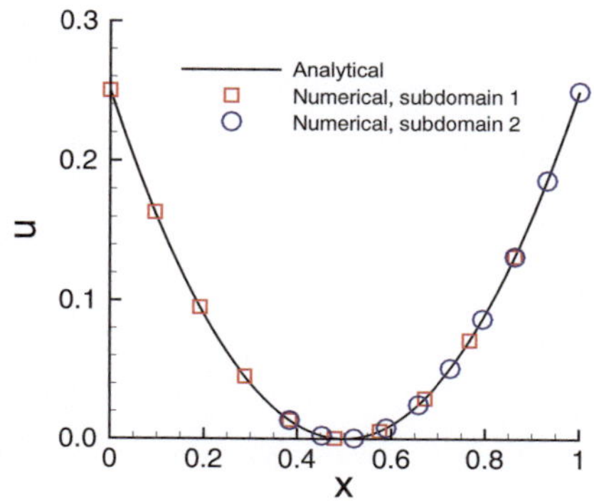

(a) $f^2(x)$

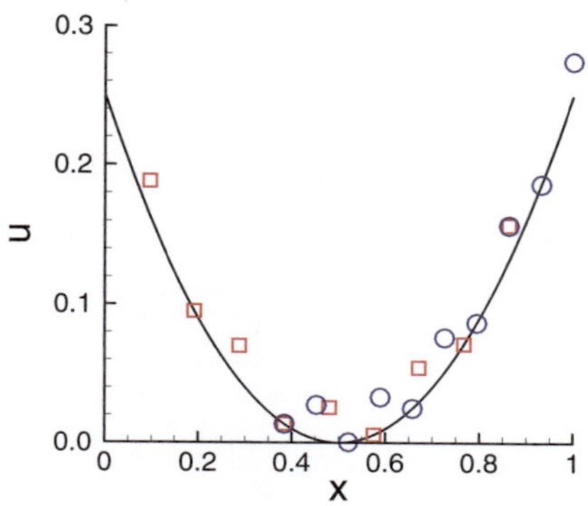

(b) $f^2(x)$, with $1.1u_L$

Fig. 6.5 Visualization of oscillation in solution (6.28) and (6.29) due to inconsistent boundary conditions

are simplified from the continuity and momentum equations, respectively, of fluid flows.

Now let us compute problem (6.30) using a multiblock algorithm:

$$\begin{cases} Subdomain\ 1: \\ v_{i+1} - v_{i-1} = \bar{f}_i,\ q_{i+1} - q_{i-1} = -({v_{i+1}}^2 - {v_{i-1}}^2)/2,\ \ i = 2, 3, \cdots, I_1 - 1 \\ v_1 = u_0,\ q_1 = p_0 \\ v_{I_1} = v_{I_1+K'},\ q_{I_1} = q_{I_1+K'} \\ Subdomain\ 2: \\ v_{I_1+i+1} - v_{I_1+i-1} \ = \bar{f}_{I_1+i} \\ q_{I_1+i+1} - q_{I_1+i-1} \ = -({v_{I_1+i+1}}^2 - {v_{I_1+i-1}}^2)/2,\ i = 2, 3, \cdots, I_2 - 1 \\ v_{I_1+1} = v_K,\ q_{I_1+1} = q_K \\ v_{I_1+I_2} - v_{I_1+I_2-1} \ = \bar{f}_{I_1+I_2},\ q_{I_1+I_2} - q_{I_1+I_2-1} = -({v_{I_1+I_2}}^2 - {v_{I_1+I_2-1}}^2)/2 \end{cases}$$

In the aforementioned, v and q represent the numerical solutions for u and p, respectively. Also, it is assumed that the numbers of the two grids' nodes, I_1 and I_2, are even, and the numbers of their grid nodes in the overlapping region are even. It is seen that discretization for the continuity equation is actually the same as discretization (6.20). Therefore, the solution to the continuity equation remains the same, i.e., (6.23) and (6.24), which may exhibit oscillation as discussed previously. Additionally, the discretization for the momentum equation is also similar to (6.20).

Noticing the Dirichlet conditions at the interfaces and

$$\sum_{k=K+1,K+3,\cdots}^{I_1\text{-}2} ({v_{k+1}}^2 - {v_{k-1}}^2) = {v_{I_1-1}}^2 - {v_K}^2, \quad subdomain\ 1$$
$$\sum_{k=I_1+3,I_1+5,\cdots}^{I_1+K'-1} ({v_{k+1}}^2 - {v_{k-1}}^2) = {v_{I_1+K'}}^2 - {v_{I_1+2}}^2, \quad subdomain\ 2$$

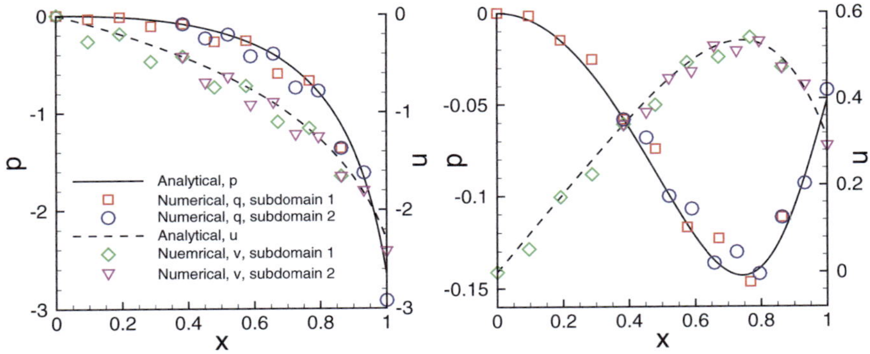

(a) $f^4(x)$ (b) $f^5(x)$

Fig. 6.6 Visualization of oscillation in the flow problem. $L = 1$. $f^4(x) = -0.9/(1 - 0.9x)$, $f^5(x) = 1 - 5x^4$, $I_1 = 10$, $I_2 = 10$, $K = 5$, and $K' = 6$

one can derive the solution for pressure. Following the procedure to derive solutions (6.23) and (6.24), after a tedious but straightforward calculation, one derives the solution for pressure as

$$q_i = p_0 - ({v_i}^2 - {u_0}^2)/2, \qquad subdomain 1,\ 2 \tag{6.31}$$

This solution indicates that the pattern of q_i depends on v_i, since u_0 and p_0 are constants with regard to i. Therefore, as v_i fluctuates, q_i will also fluctuate.

Example 6.1.5 With two functions for $f(x)$, a discussion can be made for the odd-even oscillation of v_i according to condition (6.25), and then that of q_i. The discussion is visualized in Fig. 6.6. The examples in the figure show that the solutions for both velocity and pressure exhibit the odd–even oscillation.

6.2 Coupling by Machine Learning

6.2.1 Motivation and Potential

As seen in the previous content of this book, we face difficulties in computing coupled PDEs and fluid flows via the multiblock method, despite significant progress on this topic. In the past, theoretical analysis has been mainly restricted to simplified problems, e.g., [19], while applied research deals with complex, realistic problems but lacks rigorous foundations, e.g., [20]. As a result, there is a substantial gap between theoretical and applied research, and difficulties arise in the computation of various coupling problems. These problems have been discussed and manifested by numerical examples previously in this book, such as numerical oscillations, non-physical solutions, and multiple solutions.

In computing the coupled PDEs, a key issue is how to couple solutions at the interfaces between PDEs, and it has become the bottleneck in the existing approaches. In these approaches, such as those in the previous sections, a transmission condition at the interface of a subdomain is computed directly from the solutions in the other subdomains. As a result, the coupling is realized by directly exchanging the PDEs' numerical solutions of subdomains at their interfaces according to transmission conditions. Hereafter, such an approach is called the classical coupling approach. Now, in the framework of the classical coupling approach, it has become increasingly and even forbiddingly difficult to further progress in the theory and method for computing coupled PDEs, especially distinct PDEs, e.g., the NS and GFD equations in Chap. 5. A source of the difficulty stems from the heterogeneity of the problem; it involves the coupling of different PDEs, numerical methods, and computational meshes. With the classical coupling, the overall coupled problem's well-posedness is often unclear or does not even exist. When coupling actual flow problems, the difficulties become more pronounced. In this situation, if not outright impossible, implementing advanced techniques—such as a conservative interface treatment and the Newton–Krylov–Schwarz method [21, 22]—is often far from straightforward.

The machine learning (ML) coupling approach is essentially a brand-new topic and is distinct from the classical coupling approach. In the ML coupling approach, rather than directly by the solutions in the subdomains, the transmission condition is provided by ML solutions. This idea came from the inspiration of the rapid progress in ML methods and their surprising success in applications to fluid flows. A particularly exciting result is that deep learning can predict flow fields at a later time based on their states at an earlier time [23, 24]. Additionally, theoretical analysis indicates that a deep learning solution converges to that of a PDE in terms of the number of neurons [25]. With such inspiration, we started to explore the ML coupling of PDEs and have reported preliminary results [26–28]. At about the same time, ML has also been utilized to integrate micro–macro fluid flows [29].

A basic idea for the ML coupling is to introduce an interface zone, which encloses the interface of subdomains, and construct an ML model to predict the local solution of PDEs in this zone. Unlike the classical coupling, which provides interface conditions directly from the numerical solutions in the subdomains, the ML coupling uses the ML model to predict the solutions in the interface zone, also the interface conditions, according to these numerical solutions. The ML model serves as a surrogate model for the interface zone, and it is a trained artificial neural network (ANN), providing an ML solution at the interface zone. Compared to the classical coupling, the ML coupling is conjectured to exhibit the following potential advantages in coupling PDEs and fluid flows.

(1) Capability—The ML solution for the interface zone introduces a "buffer zone" between solutions of PDEs in subdomains. It is conjectured that such a "buffer zone" may relax the rigid requirement for well-posedness for their coupling, which is necessary but could be difficult to achieve by the classical coupling.

(2) Efficiency—Schwarz iteration becomes unnecessary in a time evolution problem when an ML model can predict a solution at a later time according to numerical solutions within the subdomains. In addition, the training of the ML model may be conducted offline. These lead to the potential to reduce the computation time for the coupling significantly.

(3) Universality—It is anticipated that an ML-based coupler for two models (PDEs, numerical algorithms, meshes, etc.) could exhibit a certain degree of "universality": an ML coupler developed for the two models is directly applicable to another set of two models as long as the underlying PDEs remain the same. The classical coupling heavily depends on particular model techniques, e.g., discretization algorithms and software structures, and it lacks such "universality."

The ML coupling of PDEs is a new topic, and its advantages, performance, and limitations are yet to be explored. This section presents the ideas, concepts, and algorithms for the ML coupling, which are based on what we have proposed to couple numerical solutions of PDEs and fluid flows at subdomain interfaces [26–28]. It provides numerical examples of the feasibility and performance of the ML coupling. The section's content is exploratory and preliminary, but it shows the promise of the ML coupling and, hopefully, leads to a new paradigm for coupling PDEs and flow models.

6.2.2 Poisson Equation

Consider the computation of a boundary value problem of the 2D Poisson equation:

$$\begin{cases} u_{xx} + u_{yy} = f, & x,\ y \in \Omega \\ u = g, & x,\ y \in \partial\Omega \end{cases} \tag{6.32}$$

where u is the unknown function, and f and g are prescribed functions of x and y. Let domain Ω be divided into two overlapping subdomains Ω_1 and Ω_2, respectively with $\partial\Omega_1$ and $\partial\Omega_2$ as their boundaries, see Fig. 6.7. In this figure, Γ_1 and Γ_2 are a part of $\partial\Omega_1$ and $\partial\Omega_2$, respectively, and they are the interfaces of subdomain Ω_1 and Ω_2, respectively. Note that the interfaces are not actual boundaries but artificial boundaries for the problem.

In reference to Schwarz method (1.3), an ML Schwarz method to solve problem (6.32) is formulated as ($m = 0, 1, 2, \ldots$):

$$\begin{cases} {u_1}_{xx}^{m+1} + {u_1}_{yy}^{m+1} = f, & x,\ y \in \Omega_1 \\ u_1^{m+1} = g, & x,\ y \in \partial\Omega_1/\Gamma_1 \\ u_1^{m+1} = s^m, & x,\ y \in \Gamma_1 \end{cases}$$

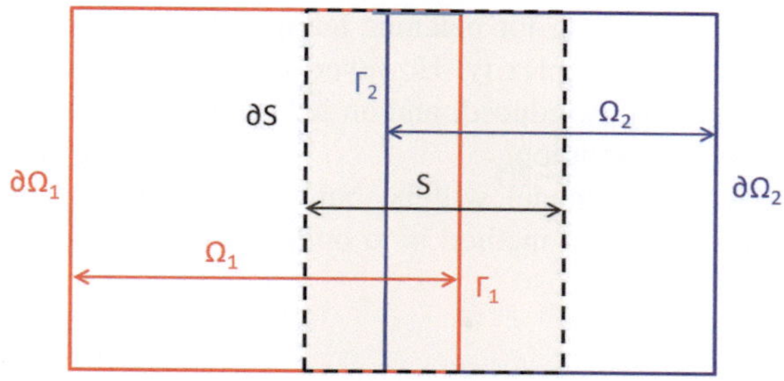

Fig. 6.7 Domain partition for ML coupling. Γ_1 and Γ_2 are the interfaces of the two subdomains. The interface zone, S, is the shadow area enclosed by the dashed line and is denoted by ∂S

$$\begin{cases} {u_2}_{xx}^{m+1} + {u_2}_{xx}^{m+1} = f, & x,\ y \in \Omega_2 \\ u_2^{m+1} = g, & x,\ y \in \partial\Omega_2/\Gamma_2 \\ u_2^{m+1} = s^m, & x,\ y \in \Gamma_2 \end{cases} \tag{6.33}$$

In the aforementioned, s is an ML solution, and the computation is conducted via the ML coupling; the ML solution s provides the interface condition, or transmission condition, on Γ_1 and Γ_2 to facilitate the computation. In a practical calculation, one may set m to a prescribed number of Schwarz iteration, M, which a number of the iteration corresponding to a prescribed iteration residual. The convergence is considered to be achieved at $m = M$.

The aforementioned ML Schwarz method is essentially the same as classical parallel Schwarz method (1.3), except that its interface condition is provided by an ML solution rather than directly from the numerical solutions in the subdomains. It is a parallel ML Schwarz method. Variants for method (6.33) may be considered. For instance, let $u_2^{m+1} = s^m$ be replaced by $u_2^{m+1} = s^{m+1}$, the method becomes an alternating ML Schwarz method.

Let us introduce an interface zone, S, which is bounded by ∂S and encloses the interfaces of the two subdomains, see Fig. 6.7. Then, consider a local problem of the Poisson equation on the interface zone:

$$\begin{cases} u_{xx} + u_{yy} = f, & x,\ y \in S \\ u = \overline{u}_0, & x,\ y \in \partial S \end{cases} \tag{6.34}$$

Here, "local" refers to the interface zone, which is a small portion of the overall domain in the original problem. Moreover, $\overline{u}_0$ is a boundary value. This local boundary value problem acts like a mapping $\mathcal{M}$: $u = \mathcal{M}(\overline{u}_0)$, with $\overline{u}_0$ and u as the input (i.e., the boundary value on ∂S) and the output, respectively.

A main task in ML coupling is to construct an ML model to mimic the mapping and provide a surrogate or approximate solution to the local boundary value problem (6.34). As a result, the task becomes: Find an ML model, $\tilde{\mathcal{M}}$, such that

$$\begin{cases} s = \tilde{\mathcal{M}}(\overline{u}_0) \approx u, & x,\ y \in S \\ s = \overline{u}_0, & x,\ y \in \partial S \end{cases} \tag{6.35}$$

$\tilde{\mathcal{M}}$ acts as an approximation to $\mathcal{M}$, and it provides an ML solution, s. The local problem and its ML model, i.e., (6.34) and (6.35), respectively, are the key to the ML coupling. The rationale of ML coupling is based on this understanding; in general, it is challenging for machine learning to directly simulate original problem (6.32) due to its complexity. However, the interface zone is small, the local problem's complexity is reduced, and an ML model is able to simulate and predict the local problem's solution.

The ML model will be built on an artificial neural network (ANN), $\mathcal{N}$. A straightforward method is to build it as an ANN, that is, $\tilde{\mathcal{M}} = \mathcal{N}(x, y, \boldsymbol{\alpha})$. As a result,

$$s = \mathcal{N}(x, y, \boldsymbol{\alpha}) \tag{6.36}$$

in which $\boldsymbol{\alpha}$ is a vector parameter, $\boldsymbol{\alpha} = (\alpha_1, \alpha_2)$. Here α_1 and α_2 are the neural network's weight and bias, respectively. A diagram and an explanation for the ANN are presented in Fig. 6.8. As seen in the figure, the ANN is a composite function; it is a composition of a series of activation functions at all neurons [30]. The ANN takes $\overline{u}_0$ as the input, and it generates $\mathcal{N}$ as the output (note here $\mathcal{N}$ stands for an ANN as well as its output) . In a more general approach, the ML model may be constructed as a function of $\mathcal{N}$ [31, 32], as shown in a numerical example in the following subsection. Moreover, an ML solution with multi-fidelity, e.g., a combination of low- and high-fidelity, may be considered [33, 34].

Since an ML model is built as an ANN, $\mathcal{N}$, a key to its construction is finding $\mathcal{N}$. The construction of $\mathcal{N}$ includes designing its structure, e.g. Fig. 6.8, and searching its best weights and biases via a minimization problem. In correspondence to (6.36), the construction of the ML model involves the following minimization problem:

$$\begin{cases} \text{Find } \mathcal{N} \text{ such that } \mathcal{N}\overline{u}_0 = s \\ \text{subject to: } \operatorname{argmin}_s ||s - u||^2 \end{cases} \tag{6.37}$$

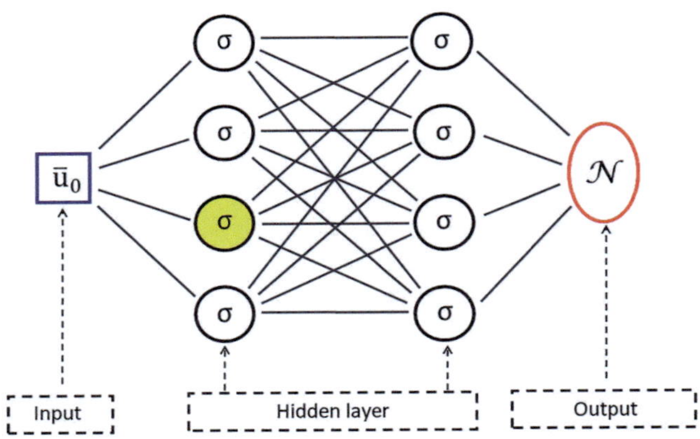

Fig. 6.8 A schematic representation of an ANN, $\mathcal{N}$. In the figure, the ANN has two hidden layers, each with four neurons. The output of a neuron, e.g., the one marked in yellow, is $\sigma(e)$, $e = \alpha_1\overline{u}_0 + \alpha_2$, σ being its activation function (e.g., Sigmoid, ReLU), $\overline{u}_0$ being its input, and α_1 and α_2 being its weight and bias, respectively

where $||\cdot||^2$ is the loss in the minimization. $u = u(x, y)$ is the data for solution to problem (6.34), and it trains the ANN, $\mathcal{N}$, which produces $s = s(x, y)$ as the ML solution. Note that a regularization term may be added to the above optimization problem to facilitate the search for the ANN. Additionally, instead of via the data, it is possible to train an ANN via learning the underlying PDEs [31].

While training the ANN, a central task is to solve minimization problem (6.37), particularly to efficiently search for the weights and biases, $\boldsymbol{\alpha}$, in the ANN that minimizes the loss function. This task is challenging when dealing with many neurons or parameters (i.e., the weights and biases). The minimization involves complicated issues, such as identifying global minimizers (i.e., the best weights and biases), speeding up the search for them, and reducing data dimensions. Thanks to the researchers who made the breakthrough and developed software packages that handle the minimization problem. Available packages include MATLAB made by MathWorks, TensorFlow by Google Brain Team, and PyTorch primarily by Facebook's AI Research lab. These packages have been proven efficient in solving the minimization problem.

Example 6.2.1 Consider the computation of problem (6.32) with

$$
\begin{aligned}
&\Omega: \ 0 < x, \ y < 1 \\
&f = e^{-x}(x - 2 + y^3 + 6y) \\
&g(0, y) = y^3 \\
&g(1, y) = (1 + y^3)e^{-1} \\
&g(x, 0) = xe^{-x} \\
&g(x, 1) = e^{-x}(x + 1)
\end{aligned}
$$

The problem has an analytical solution, $u = e^{-x}(x + y^3)$ [31]. Let domain Ω be covered by mesh $i = 1, 2, \ldots, I' - 1, I', \ldots, I, \ j = 1, 2, \ldots, J$. Then, this mesh is divided into two grids: grid 1: $i = 1, 2, \ldots, I', \ j = 1, 2, \ldots, J$, grid 2: $i = I' - 1, I', \ldots, I, \ j = 1, 2, \ldots, J$, as shown in Fig. 6.9. Additionally, let the grid of the interface zone be $i = I_{I'-k-1}, I_{I'-k}, \ldots, I_{I'+k}, \ j = 1, 2, \ldots, J$, k being an integer.

Let problem (6.33) be discretized by the central difference ($m = 0, 1, 2, \ldots$):

$$
\begin{aligned}
&\frac{v_{1,i-1,j}^{m+1} - 2v_{1,i,j}^{m+1} + v_{1,i+1,j}^{m+1}}{\Delta x^2} + \frac{v_{1,i,j-1}^{m+1} - 2v_{1,i,j}^{m+1} + v_{1,i,j+1}^{m+1}}{\Delta y^2} = f_{i,j} \\
&i = 2, \ldots, I' - 1, \ j = 2, \ldots, J - 1 \\
&\frac{v_{2,i-1,j}^{m+1} - 2v_{2,i,j}^{m+1} + v_{2,i+1,j}^{m+1}}{\Delta x^2} + \frac{v_{2,i,j-1}^{m+1} - 2v_{2,i,j}^{m+1} + v_{2,i,j+1}^{m+1}}{\Delta y^2} = f_{i,j} \\
&i = I', \ldots, I - 1, \ j = 2, \ldots, J - 1
\end{aligned}
\tag{6.38}
$$

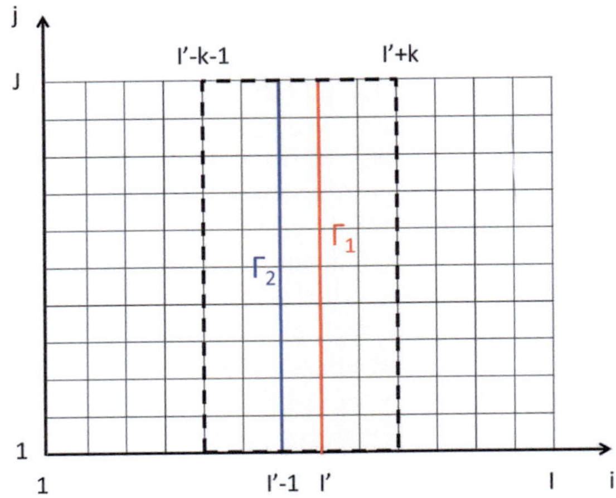

Fig. 6.9 Mesh for computation (6.38). The line at $i = I'$ and the line at $i = I' - 1$ are interfaces, and the shadow area enclosed by the dashed line is the interface zone

By (6.36), the interface algorithm becomes

$$
\begin{aligned}
v_{1,I',j}^{m+1} &= \mathcal{N}(x_{I'}, y_j, \boldsymbol{\alpha}^m), \qquad j = 2, \ldots, J-1 \\
v_{2,I'-1,j}^{m+1} &= \mathcal{N}(x_{I'-1}, y_j, \boldsymbol{\alpha}^m), \ j = 2, \ldots, J-1
\end{aligned}
\tag{6.39}
$$

Additionally, the boundary conditions are

$$
\begin{aligned}
v_{1,i,j}^{m+1} &= g_{i,j}, \ i = 1, \ldots, I', \ j = 1, J \\
v_{2,i,j}^{m+1} &= g_{i,j}, \ i = I' - 1, \ldots, I, \ j = 1, J
\end{aligned}
$$

and

$$
\begin{aligned}
v_{1,i,j}^{m+1} &= g_{i,j}, \ i = 1, \ j = 2, ..., J - 1 \\
v_{2,i,j}^{m+1} &= g_{i,j}, \ i = I, \ j = 2, ..., J - 1
\end{aligned}
$$

Note that when interface algorithm (6.39) is replaced by

$$
\begin{aligned}
v_{1,I',j}^{m+1} &= v_{2,I',j}^{m+1}, \qquad j = 2, \ldots, J-1 \\
v_{2,I'-1,j}^{m+1} &= v_{1,I'-1,j}^{m+1}, \ j = 2, \ldots, J-1
\end{aligned}
\tag{6.40}
$$

the computation becomes the classical coupling.

The ANN in (6.39) is trained as in (6.37). The iterated values produced during the classical Schwarz iteration are utilized as the training data. Particularly, let the original boundary problem be solved by the classical Schwarz method via (6.38) and (6.40). This generates a set of numerical solutions on the boundary of the interface zone, i.e., $v^m_{1,I'-k-1,j}$, $v^m_{2,I'+k,j}$ ($m = 0, 1, 2, \ldots$). With this set of solutions on the boundary of the interface zone, together with $g(x, y)$ on $i = I'-k-1, \ldots, I'+k$, $j = 1, J$, as a set of boundary values $\overline{u}_0$, local problem (6.34) is solved via the classical Schwarz method. The discretization for the local problem is the same as that for the original problem, i.e., (6.38). A set of solutions to the local problem are generated within the interface zone ($i = I' - k, \ldots, I' + k - 1$, $j = 2, \ldots, J - 1$), including at subdomain interfaces Γ_1 and Γ_2, where they are denoted as $\hat{v}^m_{1,I',j}$, $\hat{v}^m_{2,I'-1,j}$. Then, training (6.37) is conducted with the set of boundary values of the local problem as the input and the set of its corresponding solutions as the output ($m = 1, 2, \ldots$):

$$\begin{cases} \overline{u}_0 : v^m_{1,I'-k-1,j},\ v^m_{2,I'+k,j}, & j = 2, \ldots J - 1 \\ \qquad g(x_i, y_1),\ g(x_i, y_J), & i = I' - k - 1, \ldots, I' + k \\ u :\ \hat{v}^m_{1,I',j},\ \hat{v}^m_{2,I'-1,j}, & j = 2, \ldots, J - 1 \end{cases}$$

Note that in general, $\hat{v}^m_{1,I',j}$, $\hat{v}^m_{2,I'-1,j}$, which are the solution by the classical Schwarz method for local problem (6.34), are not same as $v^m_{1,I',j}$, $v^m_{2,I'-1,j}$, which are the iterated values by the classical Schwarz method for original problem (6.32). Therefore, the training does not simply repeat the iteration process of the classical Schwarz method for the original problem. As a result, the ML Schwarz method based on such training does not repeat the classical Schwarz method in solving the original problem.

In the computation of ML Schwarz method (6.33), $(I, J) = (21, 21)$, $I' = 11$, and $k = 2$. The ANN training via (6.37) is conducted using MATLAB, and one hidden layer ($N_l = 1$) and 10 neurons ($N_n = 10$) with the Sigmoid function as the activation function are adopted. Figure 6.10 presents the solution, denoted as v_{ml} ($= v_{1ml} \cup v_{2ml}$, v_{1ml} and v_{2ml} being numerical solution obtained in subdomains Ω_1 and Ω_2, respectively, with the ML coupling). It also shows the error of the solution against the analytical solution. The solution obtained with the ML Schwarz method matches the analytical solution satisfactorily.

The solution by the ML Schwarz method, via interface algorithm (6.39), is compared with that by the classical Schwarz method, via interface algorithm (6.40). Interestingly, the computation shows that, with zero as the initial value of iteration and a convergence criterion of 10^{-10} in iteration residual, the ML Schwarz method converges faster than the classical Schwarz method; the former converges in just 18 iterations, or $M = 18$, and the latter needs 121 iterations, or $M = 121$. This further manifests the fact indicated in the above: the iteration process of the ML Schwarz method is different from that of the classical Schwarz method.

In the aforementioned example, because the two grids overlap at two layers of nodes, as explained in Sect. 2.2.1, Dirichlet–Derichlet interface condition (6.40) is also an approximation to the Dirichlet–Neumann interface condition. Actually, (6.40) leads to

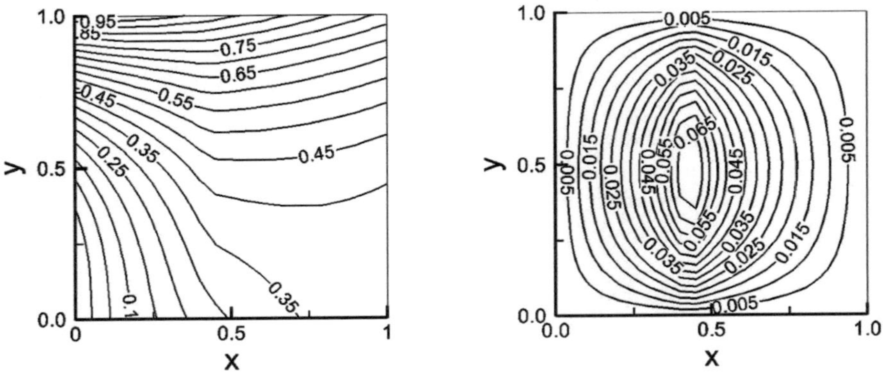

(a) (b)

Fig. 6.10 Computation of the Poisson equation. (**a**) Numerical solution with ML coupling. (**b**) Error distribution, $v_{ml} - u$

$$\begin{aligned} v_{1,I',j}^{m+1} &= v_{2,I',j}^{m}, & j = 2, \ldots, J-1 \\ \frac{v_{1,I',j}^{m+1} - v_{1,I'-1,j}^{m}}{\Delta x} &= \frac{v_{2,I',j}^{m} - v_{2,I'-1,j}^{m+1}}{\Delta x}, & j = 2, \ldots, J-1 \end{aligned}$$

which is an implementation of the Dirichlet–Neumann interface condition in (2.11).

To quantify the error of the solution with ML coupling, v_{ml}, against the analytical solution, u, in the interface zone, the following estimate is useful:

$$||v_{ml} - u|| \leq ||v_{ml} - v|| + ||v - u|| \tag{6.41}$$

in which v represents the training data obtained with the classical coupling. The above indicates that the error can be estimated via two parts, i.e., $||v_{ml} - v||$ and $||v - u||$. The first part is the error of the ML solution against the training data, and the second part is that of the training data against the analytical solution. The former error depends on the ML solution and its training method, i.e., (6.36) and (6.37), and the latter error depends on the accuracy order in the discretization of the Poisson equation, i.e., (6.38).

Example 6.2.2 Continuing from the aforementioned example, an experiment is further made using different numbers of hidden layers and neurons, and the results are shown in Table 6.5. Minimization (6.37) is solved using MATLAB. Note that a repeated MATLAB run for the minimization may not produce the same converged value for the loss, but its value is stable, and the resulting ML solution, s, remains about the same. In the training, the errors and the loss values result from the averages of three runs. The results in the table verify the estimate (6.41). Furthermore, they indicate that, in this experiment, error $||v_{ml} - v||$ is much smaller than error $||v - u||$,

Table 6.5 Error of computation for the Poisson equation. $||\cdot||$ is in the infinity norm, and the loss is the mean square error, both over S

N_l, N_n	$\|\|v_{ml} - u\|\|$	$\|\|v_{ml} - v\|\|$	$\|\|v - u\|\|$	$Loss$
1, 5	7.00×10^{-2}	1.48×10^{-2}	8.21×10^{-2}	3.49×10^{-9}
1, 10	6.97×10^{-2}	1.50×10^{-2}	8.19×10^{-2}	4.86×10^{-10}
1, 15	7.00×10^{-2}	1.51×10^{-2}	8.22×10^{-2}	1.92×10^{-10}
1, 20	7.01×10^{-2}	1.54×10^{-2}	8.22×10^{-2}	5.98×10^{-10}
2, 10	6.98×10^{-2}	1.49×10^{-2}	8.19×10^{-2}	1.00×10^{-9}
3, 10	6.97×10^{-2}	1.46×10^{-2}	8.17×10^{-2}	5.65×10^{-10}

implying that the error of the ML solution, $||v_{ml} - u||$, could primarily come from $||v - u||$, i.e., the error of the training data against the analytical solution. Moreover, when no more than 15 neurons are used, the results in the table show a decreasing trend in the loss value as the number of neurons increases, in the case of a single hidden layer. However, in general, there is no clear trend in the loss value or solution convergence with respect to the number of neurons or hidden layers.

6.2.3 Advection-Diffusion Equation

Now, let us proceed to the computation of an initial value problem of two coupled 1D advection-diffusion equations over Ω $(-\infty < x < \infty)$:

$$
\begin{cases} u_{1t} + a_1 u_{1x} = b_1 u_{1xx}, & x \in \Omega_1,\ t \in (0, T] \\ u_1 = u_2, & x = x_1,\ t \in (0, T] \\ u_1 = u_0, & x \in \Omega_1,\ t = 0 \end{cases}
$$

$$
\begin{cases} u_{2t} + a_2 u_{2x} = b_2 u_{2xx}, & x \in \Omega_2,\ t \in (0, T] \\ u_2 = u_1, & x = x_2,\ t \in (0, T] \\ u_2 = u_0, & x \in \Omega_2,\ t = 0 \end{cases} \tag{6.42}
$$

in which $a_k, b_k = const.$ $(k = 1, 2)$, and u_0 is a prescribed function of x. Ω is divided into two subdomains: Ω_1 $(-\infty < x < x_1)$ and Ω_2 $(x_2 < x < \infty)$, which overlap with each other $(x_1 > x_2)$. x_1 and x_2 are the interfaces of Ω_1 and Ω_2, respectively. Dirichlet conditions are used at the interfaces. Note that when the equations are different and the subdomains overlap, the subdomain solutions generally differ in the overlapping region.

We consider the computation of (6.42) via the conventional method as in (1.9), in which the two equations march in time simultaneously, but an ML model provides the interface conditions. Particularly, when marching from time level n to $n + 1$ $(n = 0, 1, 2, \ldots, N - 1)$, the problem is solved by ML coupling:

$$\begin{cases} u_{1t} + a_1 u_{1x} = b_1 u_{1xx}, & x \in \Omega_1 \; t \in (t_n, t_{n+1}] \\ u_1 = s, & x = x_1, \; t \in (t_n, t_{n+1}] \\ u_1 = u_1^n, & x \in \Omega_1, \; t = t_n \end{cases}$$
$$\begin{cases} u_{2t} + a_2 u_{2x} = b_2 u_{2xx}, & x \in \Omega_2, \; t \in (t_n, t_{n+1}] \\ u_2 = s, & x = x_2, t \in (t_n, t_{n+1}] \\ u_2 = u_2^n, & x \in \Omega_2, \; t = t_n \end{cases} \tag{6.43}$$

where $t_n = n\Delta t$, $\Delta t = T/N$. Superscript n indicates $t = t_n$, and $u_1^0, u_2^0 = u_0$. $s = s(x,t)$ is an ML solution that provides transmission conditions. As discussed in Sect. 2.2, the computation requires Schwarz iteration between the two subdomains when implicit discretization is adopted within the subdomains and at the interfaces. However, as a distinct feature of algorithm (6.43), such Schwarz iteration may not be necessary when marching from time level n to $n+1$ if the ML model is capable of predicting the solution on the interfaces at time level $n+1$ based on the solution at time level n. Such capability will be discussed in the content to follow in this section. Of course, if needed, Schwarz iteration may also be adopted when marching from time level n to $n+1$.

Similar to the situation of the Poisson equation, let us introduce an interface zone in the $x - t$ plane, $S: \; S_1 < x < S_2, 0 < t \le T$, S_1 and S_2 being the left and right ends, respectively, of the interface zone in the space coordinate. The interface zone has a boundary $\partial S : (t = 0) \times (S_1 < x < S_2) \cup (0 < t < T) \times (x = S_1, S_2)$. The interface zone contains the interfaces: x_1, x_2. $S_1 < x_2$, and $x_1 < S_2$. Now, consider the following local initial-boundary value problem of the advection-diffusion-reaction equations:

$$\begin{cases} u_{1t} + a_1 u_{1x} = b_1 u_{1xx}, & S_1 < x < x_1, 0 < t \le T \\ u_1 = u_0, & S_1 < x < x_1, \; t = 0 \\ u_1 = g_1, & x = S_1, \; 0 < t \le T \\ u_1 = u_2, & x = x_1, \; 0 < t \le T \end{cases}$$
$$\begin{cases} u_{2t} + a_2 u_{2x} = b_2 u_{2xx}, & x_2 < x < S_2, 0 < t \le T \\ u_2 = u_0, & S_2 < x < x_2, \; t = 0 \\ u_2 = g_2, & x = S_2, \; 0 < t \le T \\ u_2 = u_1, & x = x_2, \; 0 < t \le T \end{cases} \tag{6.44}$$

with $\bar{g}_1$ and $\bar{g}_2$ as prescribed functions of t. Here, "local" refers to the interface zone. Again, as in the previous subsection, this initial-boundary value problem acts like a mapping $\mathcal{M}$: $u = \mathcal{M}(\overline{u}_0)$. Here, $\overline{u}_0$ includes both the initial and the boundary conditions, i.e., $\overline{u}_0 = u_0 \cup g_1 \cup g_2$.

To formulate the machine learning problem, let s be the ML solution, u be the solution of local problem (6.44). Then, a task in ML coupling becomes: Find an ML model, $\tilde{\mathcal{M}}$, such that

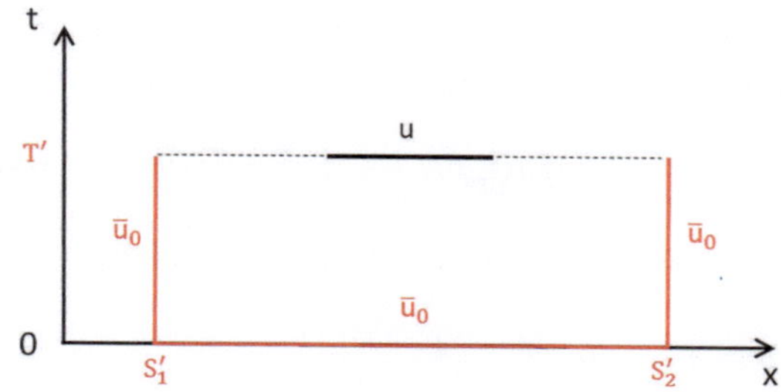

Fig. 6.11 Schematic representation of the input and output for the training

$$\begin{cases} s = \tilde{\mathcal{M}}(u_0) \approx u, & x, t \in S \\ s = \overline{u}_0, & x, t \in \partial S \end{cases} \tag{6.45}$$

Now $\tilde{\mathcal{M}}$ approximates $\mathcal{M}$, and it mimics the evolution of the solutions to the advection-diffusion equations in (6.44). In computation, the ML model for the interface zone may be constructed as an ANN:

$$s = \mathcal{N}(x, t, \boldsymbol{\alpha}) \tag{6.46}$$

Still, the training of the ML model follows (6.37). In the training, the input and output, i.e., $\overline{u}_0$ and u, and their relation are shown in Fig. 6.11. In the figure, the domain enclosed by $S_1' < x < S_2'$ and $0 < t < T'$ may be same as, or different from, the interface zone. But it should cover the domain of dependence for solution u for the sake of computational stability [35]. In the ML model, $\overline{u}_0$ resides on $(S_1' < x < S_2', t = 0)$ as an initial condition and $(x = S_1', S_2', 0 < t < T')$ as a boundary condition. The boundary condition is necessary when the PDEs are parabolic. When the PDEs are hyperbolic, no boundary condition is needed as long as the range $(S_1' < x < S_2', t = 0)$ is sufficiently large. Once trained, the ML model can predict the evolution of the solution to local problem (6.44).

With different values for S_1', S_2', and T' in Fig. 6.11, three training approaches are possible:

$$\begin{cases} \text{global}: & (-\infty < x < \infty) \times (t = 0) \\ \text{regional}: & (S_1 < x < S_2) \times (t = 0),\ (x = S_1, S_2) \times (0 < t < T) \\ \text{local}: & (S_1 < x < S_2) \times (t = 0),\ (x = S_1, S_2) \times (0 < t < \Delta) \end{cases}$$

The input data, $\overline{u}_0$, is actually the initial condition of the problem itself in the global approach (note that, by this approach, a boundary condition is also considered as a part of $\overline{u}_0$ for an initial-boundary value problem). It is the initial and boundary conditions in the regional approach (note that, in the approach, $\overline{u}_0$ also comes from the boundary of the interface zone). In the local approach, Δ is a small number, and $\overline{u}_0$ resides in a relatively small neighborhood. In practical computation, the range of (S_1, S_2) is not necessarily large since the solution data far away have little effect on the solution.

Example 6.2.3 In this problem, two subdomains have an identical advection-diffusion equation: $a_1, a_2 = a = 1,\ b_1, b_2 = b = 0.1$, together with

$$u_0 = -sin(\pi x),\ -1 < x < 1;\ u = 0,\ x = -1, 1,\ 0 \le t \le 1 \tag{6.47}$$

The problem has an analytical solution [36]:

$$\begin{aligned} u = {} & 16\pi^2 ab^3 e^{(x-at/2)a/(2b)} \\ & \times \left[sinh(\frac{a}{2b}) \sum_{m=0}^{\infty} \frac{(-1)^m\ 2m\ sin(m\pi x)\ e^{-bm^2\pi^2 t}}{a^4 + 8(ab\pi)^2(m^2+1) + 16(\pi b)^4(m^2-1)^2} \right. \\ & \left. + cosh(\frac{a}{2b}) \sum_{m=0}^{\infty} \frac{(-1)^m\ (2m+1)\ cos((2m+1)/(2\pi x))\ e^{-(2m+1)^2 b\pi^2 t/4}}{a^4 + (ab\pi)^2(8m^2+8m+10) + (\pi b)^4(4m^2+4m-3)^2} \right] \end{aligned}$$

Let the domain $-1 \le x \le 1$ be decomposed into Ω_{-1} $(-1 < x < x_{I'})$ and Ω_2 $(x_{I'-1} < x < 1)$, with their interfaces at $x_{I'}, x_{I'-1}$, respectively. Furthermore, the domain is covered by grid $(-1 = x_1, x_2, \ldots, x_{I'-1}, x_{I'}, \ldots, x_I = 1)$. The problem is discretized via implicit schemes and ML coupling as

$$\begin{cases} \dfrac{v_{1,i}^{n+1} - v_{1,i}^{n}}{\Delta t} + a_1 \dfrac{v_{1,i+1}^{n+1} - v_{1,i-1}^{n+1}}{2\Delta x} = b_1 \dfrac{v_{1,i-1}^{n+1} + v_{1,i+1}^{n+1} - 2v_{1,i}^{n+1}}{\Delta x^2}, \\ i = 2, \ldots, I'-1,\ j = 2, \ldots, J-1;\ v_{1,I'}^{n+1} = s_{I'}^{n+1} \\ \dfrac{v_{2,i}^{n+1} - v_{2,i}^{n}}{\Delta t} + a_2 \dfrac{v_{2,i+1}^{n+1} - v_{2,i-1}^{n+1}}{2\Delta x} = b_2 \dfrac{v_{2,i-1}^{n+1} + v_{2,i+1}^{n+1} - 2v_{2,i}^{n+1}}{\Delta x^2}, \\ i = I', \ldots, I-1,\ j = 2, \ldots, J-1;\ v_{2,I'-1}^{n+1} = s_{I'-1}^{n+1} \end{cases} \tag{6.48}$$

The corresponding classical coupling is to replace $s_{I'}^{n+1}$ and $s_{I'-1}^{n+1}$ at interface nodes in the above by $v_{2,I'}^{n+1}$ and $v_{1,I'-1}^{n+1}$, respectively. Particularly,

$$v_{1,I'}^{n+1} = v_{2,I'}^{n+1},\ v_{2,I'-1}^{n+1} = v_{1,I'-1}^{n+1} \tag{6.49}$$

As explained in Sect. 2.2, the aforementioned Dirichlet–Dirichlet interface condition is also an approximation of Dirichlet–Neumann interface condition (2.11) because the two grids overlap at two nodes in $i-$ direction.

The global approach is to adopt the analytical solution to train the ANN. Particularly,

$$\begin{cases} \overline{u}_0 : u_0, & x = x_1, x_2, \ldots, x_I,\ t = 0 \\ \qquad 0 & x = x_1,\ x_I,\ t = t_l,\ l = 1, 2, \ldots, L \\ u : \ u(x,t), & x = x_{I'-1}, x_{I'},\ t = t_l,\ l = 1, 2, \ldots, L \end{cases}$$

In the training, $t_l = l\Delta t'$, $\Delta t' = 1/L$. Following the early work in [31], the ML solution is constructed as a function of $\mathcal{N}$:

$$\begin{cases} s = A(x,t) + t\dfrac{x-1}{2}\dfrac{x+1}{2}\mathcal{N}(x,t,\boldsymbol{\alpha}) \\ A(x,t) = (1-t)\left(u(x,0) - \dfrac{1-x}{2}u(-1,0) - \dfrac{1+x}{2}u(1,0)\right) \\ \qquad + \dfrac{1-x}{2}u(-1,t) + \dfrac{1+x}{2}u(1,t) \end{cases} \tag{6.50}$$

in which $-1 \leq x \leq 1, 0 \leq t \leq 1$. This ML solution, s, automatically satisfies the initial and boundary conditions of the problem. This leads to an unconstrained optimization problem in training $\mathcal{N}$ via (6.37), or $\mathcal{N}$ is not restricted directly by the problem's initial and boundary conditions in (6.47). With (6.50), the boundary condition at the interfaces is given over the entire time interval, $0 \leq t \leq 1$. The computation in each subdomain uses only the initial and boundary conditions of the problem and the ML solution at the interface, without considering any information from the other subdomain. As a result, the computations in the two subdomains are independent, and their solutions are actually decoupled.

In the computation, $(I, N, L) = (41, 100, 31)$, and the Sigmoid function is chosen as the activation function. Moreover, in this example, also in the rest examples with advection-diffusion equations, $M = 1$, or no Schwarz iteration between subdomains is employed. The solution obtained with the ML coupling, v_{ml}, the analytical solution, u, and the error are shown in Fig. 6.12. The figure shows that the numerical solution associated with the ML coupling is satisfactory. Note that the training mesh, (x_i, t_l), is not the same as the mesh, (x_i, t_n), for the numerical solution. Therefore, the solution provided by the ML coupling does not simply repeat the analytical solution, the training data, in the computation.

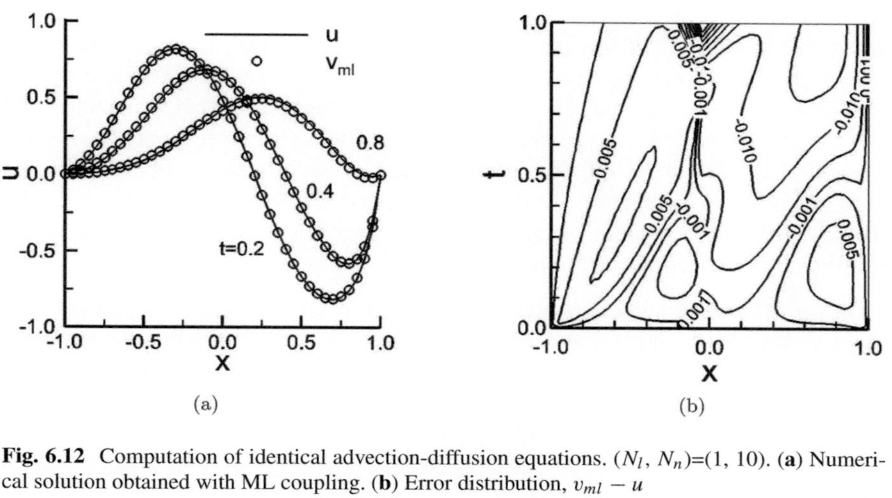

(a) (b)

Fig. 6.12 Computation of identical advection-diffusion equations. (N_l, N_n)=(1, 10). (**a**) Numerical solution obtained with ML coupling. (**b**) Error distribution, $v_{ml} - u$

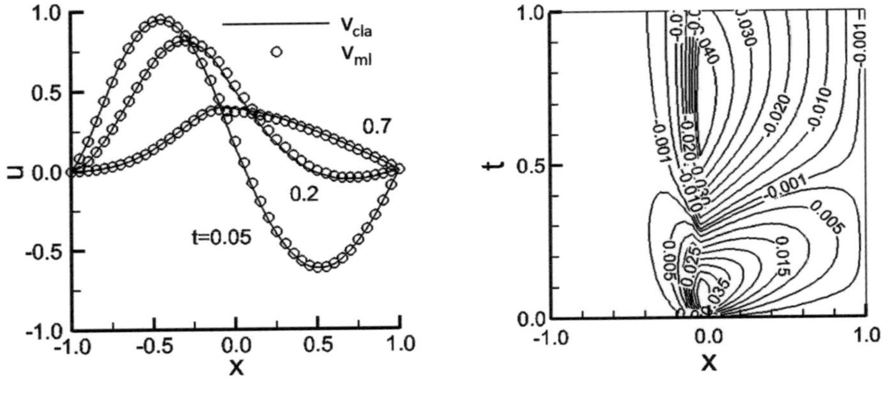

(a) (b)

Fig. 6.13 Computation of distinct advection-diffusion equations. (N_l, N_n)=(1, 4). (**a**) Numerical solutions with ML coupling and classical coupling, v_{cla}. (**b**) Error distribution, $v_{ml} - v_{cla}$

Example 6.2.4 This is a heterogeneous problem involving equations with distinct coefficients in the two subdomains: $a_1 = 1,\ a_2 = 0.1,\ b_1 = 0.1,\ b_2 = 1$. Other conditions and parameters remain the same as those in the above example, such as $(I, N, L) = (41, 100, 31)$. This problem has no analytical solution, and the ANN is trained by a numerical solution obtained with the classical coupling.

A straightforward method for the local approach is to adopt the numerical solution to train the ANN directly. The training is performed via

$$\begin{cases} \overline{u}_0 : v^l_{1,I'-k-1}, \ldots, v^l_{1,I'-1}, v^l_{2,I'}, \ldots, v^l_{2,I'+k}, & t = 0 \\ u : \ v^{l+1}_{1,I'}, v^{l+1}_{2,I'-1}, & t = \Delta t' \end{cases} \tag{6.51}$$

for all $l = 0, 1, 2, \ldots, L - 1$. Here, the training input and output, i.e., $\overline{u}_0$ and u, are the numerical solutions at different time levels l obtained with the classical coupling (6.49). With all the parameters as before, the computation of (6.48) is made, and the computed solution is presented in Fig. 6.13. The figure shows that the solution computed by the ML coupling is close to the numerical solution obtained by the classical coupling, and a larger error occurs around the interfaces. Again, since the mesh for the training and that for the computation are not the same, the ML-coupling solution does not repeat the training.

In the aforementioned two examples, as mentioned previously, since the training's output u is at a later time than the input $\overline{u}_0$, the ML model predicts the temporal evolution of the solution to the corresponding local problem to a certain extent. As a result, the interface conditions $v^{n+1}_{1,I'}$ and $v^{n+1}_{2,I'-1}$ in (6.48) are explicitly predicted by the ML model according to the initial value of the problem or the solution at time level n in the interface zone, and Schwarz iteration becomes unnecessary.

The local approach of training in (6.51) generates an ML solution $s = \mathcal{N}(x, t, \boldsymbol{\alpha})$. Instead of $\mathcal{N}(x, t, \boldsymbol{\alpha})$, the local approach to train the ANN may generate a function of it via (6.50). Particularly, with a transformation in coordinates in (6.50), the ML solution is constructed as

$$\begin{cases} s = A(x', t') + t\dfrac{x'-1}{2}\dfrac{x'+1}{2}\mathcal{N}(x', t', \boldsymbol{\alpha}) \\ A(x', t') = (1-t')\left(u(x, t-t_l) - \dfrac{1-x'}{2}u(S_1, t-t_l) - \dfrac{1+x'}{2}u(S_2, t-t_l)\right) \\ \quad + \dfrac{1-x'}{2}u(S_1, t-t_l) + \dfrac{1+x'}{2}u(S_2, t-t_l) \end{cases}$$

in which $t' = (t - t_l)/\Delta t'$, $x' = (2x - S_1 - S_2)/(S_2 - S_1)$, $-1 \le x' \le 1$, and $0 \le t' \le 1$. Additionally, $S_1 \le x \le S_2$, and $t_l \le t \le t_l + \Delta t'$.

Example 6.2.5 This example examines the convergence behaviors of the numerical solutions obtained with ML coupling. Continued from Example 6.2.3, an experiment is made using different numbers of hidden layers and neurons, see Table 6.6. The table shows that the computation with one hidden layer exhibits convergence concerning the number of neurons; the error of the numerical solution and the magnitude of the loss decrease when more neurons are utilized. Such observation may be relevant to an earlier analysis [25], which proves that the ML solution to a time-dependent PDE converges with respect to the number of neurons in the case of a single hidden layer. For Example 6.2.4, an experiment is also made as in Table 6.7. Still, a convergence trend is observed in terms of the number of neurons.

Table 6.6 Error of computation for identical advection-diffusion equations. $||\cdot||_\infty$ is over $\Omega \times [0, T]$

N_l, N_n	$\|\|v_{ml} - u\|\|_\infty$	*Loss*
1, 5	1.50×10^{-1}	6.16×10^{-1}
1, 10	3.78×10^{-2}	6.46×10^{-2}
1, 15	3.42×10^{-2}	2.42×10^{-2}
1, 20	2.97×10^{-2}	3.73×10^{-3}
2, 10	1.67×10^{-2}	3.67×10^{-7}
3, 10	1.67×10^{-2}	1.04×10^{-8}

Table 6.7 Error of computation for different advection-diffusion equations. $||\cdot||_\infty$ is over $\Omega \times [0, T]$

N_l, N_n	$\|\|v_{ml} - v_{cla}\|\|_\infty$	*Loss*
1, 2	6.90×10^{-2}	1.01×10^{-10}
1, 4	5.02×10^{-2}	6.95×10^{-11}
1, 8	4.60×10^{-2}	3.33×10^{-11}
1, 16	4.35×10^{-2}	3.02×10^{-11}
2, 4	4.88×10^{-2}	1.90×10^{-11}
3, 4	4.50×10^{-2}	1.71×10^{-11}

6.2.4 ML Godunov Scheme and Interface Algorithm

Initial Value Problem and Local Solution This subsection discusses ML coupling for general advection-diffusion equations. Let $\mathbf{u}_k$ be the unknown variables in overlapping subdomain Ω_k, and L_k and N_k be the linear operator and the nonlinear operator in the equations, respectively ($k = 1, 2$). Particularly, L_k represents the diffusion term, and N_k stands for the advection term in the equations. Such equations may be considered model equations for Navier–Stokes equations (4.2) or shallow water equations (5.24). Consider an initial value problem of two coupled general advection-diffusion equations as follows:

$$\begin{cases} \dfrac{\partial \mathbf{u}_1}{\partial t} = L_1\mathbf{u}_1 + N_1(\mathbf{u}_1), & \mathbf{x} \in \Omega_1,\ t \in (0, T] \\ \mathbf{u}_1 = \mathbf{u}_2, & \mathbf{x} \in \Gamma_1,\ t \in (0, T] \\ \mathbf{u}_1 = \mathbf{u}_0, & \mathbf{x} \in \Omega_1,\ t = 0 \end{cases}$$
$$\begin{cases} \dfrac{\partial \mathbf{u}_2}{\partial t} = L_2\mathbf{u}_2 + N_2(\mathbf{u}_2), & \mathbf{x} \in \Omega_2,\ t \in (0, T] \\ \mathbf{u}_2 = \mathbf{u}_1, & \mathbf{x} \in \Gamma_2, t \in (0, T] \\ \mathbf{u}_2 = \mathbf{u}_0, & \mathbf{x} \in \Omega_2,\ t = 0 \end{cases} \tag{6.52}$$

Here bold indicates a vector, and $\mathbf{u}_0$ is a prescribed function. Let the aforementioned problem be computed via the conventional method for temporal integration within the subdomains and ML coupling at the interfaces. Particularly, when marching from time level n to $n + 1$ ($n = 0, 1, 2, \ldots, N - 1$), the computation is formulated as

$$\begin{cases} \dfrac{\partial \mathbf{u}_1}{\partial t} = L_1\mathbf{u}_1 + N_1(\mathbf{u}_1), & \mathbf{x} \in \Omega_1\ t \in (t_n, t_{n+1}] \\ \mathbf{u}_1 = \mathbf{s}, & \mathbf{x} \in \Gamma_1,\ t \in (t_n, t_{n+1}] \\ \mathbf{u}_1 = \mathbf{u}_1^n, & \mathbf{x} \in \Omega_1,\ t = t_n \end{cases}$$
$$\begin{cases} \dfrac{\partial \mathbf{u}_2}{\partial t} = L_2\mathbf{u}_2 + N_2(\mathbf{u}_2), & \mathbf{x} \in \Omega_2,\ t \in (t_n, t_{n+1}] \\ \mathbf{u}_2 = \mathbf{s}, & \mathbf{x} \in \Gamma_2, t \in (t_n, t_{n+1}] \\ \mathbf{u}_2 = \mathbf{u}_2^n, & \mathbf{x} \in \Omega_2,\ t = t_n \end{cases} \tag{6.53}$$

in which $\mathbf{u}_1^0, \mathbf{u}_2^0 = \mathbf{u}_0$.

Now, consider a 1D initial value problem, or the Riemann problem, of a general advection-diffusion equation:

$$\begin{aligned} &\frac{\partial \mathbf{u}}{\partial t} = L\mathbf{u} + N(\mathbf{u}), \quad -\infty < x < +\infty, \quad t > 0 \\ &\mathbf{u} = \begin{cases} \mathbf{c}_l, & x < 0, \quad t = 0 \\ \mathbf{c}_r, & x > 0, \quad t = 0 \end{cases} \end{aligned} \tag{6.54}$$

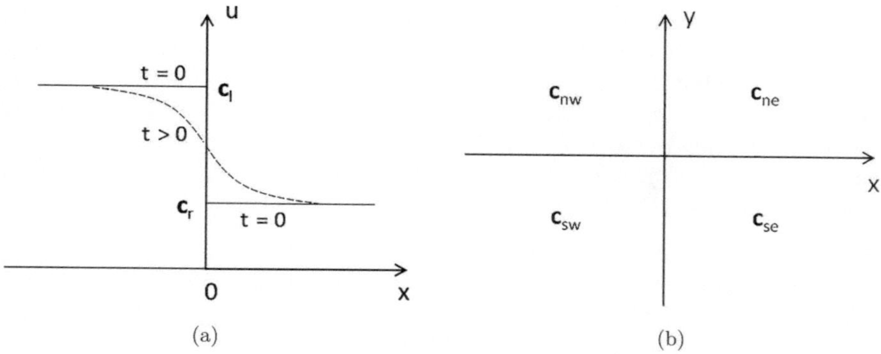

Fig. 6.14 A schematic representation of Riemann problems. (**a**) 1D problem, initial condition ($t = 0$), and solution ($t > 0$). $\mathbf{c_l}, \mathbf{c_r} = const$. (**b**) 2D problem, with four piecewise constants in the initial condition. $\mathbf{c}_{ne}, \mathbf{c}_{nw}, \mathbf{c}_{se}, \mathbf{c}_{sw} = const$

where subscript l and r mean the left and right sides of $x = 0$, respectively, and $\mathbf{c} = const$. In general, $\mathbf{c}_l \neq \mathbf{c}_r$, leading to a discontinuity in the initial condition at $x = 0$, see Fig. 6.14a. A Riemann problem usually refers to an initial value problem consisting of a conservation law and a piecewise constant initial value. The aforementioned problem includes a general advection-diffusion equation, and it may be considered as a generalized Riemann problem. Moreover, a Riemann problem may also be formulated for higher dimensions in space [37], as illustrated in Fig. 6.14b.

There is rich literature on local solutions for the general advection-diffusion equation, including exact, approximate, and numerical solutions in various scenarios (e.g., high dimensions in space) for Riemann problems of the PDEs. Here and hereafter, "local" refers to a small neighborhood of (x, t)=(0, 0). For instance, for a 1D, scalar diffusion equation, a local solution for the Riemann problem is [38, 39]

$$u^* = \int_{-\infty}^{0} \frac{c_l}{\sqrt{4\pi bt}} e^{-(x-\xi)^2/(4bt)} d\xi + \int_{0}^{+\infty} \frac{c_r}{\sqrt{4\pi bt}} e^{-(x-\xi)^2/(4bt)} d\xi \tag{6.55}$$

where b is the diffusion coefficient. Here, superscript $*$ represents the Riemann resolution. In the case of the 1D shallow water equations, the Riemann problem's analytical solution consists of a left-moving wave (a rarefaction wave or a shock/surge bore) and a right-moving wave (e.g., a rarefaction wave or a shock/surge bore) [40, 41]. The key to analytically solving the Riemann problem is to solve a nonlinear algebraic equation for water surface elevation, η^*, between the left- and right-moving waves:

$$f(\eta^*, \eta_l, u_l, \eta_r, u_r) = 0 \tag{6.56}$$

Subsequently, the other variables are determined analytically using η^* [41].

For some complex Riemann problems, such as those for GFD equations, (5.6), and NS equations, (4.1) and (4.2), analytical solutions become difficult. However, it has become mature to solve such problems numerically [35, 42, 43], and various software packages are available, such as ANYSYS FLUENT and OpenFOAM [44, 45]. Additionally, solutions may be derived analytically (e.g., by characteristic analysis) or obtained numerically for Riemann problems with different sets of PDEs in subdomains [46, 47].

Machine Learning Local Solution Let $\mathbf{u}_0 = \mathbf{c}_l \cup \mathbf{c}_r$, being the initial value, $\mathbf{u}^*(\mathbf{x}, t)$ be the local solution near $(x, t) = (0, 0)$ to Riemann problem (6.54), and $\mathbf{s}$ be the prediction of ANN. Here, the local solution is within a short time after $t = 0$ and a small neighborhood of $x = 0$. ANNs can be constructed by learning/training with $\mathbf{u}_0$ as input and $\mathbf{u}^*(\mathbf{x}, t)$ as the output. Note that a Riemann problem has no boundary condition, and $\bar{\mathbf{u}}_0$ is the initial value $\mathbf{u}_0$. Additionally, $\mathbf{u}^*(\mathbf{x}, t)$ either comes from analytical or numerical solutions. In this situation, the construction of the ANN involves the following minimization problem:

$$\begin{cases} \text{Find } \mathcal{N} \text{ such that } \mathcal{N}\mathbf{u}_0 = \mathbf{s}(\mathbf{x}, t) \\ \text{subject to: } \text{argmin}_{\mathbf{s}} ||\mathbf{s}(\mathbf{x}, t) - \mathbf{u}^*(\mathbf{x}, t)||_2^2 \end{cases} \tag{6.57}$$

Machine learning and predicting the local solution are innovative, and they are the cornerstone of the ML coupling. In general, the local solution to Riemann problem (6.54) is relatively simple in terms of its behaviors and structures, and it is conjectured that an ML model can learn and then predict such a solution well. Such a conjecture is based on machine learning's tremendous success, although its mechanism and foundation are still research topics. Direct evidence for such learning and prediction is available. For instance, learning and prediction are made for the Sod shock tube problem, a Riemann problem of the Euler equations, which somewhat resemble the shallow water equations. The computation shows that ANN predictions are even sharper around the shock wave and the contact discontinuity than those computed from a state-of-the-art, high-resolution numerical scheme, see Fig. 6.15.

A key issue in ANNs' learning/training is letting them experience all possible initial value, $\mathbf{u}_0$, which is the input, and the corresponding solution, $\mathbf{u}^*$, which is the output. One may evenly divide the range of $\mathbf{u}_0$ $(= \mathbf{c}_l \cup \mathbf{c}_r)$ into small intervals (thanks to the piecewise constancy in $\mathbf{u}_0$ in the Riemann problem), each of which approximates a possible initial value. The training is formulated as

$$\begin{cases} Do\ i = 1, \ldots, I, \quad j = 1, \ldots, J \\ \quad \mathbf{c}_l = \mathbf{c}_l^{min} + (i-1)\Delta\mathbf{c}_l, \ \Delta\mathbf{c}_r = (\mathbf{c}_l^{max} - \mathbf{c}_l^{min})/(I-1) \\ \quad \mathbf{c}_r = \mathbf{c}_r^{min} + (j-1)\Delta\mathbf{c}_r, \ \Delta\mathbf{c}_r = (\mathbf{c}_r^{max} - \mathbf{c}_r^{min})/(J-1) \\ \quad ANN\ training,\ e.g.,\ Eq.\ (6.57) \\ End\ Do \end{cases} \tag{6.58}$$

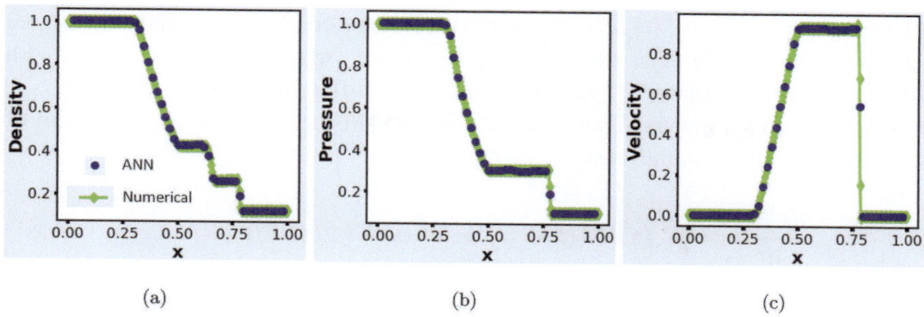

Fig. 6.15 Example ANN prediction and high-resolution numerical solution of a Riemann problem (courtesy of Huang et al. [48])

Here, superscripts "min" and "max" mean the lower and upper bounds, respectively. These bounds may be determined with prior knowledge of the initial value. Three types of local solution data can be used to train the ANNs (e.g., to predict fluid flows): (1) low-fidelity data (e.g., from the SW equations), (2) high-fidelity data (e.g., from the NS equations), and (3) a combination of the low- and high-fidelity data, leading to ANNs with multi-fidelity [33, 34].

Another issue is that the data for the local solution can bear structural complexity and high dimensionality. Such high complexity and dimensionality present difficulties for an ANN to learn the data. When available, e.g., formulated as an analytical solution, the data structures may be utilized to mitigate the difficulties while learning the data. For instance, in learning the data for the Riemann problem of shallow water equations, one can divide the range of initial value into many intervals as in Eq. (6.58), and then find the corresponding water surface elevation, η^*, according to solution/structure (6.56). The construction of the ANN reads as the following minimization problem:

$$\begin{cases} \text{Find } \mathcal{N} \text{ such that } \mathcal{N}\mathbf{u}_0 = s(\mathbf{x}, t) \\ \text{subject to: } \operatorname{argmin}_s ||s(\mathbf{x}, t) - \eta^*(\mathbf{x}, t)||_2^2 \end{cases} \tag{6.59}$$

Once the ANN has learned the water surface elevation, other variables, such as velocity, can be determined using relevant analytical solutions [41]. As a result, the ANN learns only one variable, i.e., the water surface elevation, and this reduces the complexity and expense for ANN learning.

As a general approach to overcome the difficulties due to the data's high dimensionality, let us consider the Galerkin projection for constructing a local solution. By the projection, the local solution reads as

$$\mathbf{u}^*(\mathbf{x}, t) = \sum_{l=1}^{L} c_l^*(t)\phi_l(\mathbf{x}) = \Phi(\mathbf{x})\mathbf{c}^*(t) \tag{6.60}$$

where $\mathbf{c}^*(t) = (c_1^*(t), c_2^*(t), \ldots, c_L^*(t))^T$ are coefficients, and $\Phi(\mathbf{x}) = (\phi_1(\mathbf{x}), \phi_2(\mathbf{x}), \ldots, \phi_L(\mathbf{x}))$ are orthogonal bases spanning a subspace with dimension L. Note that $\mathbf{c}^*(t)$ may sample data at different time, t_j ($j = 1, 2, \ldots$), and it also forms a matrix. Then, the PDE in (6.54) is transformed into a system of ordinary differential equations:

$$\frac{d\mathbf{c}^*(t)}{dt} = \Phi^T(\mathbf{x})L\Phi(\mathbf{x})\mathbf{c}^*(t) + \Phi^T(\mathbf{x})N(\Phi(\mathbf{x})\mathbf{c}^*(t)) \tag{6.61}$$

with the initial condition $\mathbf{c}^*(0)$. The problem of learning the local solution becomes

$$\begin{cases} \text{Find } \mathcal{N} \text{ such that } \mathcal{N}\mathbf{c}^*(0) = \mathbf{s}(t) \\ \text{subject to: } \text{argmin}_{\mathbf{s}}||\mathbf{s}(t) - \mathbf{c}^*(t)||_2^2 \end{cases} \tag{6.62}$$

in which $\mathbf{s}(t)$ is the ML solution.

To lower the high cost in computation (6.62) due to the high dimensions of the local solution data, reduced order modeling may be used to remove trivial modes in (6.60). This is only to keep the first few dominant modes:

$$\mathbf{u}^*(\mathbf{x}, t) = \sum_{l=1}^{r} c_l^*(t)\phi_l(\mathbf{x}) = \Phi^r(\mathbf{x})\mathbf{c}^{*r}(t) \tag{6.63}$$

in which $r < L$, r being the reduced dimension number. For this purpose, snapshots are collected for the Riemann problems' local solutions at different times and locations to form a data matrix: $U = \mathbf{u}^*(\mathbf{x}_i, t_j)$ ($i = 1, 2, \ldots, I$, $j = 1, 2, \ldots, J$). Then, the matrix is decomposed via the singular value decomposition (SVD) technique into three matrices [49, 50], as illustrated in Fig. 6.16. The value of r may be determined via techniques such as the principal component analysis (PCA) in association with the data matrix U [51]. When the data, e.g., if containing discontinuities (e.g., hydraulic jump), cannot be approximated well

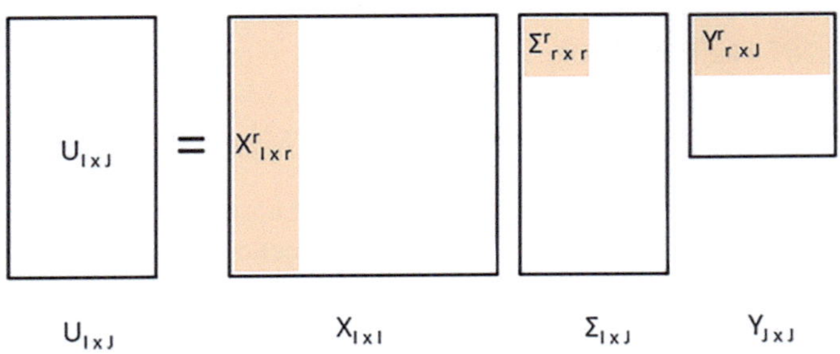

Fig. 6.16 A schematic representation of SVD. X and Y are unitary matrices, and $\sum_{r\times r}^{r}$ is a diagonal matrix. I and J indicate the data matrix dimensions, and r represents a reduced dimension. The orange shadow areas correspond to the modes with the reduced dimension

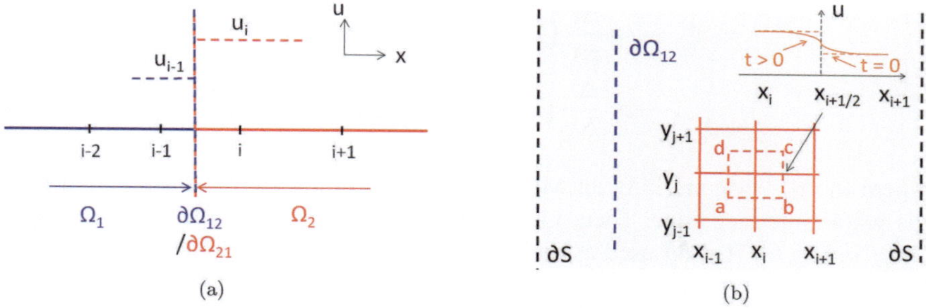

Fig. 6.17 A schematic representation of an interface zone, ML Godunov scheme, and ML interface algorithm. (**a**) Two 1D subdomains patch at an interface at $i-1/2$. (**b**) A 2D interface zone (depicted as the shadow area within the dashed lines), its boundaries, ∂S, a subdomain interface enclosed by the interface zone, $\partial\Omega_{12}$, a grid cell, and Riemann resolution

in low dimensions in a linear space, nonlinear dimensionality reduction may be considered [52].

ML Godunov Scheme and ML Interface Algorithm The aforementioned ANNs' local solutions may be utilized to develop a Godunov scheme, serving as an interface algorithm to compute interface conditions. Such a Godunov scheme is referred to as the ML Godunov scheme, which is a conservative scheme. Recall that the Godunov scheme is used as an interface algorithm in the computation of shock reflection in chap. 3. Here, the interface algorithm based on the ML Godunov scheme may be viewed as an extension of that approach.

For illustration, consider a problem of 1D advection equations (2.1), without diffusion and reaction, on Ω_1 and Ω_2, whose grids patch at their interface at $i-1/2$, see Fig. 6.17a. The application of the Green formula to the equations leads to an ML Godunov scheme, or an ML interface algorithm, for interface grid elements $(x_{i-3/2}, x_{i-1/2})$ and $(x_{i-1/2}, x_{i+1/2})$:

$$
\begin{aligned}
v_{i-1}^{n+1} &= v_{i-1}^{n} - a\frac{\Delta t}{\Delta x}\left(\langle s\rangle_{(i-1/2)^-} - \langle u\rangle_{(i-3/2)^+}\right) \\
v_{i}^{n+1} &= v_{i}^{n} - a\frac{\Delta t}{\Delta x}\left(\langle u\rangle_{(i+1/2)^-} - \langle s\rangle_{(i-1/2)^+}\right)
\end{aligned}
\tag{6.64}
$$

in which $\langle s\rangle$ and $\langle u\rangle$ are respectively the ANN local solution and the analytical local solution for the Riemann problems. Other terms, such as diffusion and source terms, if included, may be treated via operator splitting techniques, e.g., [35].

A more general approach is to introduce an interface zone that encloses subdomain interfaces, and an ML Godunov scheme is developed for this zone. Consider a scalar, 2D advection equation, the interface zone, and its grid (Fig. 6.17b). Between every two adjacent grid cells in the interface zone, a Riemann problem (e.g., 1D), a local initial value problem again, can be formulated. Built on ML local solutions for these problems, an ML Godunov scheme is

$$v_{i,j}^{n+1} = v_{i,j}^{n} - a_x \frac{\Delta t}{\Delta x}\left(\langle s\rangle_{(i+1/2)^-,j} - \langle s\rangle_{(i-1/2)^+,j}\right) \\ -a_y \frac{\Delta t}{\Delta y}\left(\langle s\rangle_{i,(j+1/2)^-} - \langle s\rangle_{i,(j-1/2)^+}\right) \tag{6.65}$$

where a_x, a_y $=$ $const$. As an ML interface algorithm, (6.65) is computed over the whole interface zone. Then, its solution is transferred onto the interfaces, e.g., $\partial\Omega_{12}$ in Fig. 6.17b, and used as the interface conditions. The solution transfer can be done directly if the two grids of the subdomains share same elements/nodes at the interfaces. Otherwise, it may be conducted indirectly via techniques such as interpolation.

The ML Godunov scheme may utilize mature techniques for conventional Godunov schemes and their variants. Available techniques are multi-dimensional methods [35], implicit algorithms [53], high-order methods [43], smooth techniques [54], and Riemann resolution for multiple fluids [46]. However, building a Godunov scheme on machine learning is a largely untouched topic, and its performance and pros and cons compared to the conventional Godunov schemes are yet to be explored.

6.2.5 *Predictive Capability and Application to Fluid Flow*

Predictive Capability of Machine Learning In the numerical experiments in the previous two subsections, the predictions of ML models, or surrogate models, for the interface zones are not exactly the same as the data used in their training. Still, the solutions by the ML coupling exhibit reasonable accuracy. Additionally, it has been shown that, in a time evolution problem, an ANN trained with data from the past time can predict the solution at a future time [23, 24]. All of these suggest that ML solutions are not merely repeating and fitting their training data. Instead, they could have a certain degree of predictive capability. Two examples will be presented to further illustrate and explore such capability. In these examples, ANN models trained by solutions in association with a set of initial and boundary conditions are applied to compute solutions with another set of initial and boundary conditions.

Example 6.2.6 This example computes a problem that is a perturbation of the problem in Example 6.2.1. Let us keep all others, i.e., algorithms and parameters, as in that example but introduce a perturbation in its boundary condition: $g(0, y) = y^3 + asin(\pi y)$, $g(x, 0) = xe^{-x} + asin(\pi x)$, with a being a coefficient. With $a = 0.2$, the perturbed problem is computed with the interface conditions provided by the ML model trained by the analytical solution with $a = 0$ in Example 6.2.1. Note that this perturbed problem has no analytical solution.

The computed solution for the perturbed problem and its error, particularly its difference to the numerical solution obtained with the classical coupling, v_{cla},

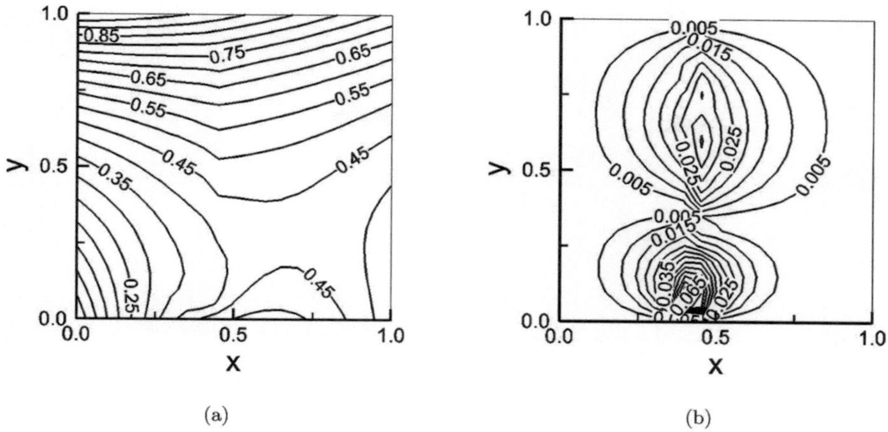

Fig. 6.18 Computation of the Poisson equation with a perturbed boundary condition. (**a**) Numerical solution with ML coupling. (**b**) Error distribution, $v_{ml} - v_{cla}$

are shown in Fig. 6.18. Figure 6.18b shows that the accuracy of the ML-coupling associated solution is reasonably well against that of the classical coupling. A comparison of this figure with Fig. 6.10b shows that its error remains similar in magnitude to that of the original problem. Note that the ML coupling solution works reasonably well until the perturbation in the boundary condition becomes too big, say when $a = 1$.

Example 6.2.7 Now, compute an initial-boundary value problem modified from that in Example 6.2.4. It is the same as the previous example, except for a different initial condition: $u_0 = sin(\pi x)$. Again, let us keep all the methods and parameters in Example 6.2.4 and again use the ML model trained in that example. Interestingly, the computed solution is still satisfactory in accuracy, and its error distribution is similar to that of Example 6.2.4 in terms of magnitude, see Figs. 6.13 and 6.19. As explained before, no Schwarz iteration is used for the computation.

In view of the two examples above, one would conjecture that, beyond merely recovering the training data, an ML model may also have the capacity to learn and predict deeper aspects of a PDE, particularly its structure and behavior.

ML Prediction of Fluid Flow The previous content has shown that ML models can learn and then predict PDE solutions in the interface zones, and, on this basis, they serve as interface algorithms for multiblock methods to compute the coupling of these PDEs. As a further move from the ML coupling of these PDEs toward ML coupling for fluid flows, here is a discussion on machine learning fluid flows at interface zones in our efforts [28]. ANNs in this subsection directly learn from the data of fluid flows obtained with numerical solutions.

Example 6.2.8 This example shows the capability of an ANN to simulate the oscillatory cavity flow in Sect. 4.3.2. The Reynolds number of the flow is $Re = 400$,

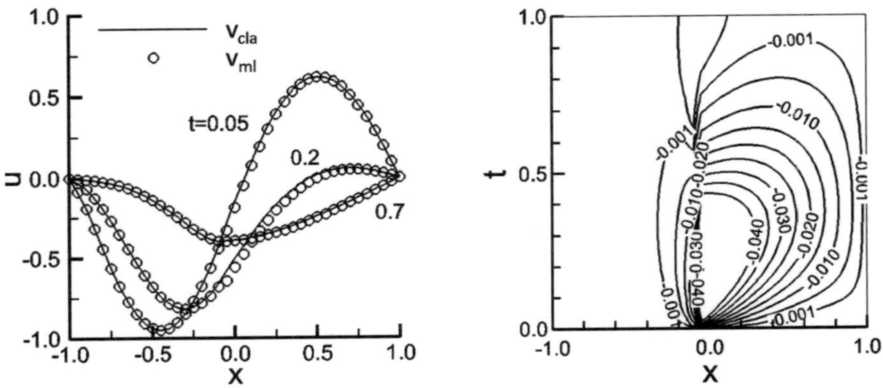

(a) (b)

Fig. 6.19 Computation of distinct advection-diffusion equations with a different initial condition: $u_0 = sin(\pi x)$. (**a**) Numerical solutions with ML coupling and classical coupling. (**b**) Error distribution, $v_{ml} - v_{cla}$

and simulation data are obtained with the runs of the NS model on a single grid of 25×25×25, as described in Sect. 4.3.2. 61 images for pressure on the plane $y = 0$ are produced by the model runs at different time instants: $t = 0, 0.01, \ldots, 0.60$. Among the 61 images, three images at $t = 0.31$, 0.41, and 0.51 are taken as test images, and the remaining 58 images are used to train the ANN, with t being the input and flow pressure on the plane being the output. The training is conducted with the aid of TensorFlow. Three hidden layers are used; the first and second layers have 20 neurons, the third layer has 10 neurons. These neurons have ReLU as their activation function. Then, the trained ANN predicts the three images of pressure. An example prediction on the pressure image, plus pressure distribution on the bottom wall, is shown in Fig. 6.20. The prediction in the figure shows that the ML prediction captures the flow fields and the pressure distribution.

Example 6.2.9 A numerical simulation on the oscillatory cavity flow detailed in the aforementioned example is made at 21 Reynolds numbers $Re = 300, 310, \ldots,$ and 500, all at $t = 0.25$. Then, the cavity is divided into three parts: upper zone, interface zone, and lower zone (Fig. 6.21a). Among the 21 images of the velocity field (vertical velocity, w), 20, without that at $Re = 400$, are adopted as the data to train the ANN. In the training, the velocity at the upper and lower zones is the input, and the velocity in the interface zone is the output. Again, the training is conducted with the aid of TensorFlow. Two hidden layers are used; both have 20 neurons with ReLU as their activation function. The prediction by the trained ANN is shown in Fig. 6.21b. It is seen from the figure that the trained network satisfactorily predicts the velocity at $Re = 400$ in the interface zone.

The aforementioned two examples are preliminary but encouraging; they demonstrate that the ML coupling is promising in serving as interface models in the

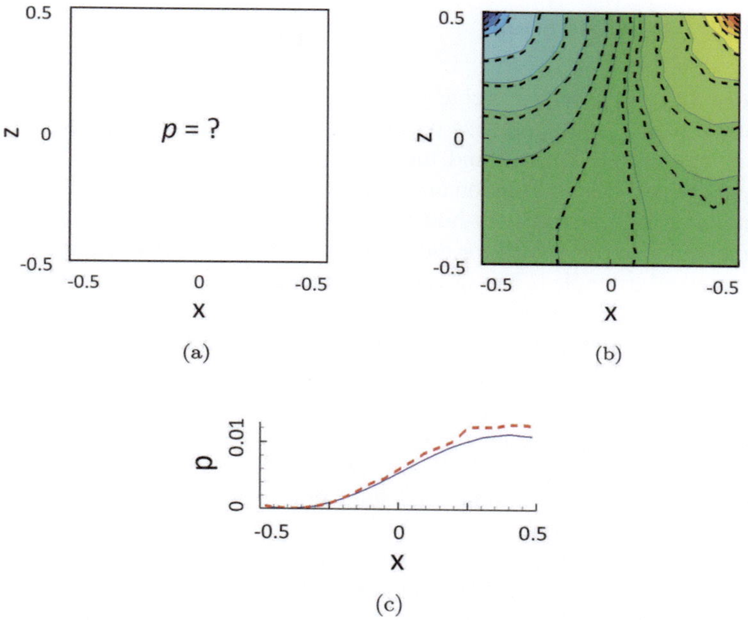

Fig. 6.20 ML prediction of cavity flow. $y = 0$, and $t = 0.31$. Dash line—prediction, solid line—data. (**a**) Flow configuration. (**b**) Pressure distribution. (**c**) Pressure load at the bottom, $z = -0.5$

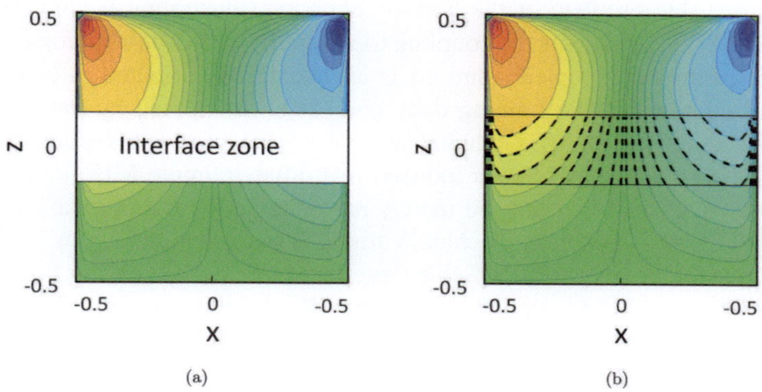

Fig. 6.21 ML prediction on cavity flow in the interface zone. $Re = 400$, $y = 0$, and $t = 0.25$. Dash line—prediction, solid line—data. (**a**) Training data for velocity. (**b**) ML prediction of velocity in the interface zone

multiblock method for fluid flows. As a further note, building on Example 6.2.8 and utilizing an ML model developed for the flow in its interface zone, a preliminary effort has been made to simulate the cavity flow using the multiblock method, yielding reasonable results while also raising issues for further investigation [55].

Summary and Question

Summary In its first part, this chapter begins by defining the concept of odd-even oscillation. It first investigates the oscillation in the numerical solution of a model problem on a single domain and then extends the analysis to two subdomains. Two types of odd-even oscillation are considered: (1) dual-mode-solution odd-even oscillation, and (2) inconsistent-boundary-conditions odd-even oscillation. Necessary and sufficient conditions for the onset of each type are derived and illustrated with numerical examples. The inconsistent-boundary-conditions oscillation arises from a decoupling of solutions at odd and even nodes and is consistent with what is described in the literature. In contrast, the dual-mode-solution oscillation does not involve such decoupling; instead, it is caused by the presence of two solution modes. These two types of oscillation behave differently: the magnitude of the type (1) decreases as the grid spacing is reduced. However, that of the type (2) remains unaffected with grid refinement. Finally, the analysis and conclusions are applied to a more complex case: a simplified fluid flow problem that exhibits odd-even oscillations.

In its second part, the chapter explores a novel idea: solving PDEs coupled at their subdomain interface via ML solutions at the interface. This idea is motivated by the limitations of the classical coupling, the success of ML in applications, and the potential of the ML coupling in terms of capability, efficiency, and universality. The chapter formulates the ML coupling using the concepts of interface zones and ML models for local solutions. The discussion begins with the ML coupling for a boundary value problem of the Poisson equation, illustrating it with numerical experiments. It then extends the coupling to advection-diffusion-reaction equations. The error in the ML-coupled solutions is analyzed and attributed to two sources: (1) errors inherited in the training data, and (2) errors caused by the ML model itself. The concept of an ML Godunov scheme is introduced and applied to the ML coupling. Numerical experiments indicate that, after training, ML models exhibit some predictive capability beyond merely reproducing the training data. As a step toward fluid flow simulation, machine learning is used to simulate the cavity flow and generate the flow solution within an interface zone.

Both topics, odd-even oscillation and the ML coupling, warrant further study. Since the analysis of odd-even oscillation is based on a simplified model, it does not yet directly explain more complex phenomena, such as the oscillation observed near shocks in compressible flows (Chap. 3) and checkerboard pressure patterns in incompressible flows (Chap. 4). The ML coupling approach has shown promise in solving PDEs, but its full potential requires additional exploration, particularly with respect to the proposed ML Godunov scheme and its effectiveness in coupling. A more systematic investigation is needed to establish the theoretical foundations and test the practical potential of the ML coupling in delivering the conjectured capabilities, efficiency, and universality, especially for fluid flows.

Questions The following questions are provided to aid in understanding, discussing, and investigating the content. Their answers and clues are embedded within the chapter.

1. What is the odd-even oscillation described in the literature, and how is it defined in this chapter?
2. This chapter identifies two types of odd-even oscillation. What are they?
3. The discussion of oscillation due to inconsistent boundary conditions is under certain constraints on the number of grid nodes. What are these constraints?
4. The analyses and conclusions on the oscillation are based on a model equation, Eq. (6.1), and are subsequently applied to model flows with more complex equations. Consider whether they can be extended to other equations.
5. This chapter presents necessary and sufficient conditions for the onset of odd-even oscillation. Using these conditions, discuss the effect of the right-hand side term of the PDE, i.e., $f(x)$ in Eq. (6.1), on computation of (6.20), such as its the onset of oscillation.
6. Explore potential approaches for further analyzing the odd-even oscillation, particularly in more realistic scenarios, such as the cavity flows discussed in Chap. 4.
7. Explain the classical coupling and the ML coupling, and their procedure.
8. The ML coupling may be considered to be a modification of the classical coupling: rather than directly using the subdomain solution as the interface condition, the ML coupling introduces an extra step that generates an ML-based solution from the numerical solution to serve as the interface condition. What are the potential advantages and disadvantages of this added step?
9. How is an interface zone constructed for the ML coupling in the computation of a boundary value problem?
10. Discuss the relationship between the domain of dependence and the interface zone when computing the convection-diffusion equations.
11. In Example 6.2.3, particularly in Eq. (6.50), the ML solution is constructed as a function of an ANN, rather than the ANN itself. Discuss the advantages of such a function.
12. When training an ML model, one may face a dataset in high dimensions. What is the dimensionality of the dataset, and how can it be reduced?
13. Is the interface treatment using the ML Godunov scheme conservative in the 1D case? What about in 2D? If not, how might a conservative interface treatment be achieved?
14. With the result of the ML learning for the cavity flow and its interface zone in this chapter, discuss how to compute the cavity flow using the ML Schwarz method.
15. Design tests to validate or refute the conjecture about the potential of ML coupling in terms of capability, efficiency, and universality.

References

1. J. von Neumann, R.D. Richtmyer, A method for the numerical calculation of hydrodynamic shocks. J. Appl. Phys. **21**, 232–237 (1950)
2. R.D. Richtmyer, K.W. Morton, *Difference Methods for Initial-Value Problems* (Interscience Publishers, 1957)
3. S. Patankar, *Numerical Heat Transfer and Fluid Flow* (McGraw-Hill, New York, 1980)
4. O.P. Jacquotte, J.T. Oden, Analysis of hourglass instabilities and control in underintegrated finite element methods. Comput. Methods Appl. Mech. Eng. **44**, 339–363 (1984)
5. S.W. Armfield, Finite difference solutions of the Navier-Stokes equations on staggered and non-staggered grids. Comput. Fluids **20**, 1–17 (1990)
6. J. Reiss, A family of energy stable, skew-symmetric finite difference schemes on collocated grids: A simple way to avoid odd-even decoupling. J. Sci. Comput. **65**, 821–838 (2015)
7. T.E. Tezduyar, S. Mittal, S.E. Ray, R. Shih, Incompressible flow computations with stabilized bilinear and linear equal-order-interpolation velocity-pressure elements. Comput. Methods Appl. Mech. Eng. **95**, 221–242 (1992)
8. B. van Leer, Towards the ultimate conservative difference scheme. V - A second-order sequel to Godunov's method. J. Comput. Phys. **32**, 101–136 (1979)
9. A. Harten, High resolution schemes for hyperbolic conservation laws. J. Comput. Phys. **49**, 357–393 (1983)
10. C.W. Shu, S. Osher, Efficient implementation of essentially non-oscillatory shock-capturing schemes. J. Comput. Phys. **77**, 439–471 (1988)
11. R. Peyret, T.D. Taylor, *Computational Methods for Fluid Flow* (Springer, New York, 1983)
12. W. Zhang, Z.G. Zhang, X.F. Chen, Three-dimensional elastic wave numerical modelling in the presence of surface topography by a collocated-grid finite-difference method on curvilinear grids. Geophys. J. Int. **190**, 358–378 (2012)
13. H.S. Tang, D.L. Zhang, C.H. Lee, Comments on algorithms for grid interfaces in simulating Euler flows. Comm. Nonlinear Sci. Numer. Simul. **1**, 50–54 (1996)
14. H.S. Tang, T. Zhou, On nonconservative algorithms for grid interfaces. SIAM J. Numer. Anal. **37**, 173–193 (1999)
15. B. Hubbard, H.-C. Chen, A Chimera scheme for incompressible viscous flows with application to submarine hydrodynamics. AIAA paper 94-2210, 1994
16. H.S. Tang, C. Jones, F. Sotiropoulos, An overset-grid method for 3D unsteady incompressible flows. J. Comput. Phys. **191**, 567–600 (2003)
17. C. Runge, Über empirische Funktionen und die interpolation zwischen äquidistanten Ordinaten. Zeitschrift für Mathematik und Physik **46**, 224–243 (1901)
18. H.S. Tang, W.B. Dong, A. Agrawal, A phenomenon of artificial odd-even grid oscillation and its presence in domain decomposition computation: Algebraic analysis and numerical illustration. J. Comput. Appl. Math. **333**, 404–427 (2018)
19. M. Eisenmann, E. Hansen, Convergence analysis of domain decomposition based time integrators for degenerate parabolic equations. Numerische Mathematik **140**, 913–938 (2018)
20. K. Qu, H.S. Tang, A. Agrawal, Integration of fully 3D fluid dynamics and geophysical fluid dynamics models for multiphysics coastal ocean flows: Simulation of local complex free-surface phenomena. Ocean Modell. **135**, 14–30 (2019)
21. Z.J. Wang, A conservative interface algorithm for moving Chimera (overlapped) grids. Int. J. Comput. Fluid Dynam. **10**, 255–265 (1998)
22. R.L. Chen, Y.Q. Wu, Z.Z. Yan, Y. Zhao, X.-C. Cai, A parallel domain decomposition method for 3D unsteady incompressible flows at high Reynolds number. J. Sci. Comput. **58**, 275–289 (2014)
23. S. Lee, D. You, Prediction of laminar vortex shedding over a cylinder using deep learning. arXiv:1712.07854 v1 [physics.flu-dyn], 2017
24. S. Wiewel, M. Becher, N. Thuerey, Latent space physics: Towards learning the temporal evolution of fluid flow. Eurographics **38**(2), 71–82 (2019)

25. J. Sirignan, K. Spiliopoulos, DGM: A deep learning algorithm for solving partial differential equations. J. Comput. Phys. **375**, 1339–1364 (2018)
26. H.S. Tang, Y.J. Liu, M. Grossberg, Research proposals. 1) collaborative research: CDS& E: A modeling system and simulation for emerging complex flow problems in coastal waters. NSF, Sept., 2019; 2) A machine-learning framework to couple numerical solutions of partial differential equations towards flow simulations, DOE, Apr. 2020
27. H.S. Tang, L. Li, M. Grossberg, Y.J. Liu, Y.M. Jia, S.S. Li, W.B. Dong, An exploratory study on machine learning to couple numerical solutions of partial differential equations. Comm. Nonlin. Sci. Numer. Simul. **97**, 105729 (2021)
28. H.S. Tang, C.R. Nichols, L.D. Wright, D. Resio, Modeling multiscale and multiphysics coastal ocean processes: A discussion on necessity, status, and advances, J. Marine Sci. Eng. **9**, 847 (2021)
29. S. Pawar, S.E. Ahmed, M. San, Interface learning in fluid dynamics: Statistical inference of closures within micro-macro coupling model. arXiv:2008.04490v1 [physics.comp-ph], 2020
30. M. Taylor, *The Math of Neural Networks* (Blue Windmill Media, 2017)
31. I.E. Lagaris, A. Likas, D.I. Fotiadis, Artificial neural networks for solving ordinary and partial differential equations. IEEE Trans. Neural Networks **9**, 987–1000 (1998)
32. M. Baymani, S. Effati, H. Niazmand, A. Kerayechian, Artificial neural network method for solving the Navier–Stokes equations. Neural Comput. Appl. **26**, 765–773 (2015)
33. J. Kou, W. Zhang, Multi-fidelity modeling framework for nonlinear unsteady aerodynamics of airfoils. Appl. Math. Modell. **76**, 832–855 (2019)
34. D. Liu, Y. Wang, Multi-fidelity physics-constrained neural network and its application in materials modeling. J. Mech. Design **141**(12), 121403 (2019)
35. G.A. Sod, *Numerical Methods in Fluid Dynamics – Initial and Initial Boundary-Value Problems* (Cambridge University Press, 1985)
36. A. Mojtabi, M. Deville, One-dimensional linear advection–diffusion equation: Analytical and finite element solutions. Comput. Fluids **107**, 189–195 (2015)
37. W.B. Lindquist, *Construction of Solutions for Two Dimensional Riemann Problems* (Palala Press, 2015)
38. L.C. Evans, *Partial Differential Equations, Graduate Studies in Mathematics* (American Math Society, 1991)
39. E. Kreyszig, *Advanced Engineering Mathematics, 9th edition* (Wiley, 2006)
40. J.Y. Wang, H.S. Tang, H.W. Fang, A fully coupled method for simulation of wave-current-seabed systems. Comm. Nonlin. Sci. Numer. Simul. **18**, 1694–1709 (2013)
41. J. Hudson, P.K. Sweby, Formulations for numerically approximating hyperbolic systems governing sediment transport. J. Sci. Comput. **19**, 225–252 (2003)
42. P.D. Lax, X.D. Liu, Solution of two-dimensional Riemann problems of gas dynamics by positive schemes. SIAM J. Sci. Comput. **19**, 319–340 (1998)
43. E.F. Toro, L.A.M. Millington, R.C. Nejad, *Towards Very High Order Godunov Schemes*. In Godunov Methods, eds by E. F. Toro (Springer, New York, 2001)
44. ANSYS, Inc., ANSYS FLUENT User's Guide, release 14.0, 2011
45. OpenFOAM, The Open Source CFD Toolbox, version 2312, 2023
46. H.S. Tang, D. Huang, A second-order accurate capturing scheme for 1D inviscid flows of gas and water with vacuum zones. J. Comput. Phys. **128**, 301–318 (1996)
47. H.S. Tang, F. Sotiropoulos, A second-order Godunov method for wave problems in coupled solid-water-gas systems. J. Comput. Phys. **151**, 790–815 (1999)
48. H.X. Huang, Y.J. Liu, V. Yang, Neural networks with inputs based on domain of dependence and a converging sequence for solving conservation laws, part I: 1D Riemann problems. arXiv:2109.09316 v1 [math.NA] 2021
49. G. Strang, *Introduction to Linear Algebra, 3rd edition* (Wellesley-Cambridge Press, 2003)
50. H. Dym, *Linear Algebra in Action* (American Mathematical Society, 2007)
51. I.T. Jolliffe, *Principal Component Analysis, 2nd edition* (Springer, 2002)
52. K. Lee, K.T. Carlberg, Model reduction of dynamical systems on nonlinear manifolds using deep convolutional autoencoders. J. Comput. Phys. **404**, 108973 (2020)

53. J.P. Collins, P. Colella, H.M Glaz, An implicit-explicit Eulerian Godunov scheme for compressible flow. J. Comput. Phys. **116**, 195–211 (1995)
54. Y.J. Liu, C.-W. Shu, E. Tadmor, M.P. Zhang, Central discontinuous Galerkin methods on overlapping cells with a non-oscillatory hierarchical reconstruction. SIAM J. Numer. Anal. **45**, 2442–2467 (2007)
55. W. Dong, *Towards simulation of ocean flows: Analysis and algorithm for computation of coupled partial differential equations*. Ph.D. thesis, Civil Eng. Dept., City College of New York, 2022

Appendix A
Metrics of Geometric Transformation

The details of the geometric transformation from Cartesian coordinates to general, curvilinear coordinates $\left(x^1, x^2, x^3\right) \rightarrow \left(\xi^1, \xi^2, \xi^3\right)$ are contained in the so-called metrics of the transformation. These metrics are computed according to the equations presented in this appendix.

The Jacobian J is computed from

$$J = \frac{1}{G}$$

where G is

$$G = \det \begin{pmatrix} \left(x^1\right)_{\xi^1} & \left(x^1\right)_{\xi^2} & \left(x^1\right)_{\xi^3} \\ \left(x^2\right)_{\xi^1} & \left(x^2\right)_{\xi^2} & \left(x^2\right)_{\xi^3} \\ \left(x^3\right)_{\xi^1} & \left(x^3\right)_{\xi^2} & \left(x^3\right)_{\xi^3} \end{pmatrix}$$

in which

$$\left(x^m\right)_{\xi^l} \equiv \frac{\partial x^m}{\partial \xi^l}$$

for $m, l = 1, 2, 3$. The terms $(x^m)_{\xi^l}$ are computed on interior nodes of the computational domain using the second-order, three-point central difference. On the domain's boundaries, the one-sided, first-order upwind difference is used.

H. Tang, *Multiblock Method for Fluid Flow*, Nonlinear Systems and Complexity 40,
https://doi.org/10.1007/978-3-031-78568-9

Once the $(x^m)_{\xi^l}$ terms are computed, the metrics are computed as

$$
\begin{aligned}
\xi^1_{x^1} &= \frac{1}{J}\left[\left(x^2\right)_{\xi^2}\left(x^3\right)_{\xi^3} - \left(x^2\right)_{\xi^3}\left(x^3\right)_{\xi^2}\right]\\
\xi^1_{x^2} &= \frac{1}{J}\left[\left(x^1\right)_{\xi^3}\left(x^3\right)_{\xi^2} - \left(x^1\right)_{\xi^2}\left(x^3\right)_{\xi^3}\right]\\
\xi^1_{x^3} &= \frac{1}{J}\left[\left(x^1\right)_{\xi^2}\left(x^2\right)_{\xi^3} - \left(x^1\right)_{\xi^3}\left(x^2\right)_{\xi^2}\right]
\end{aligned}
$$

$$
\begin{aligned}
\xi^2_{x^1} &= \frac{1}{J}\left[\left(x^2\right)_{\xi^3}\left(x^3\right)_{\xi^1} - \left(x^2\right)_{\xi^1}\left(x^3\right)_{\xi^3}\right]\\
\xi^2_{x^2} &= \frac{1}{J}\left[\left(x^1\right)_{\xi^1}\left(x^3\right)_{\xi^3} - \left(x^1\right)_{\xi^3}\left(x^3\right)_{\xi^1}\right]\\
\xi^2_{x^3} &= \frac{1}{J}\left[\left(x^1\right)_{\xi^3}\left(x^2\right)_{\xi^1} - \left(x^1\right)_{\xi^1}\left(x^2\right)_{\xi^3}\right]
\end{aligned}
$$

$$
\begin{aligned}
\xi^3_{x^1} &= \frac{1}{J}\left[\left(x^2\right)_{\xi^1}\left(x^3\right)_{\xi^2} - \left(x^2\right)_{\xi^2}\left(x^3\right)_{\xi^1}\right]\\
\xi^3_{x^2} &= \frac{1}{J}\left[\left(x^1\right)_{\xi^2}\left(x^3\right)_{\xi^1} - \left(x^1\right)_{\xi^1}\left(x^3\right)_{\xi^2}\right]\\
\xi^3_{x^3} &= \frac{1}{J}\left[\left(x^1\right)_{\xi^1}\left(x^2\right)_{\xi^2} - \left(x^1\right)_{\xi^2}\left(x^2\right)_{\xi^1}\right]
\end{aligned}
$$

The contravariant metric tensor g^{ml} and the covariant metric tensor g_{ml} are ($k = 1, 2, 3$)

$$
\begin{aligned}
g^{ml} &= \xi^m_{x^k}\xi^l_{x^k}\\
g_{ml} &= \left(x^m\right)_{\xi^k}\left(x^l\right)_{\xi^k}
\end{aligned}
\tag{A.1}
$$

Appendix B
Matrix-Valued Artificial Dissipation

The matrix-valued artificial dissipation scheme uses the absolute value of the Jacobian matrix $\left|A^l\right|$ $(l = 1, 2, 3)$ to scale the dissipation [1]. The absolute value of this matrix is

$$\left|A^l\right| = \frac{1}{2JC_l^2}\begin{pmatrix} a_{11} & a_{12} & a_{13} & a_{14} \\ a_{21} & a_{22} & a_{23} & a_{24} \\ a_{31} & a_{32} & a_{33} & a_{34} \\ a_{41} & a_{42} & a_{43} & a_{44} \end{pmatrix}$$

where C_l is defined in Eq. (C.1). The components of $\left|A^l\right|$ are

$$\begin{aligned}
a_{11} &= C_l(|\lambda_2^l|\lambda_1^l - |\lambda_1^l|\lambda_2^l) \\
a_{12} &= \xi^l_{x1} C_l(|\lambda_1^l| - |\lambda_2^l|) \\
a_{13} &= \xi^l_{x2} C_l(|\lambda_1^l| - |\lambda_2^l|) \\
a_{14} &= \xi^l_{x3} C_l(|\lambda_1^l| - |\lambda_2^l|)
\end{aligned}$$

$$\begin{aligned}
a_{21} &= |\lambda_1^l|(u^1 g^{ll} - \xi^l_{x1}\lambda_1^l) + |\lambda_2^l|(u^1 g^{ll} - \xi^l_{x1}\lambda_1^l) - 2|\lambda_3^l|(u^1 g^{ll} - \xi^l_{x1} U^l) \\
a_{22} &= |\lambda_1^l|(u^1 \xi^l_{x1}\lambda_1^l + (\xi^l_{x1})^2) + |\lambda_2^l|(u^1 \xi^l_{x1}\lambda_2^l + (\xi^l_{x1})^2) \\
&\quad +2|\lambda_3^l|((U^l)^2 - u^1 U^l \xi^l_{x1} + (\xi^l_{x2})^2 + (\xi^l_{x3})^2) \\
a_{23} &= \xi^l_{x2}(|\lambda_1^l|(u^1\lambda_1^l + \xi^l_{x1}) + |\lambda_2^l|(u^1\lambda_2^l + \xi^l_{x1}) - 2|\lambda_3^l|(u^1 U^l + \xi^l_{x1})) \\
a_{24} &= \xi^l_{x3}(|\lambda_1^l|(u^1\lambda_1^l + \xi^l_{x1}) + |\lambda_2^l|(u^1\lambda_2^l + \xi^l_{x1}) - 2|\lambda_3^l|(u^1 U^l + \xi^l_{x1}))
\end{aligned}$$

$$\begin{aligned}
a_{31} &= |\lambda_1^l|(u^2 g^{ll} - \xi^l_{x2}\lambda_2^l) + |\lambda_2^l|(u^2 g^{ll} - \xi^l_{x2}\lambda_1^l) - 2|\lambda_3^l|(u^2 g^{ll} - \xi^l_{x2} U^l) \\
a_{32} &= \xi^l_{x1}(|\lambda_1^l|(u^2\lambda_1^l + \xi^l_{x2}) + |\lambda_2^l|(u^2\lambda_2^l + \xi^l_{x2}) - 2|\lambda_3^l|(u^2 U^l + \xi^l_{x2})) \\
a_{33} &= |\lambda_1^l|(u^2 \xi^l_{x2}\lambda_1^l + (\xi^l_{x2})^2) + |\lambda_2^l|(u^2 \xi^l_{x2}\lambda_2^l + (\xi^l_{x2})^2) \\
&\quad +2|\lambda_3^l|((U^l)^2 - u^2 U^l \xi^l_{x2} + (\xi^l_{x1})^2 + (\xi^l_{x3})^2) \\
a_{34} &= \xi^l_{x3}(|\lambda_1^l|(u^2\lambda_1^l + \xi^l_{x2}) + |\lambda_2^l|(u^2\lambda_2^l + \xi^l_{x2}) - 2|\lambda_3^l|(u^2 U^l + \xi^l_{x2}))
\end{aligned}$$

H. Tang, *Multiblock Method for Fluid Flow*, Nonlinear Systems and Complexity 40,
https://doi.org/10.1007/978-3-031-78568-9

$$
\begin{aligned}
a_{41} &= |\lambda_1^l|(u^3 g^{ll} - \xi_{x^3}^l \lambda_3^l) + |\lambda_2^l|(u^3 g^{ll} - \xi_{x^3}^l \lambda_1^l) - 2|\lambda_3^l|(u^3 g^{ll} - \xi_{x^3}^l U^l) \\
a_{42} &= \xi_{x^1}^l (|\lambda_1^l|(u^3 \lambda_1^l + \xi_{x^3}^l) + |\lambda_2^l|(u^3 \lambda_2^l + \xi_{x^3}^l) - 2|\lambda_3^l|(u^3 U^l + \xi_{x^3}^l)) \\
a_{43} &= \xi_{x^2}^l (|\lambda_1^l|(u^3 \lambda_1^l + \xi_{x^3}^l) + |\lambda_2^l|(u^3 \lambda_2^l + \xi_{x^3}^l) - 2|\lambda_3^l|(u^3 U^l + \xi_{x^3}^l)) \\
a_{44} &= |\lambda_1^l|(u^3 \xi_{x^3}^l \lambda_1^l + (\xi_{x^3}^l)^2) + |\lambda_2^l|(u^3 \xi_{x^3}^l \lambda_2^l + (\xi_{x^3}^l)^2) \\
&\quad + 2|\lambda_3^l|((U^l)^2 - u^3 U^l \xi_{x^3}^l + (\xi_{x^1}^l)^2 + (\xi_{x^2}^l)^2)
\end{aligned}
$$

References

1. F.B. Lin, F. Sotiropoulos, Assessment of artificial dissipation models for three-dimensional incompressible flows. Trans. ASME J. Fluids Eng. **119**, 331–340 (1997)

Appendix C
Local Time Stepping

Local time stepping maximizes the error damping of an iterative algorithm at every node of the computational mesh [1, 2]. Both hyperbolic and parabolic stability criteria are used. A numerical method's stability depends on the amplification factor's magnitude, which is a function of dimensionless numbers such as the Courant–Friedrich–Lewy (CFL) number and the von Neumann number (VN). Local time stepping maintains these numbers constant at every node. Therefore, the stepping can vary depending on the stability conditions at each node. Strictly speaking, local time stepping is only valid for steady flows; however, it can also be used with a dual time stepping technique for unsteady flows.

The time step $\Delta\tau_{i,j,k}$ is calculated and stored at every node following

$$\Delta\tau_{i,j,k} = \min\left\{\frac{\text{CFL}}{\max\left(\lambda_{\xi^1}, \lambda_{\xi^2}, \lambda_{\xi^3}\right)_{i,j,k}}, \text{Re}\left[\frac{\text{VN}}{\max\left(g^{11}, g^{22}, g^{33}\right)_{i,j,k}}\right]\right\}$$

where λ_{ξ^l} is calculated as ($l = 1,2,3$)

$$\lambda_{\xi^l} = \left|U^l\right| + \sqrt{\left(U^l\right)^2 + g^{ll}}$$

and it is the spectral radius of the Jacobian matrix

$$A^l = \frac{\partial F^l}{\partial Q} = \frac{1}{J}\begin{pmatrix} 0 & \xi^l_{x^1} & \xi^l_{x^2} & \xi^l_{x^3} \\ \xi^l_{x^1} & U^l + u^1\xi^l_{x^1} & u^1\xi^l_{x^2} & u^1\xi^l_{x^3} \\ \xi^l_{x^2} & u^2\xi^l_{x^1} & U^l + u^2\xi^l_{x^2} & u^2\xi^l_{x^3} \\ \xi^l_{x^3} & u^3\xi^l_{x^1} & u^3\xi^l_{x^2} & U^l + u^3\xi^l_{x^3} \end{pmatrix}$$

H. Tang, *Multiblock Method for Fluid Flow*, Nonlinear Systems and Complexity 40,
https://doi.org/10.1007/978-3-031-78568-9

whose eigenvalues are

$$
\begin{aligned}
\lambda_1^l &= U^l - C_l \\
\lambda_2^l &= U^l + C_l \\
\lambda_3^l &= \lambda_4^l = U^l \\
C_l &= \sqrt{\left(U^l\right)^2 + g^{ll}}
\end{aligned}
\tag{C.1}
$$

The metrics g^{ll} are given by (A.1).

References

1. A. Jameson, W. Schmit, E. Turkel, Numerical solutions of the Euler equations by finite volume methods using Runge-Kutta time-stepping schemes. AIAA paper 81–1259, 1981
2. F.B. Lin, F. Sotiropoulos, Strongly-coupled multigrid method for 3-D incompressible flows using near-wall turbulence closures. ASME J. Fluid Eng. **119**(2), 314–324 (1997)

Appendix D
Particle Tracking

The particle-tracking equation is

$$\frac{dx^i}{dt} = u^i(x^1, x^2, x^3, t), \ i = 1, 2, 3 \tag{D.1}$$

Here, x^1, x^2, and x^3 are respectively particle coordinates x, y, and z, and u^1, u^2, and u^3 are respectively particle speeds u, v, and w. We use a 2-stage Runge-Kutta algorithm

$$\begin{aligned}
x^{i,n+1} &= x^{i,n} + \frac{1}{2}(a_i + b_i) \\
a_i &= \Delta t u^i(x^{1,n}, x^{2,n}, x^{3,n}, t_n) \\
b_i &= \Delta t u^i(x^{1,n} + a_1, x^{2,n} + a_2, x^{3,n} + a_3, t_n + \Delta t)
\end{aligned}$$

Δt is the timer step, and index n refers to the time t_n. In Chimera grids, the following is the logic of computation:

Step 1 Suppose at time t_n, a particle is inside a grid and its location is (x^n, y^n, z^n).

Step 2 Use Eq. (D.1) to predict its position at time t_{n+1}.

Step 3 If **Step 2** is finished and the predicted position is still in the grid, the predicted position is the new position $(x^{n+1}, y^{n+1}, z^{n+1})$ at time t_{n+1}. Then return to **Step 1**; Otherwise, search and see if (x^n, y^n, z^n) is located in another grid. If it is, assign the particle to the grid and go to **Step 1**. If it is not, stop tracking the particle since this means the particle has exited the whole computational domain.

H. Tang, *Multiblock Method for Fluid Flow*, Nonlinear Systems and Complexity 40,
https://doi.org/10.1007/978-3-031-78568-9

Index

H. Tang, *Multiblock Method for Fluid Flow*, Nonlinear Systems and Complexity 40,
https://doi.org/10.1007/978-3-031-78568-9

F

G

H

I

J

L

M

MIX
Papier aus verantwortungsvollen Quellen
Paper from responsible sources
FSC® C105338

If you have any concerns about our products,
you can contact us on
ProductSafety@springernature.com

In case Publisher is established outside the EU,
the EU authorized representative is:
Springer Nature Customer Service Center GmbH
Europaplatz 3, 69115 Heidelberg, Germany

Printed by Libri Plureos GmbH
in Hamburg, Germany